Advanced Textbooks in Control and Signal Processing

Springer
London
Berlin
Heidelberg
New York
Barcelona
Hong Kong
Milan
Paris
Santa Clara
Singapore
Tokyo

Eduardo F. Camacho and Carlos Bordons

Model Predictive Control

With 90 Figures

Springer

Dr Eduardo F. Camacho, PhD
Dr Carlos Bordons, PhD
Escuela Superior de Ingenieros, Universidad de Sevilla, Camino de los Descubrimientos s/n, 41092 Sevilla, Spain

ISBN 978-3-540-76241-6

British Library Cataloguing in Publication Data
Camacho, E. F.
Model predictive control
1.Predictive control
I.Title II.Bordons, C. (Carlos)
629.8
ISBN 978-3-540-76241-6

Library of Congress Cataloging-in-Publication Data
Camacho, E.F..
Model predictive control / Eduardo F. Camacho and Carlos Bordons.
p. cm.
Includes bibliographical references.
ISBN 978-3-540-76241-6 **ISBN 978-3-4471-3398-8 (eBook)**
DOI: 10.1007/978-3-4471-3398-8
1. Predictive control. I. Bordons, C. (Carlos), 1962-
II. Title.
TJ217.6.C35 1998 98-4205
629.8--dc21 CIP

2nd printing 2000
3rd printing 2002

Typesetting: Camera ready by authors

69/3830-5432 Printed on acid-free paper SPIN 10894401

to Janet
E.F.C.

to Ana
C.B.

Preface

Model Predictive Control has developed considerably over the last few years, both within the research control community and in industry. The reason for this success can be attributed to the fact that Model Predictive Control is, perhaps, the most general way of posing the process control problem in the time domain. Model Predictive Control formulation integrates optimal control, stochastic control, control of processes with dead time, multivariable control and future references when available. Another advantage of Model Predictive Control is that because of the finite control horizon used, constraints and in general non-linear processes which are frequently found in industry, can be handled. Although Model Predictive Control has been found to be quite a robust type of control in most reported applications, stability and robustness proofs have been difficult to obtain because of the finite horizon used. This has been a drawback for a wider dissemination of Model Predictive Control in the control research community. Some new and very promising results in this context allow one to think that this control technique will experience greater expansion within this community in the near future. On the other hand, although a number of applications have been reported both in industry and research institutions, Model Predictive Control has not yet reached in industry the popularity that its potential would suggest. One of the reason for this is that its implementation requires some mathematical complexities which are not a problem in general for the research control community, where mathematical packages are normally fully available, but represent a drawback for the use of the technique by control engineers in practice.

One of the goals of this text is to contribute to filling the gap between the empirical way in which practitioners tend to use control algorithms and the powerful but sometimes abstractly formulated techniques developed by the control researchers. The book focuses on implementation issues for Model Predictive Controllers and intends to present easy ways of implementing them in industry. The book also aims to serve as a guideline of how to implement Model Predictive Control and as a motivation for doing so by showing that using such a powerful control technique does not require complex control algorithms.

The book is aimed mainly at practitioners, although it can be followed by a wide range of readers, as only basic knowledge of control theory and sample data systems are required. The general survey of the field, guidance of the choice of appropriate implementation techniques, as well as many illustrative examples are explained for practising engineers and senior undergraduate and graduate students. The book is developed mainly around Generalized Predictive Control, one of the most popular Model Predictive Control methods, and we have not tried to give a full description of all Model Predictive Control algorithms and their properties, although some of them and their main properties are described. Neither do we claim this technique to be the best choice for the control of every process, although we feel that it has many advantages, and therefore, we have not tried to make a comparative study of different Model Predictive Control algorithms amongst themselves and versus other control strategies.

The text is composed of material collected from lectures given to senior undergraduate students and articles written by the authors and is based in a previous book (*Model Predictive Control in the Process Industry*, Springer, 1995), written by the authors to which more recent developments and applications have been added.

Acknowledgements

The authors would like to thank a number of people who in various ways have made this book possible. Firstly we thank Janet Buckley who translated part of the book from our native language to English and corrected and polished the style of the rest. Our thanks also to Manuel Berenguel who implemented and tested the controller on the solar power plant and revised the manuscript, to Daniel Limón and Daniel Rodriguez who helped us with some of the examples given and to Julio Normey who helped with the analysis of the effects of predictions on robustness.

Our thanks to Javier Aracil who introduced us to the exciting world of Control and to many other colleagues and friends from the Department, especially Francisco R. Rubio, whose previous works on the solar power plant were of great help. Part of the material included in the book is the result of research work funded by CICYT and CIEMAT. We gratefully acknowledge these institutions for their support.

Finally both authors thank their families for their support, patience and understanding of family time lost during the writing of the book.

Sevilla, November 1998
Eduardo F. Camacho and Carlos Bordons

Glossary

Notation

$\mathbf{A}(\cdot)$	boldface upper case letters denote polynomial matrices.
$A(\cdot)$	italic and upper case letters denote polynomials.
$\mathbf{M}$	bolface upper case letters denote real matrices.
M	italic upper case letters denote real matrices in chapter 6.
$\mathbf{b}$	boldface lower letters indicate real vectors.

Symbols

s	complex variable used in Laplace transform
z^{-1}	backward shift operator
z	forward shift operator and complex variable used in z – transform
$(M)_{ij}$	element ij of matrix M
$(v)_i$	i^{th} – element of vector v
$(\cdot)^T$	transpose of $(\cdot)$
$diag(x_1, \cdots, x_n)$	diagonal matrix with diagonal elements equal to $x_1, \cdots, x_n$
$\|(\cdot)\|$	absolute value of $(\cdot)$
$\\|\mathbf{v}\\|_Q^2$	$\mathbf{v}^T Q \mathbf{v}$
$\\|\mathbf{v}\\|_l$	l – norm of $\mathbf{v}$
$\\|\mathbf{v}\\|_\infty$	infinity norm of $\mathbf{v}$
$I_{n\times n}$	$(n \times n)$ identity matrix
I	identity matrix of appropriate dimensions
$\mathbf{0}_{p\times q}$	$(p \times q)$ matrix with all entries equal to zero
$\mathbf{0}$	matrix of appropriate dimensions with all entries equal to zero

$\mathbf{1}$	column vector with all entries equal to one
$< x, z >$	dot product of vectors x and z
$E[\cdot]$	expectation operator
$\hat{\cdot}$	expected value
$\hat{x}(t+j\|t)$	expected value of $x(t+j)$ with available information at instant t
$\delta(P(\cdot))$	degree of polynomial $P(\cdot)$
$\triangle$	$1 - z^{-1}$
$det(M)$	determinant of matrix M
$\min_{x \in \mathbf{X}} J(x)$	the minimum value of $J(x)$ for all values of $x \in \mathbf{X}$

Model parameters and variables

m	number of input variables
n	number of output variables
$u(t)$	input variables at instant t
$y(t)$	output variables at instant t
$e(t)$	discrete white noise with zero mean
d	dead time of the process expressed in sampling time units
$\mathbf{A}(z^{-1})$	process left polynomial matrix for the LMFD
$\mathbf{B}(z^{-1})$	process right polynomial matrix for the LMFD
$\mathbf{C}(z^{-1})$	colouring polynomial matrix
M, N, Q	process state-space description matrices
P	noise matrix for state-space description
$v(t)$	state noise for the state-space description
$w(t)$	output noise for state-space description

Controller parameters and variables

N_1	lower value of predicting horizon
N_2	higher value of predicting horizon
N	number of points of predicting horizon ($N = N_2 - N_1$)
N_3	control horizon (N_u)
λ	weighting factor for control increments
$\mathbf{G}_{N_{123}}$	GPC prediction matrix with horizons (N_1, N_2, N_3)
$\mathbf{u}$	vector of future control increments for the control horizon
$\mathbf{y}$	vector of predicted outputs for prediction horizon
$\mathbf{f}$	vector of predicted free response
$\mathbf{w}$	vector of future references

$\overline{U}$	vector of maximum allowed values of manipulated variables
$\underline{U}$	vector of minimum allowed values of manipulated variables
$\overline{u}$	vector of maximum allowed values of manipulated variable slew rates
$\underline{u}$	vector of minimum allowed values of manipulated variable slew rates
$\overline{y}$	vector of maximum allowed values of output variables
$\underline{y}$	vector of minimum allowed values of output variables
$\tilde{\mathbf{A}}(z^{-1})$	polynomial $\mathbf{A}(z^{-1})$ multiplied by $\triangle$

Acronyms

CARIMA	Controlled Autoregressive Integrated Moving Average
CARMA	Controlled Autoregressive Moving Average
CRHPC	Constrained Receding Horizon Predictive Control
DMC	Dynamic Matrix Control
EHAC	Extended Horizon Adaptive Control
EPSAC	Extended Prediction Self-Adaptive Control
FIR	Finite Impulse Response
FLOP	Floating Point Operation
GMV	Generalized Minimum Variance
GPC	Generalized Predictive Control
HIECON	Hierarchical Constraint Control
IDCOM	Identification and Command
LCP	Linear Complementary Problem
LMFD	Left Matrix Fraction Description
LMI	Linear Matrix Inequalities
LP	Linear Programming
LQ	Linear Quadratic
LQG	Linear Quadratic Gaussian
LRPC	Long Range Predictive Control
LTR	Loop Transfer Recovery
MAC	Model Algorithmic Control
MILP	Mixed Integer Linear Programming
MIMO	Multi-Input Multi-Output
MIQP	Mixed Integer Quadratic Programming
MPC	Model Predictive Control
MPHC	Model Predictive Heuristic Control
MUSMAR	Multi-Step Multivariable Adaptive Control
MURHAC	Multipredictor Receding Horizon Adaptive Control
NLP	Nonlinear Programming
OPC	Optimum Predictive Control
PCT	Predictive Control Technology

PFC	Predictive Functional Control
PID	Proportional Integral Derivative
QP	Quadratic Programming
RMPCT	Robust Model Predictive Control Technology
SCADA	Supervisory Control and Data Acquisition
SCAP	Adaptive Predictive Control System
SGPC	Stable Generalized Predictive Control
SISO	Single-Input Single-Output
SMCA	Setpoint Multivariable Control Architecture
UPC	Unified Predictive Control

Contents

Chapter 1

Introduction to Model Based Predictive Control

Model (Based) Predictive Control (MBPC or MPC) originated in the late seventies and has developed considerably since then. The term Model Predictive Control does not designate a specific control strategy but a very ample range of control methods which make an explicit use of a model of the process to obtain the control signal by minimizing an objective function. These design methods lead to linear controllers which have practically the same structure and present adequate degrees of freedom. The ideas appearing in greater or lesser degree in all the predictive control family are basically:

- Explicit use of a model to predict the process output at future time instants (horizon).
- Calculation of a control sequence minimizing an objective function.
- Receding strategy, so that at each instant the horizon is displaced towards the future, which involves the application of the first control signal of the sequence calculated at each step.

The various MPC algorithms (also called long-range Predictive Control or LRPC) only differ amongst themselves in the model used to represent the process and the noises and the cost function to be minimized. This type of control is of an open nature within which many works have been developed, being widely received by the academic world and by industry. There are many applications of predictive control successfully in use at the present time, not only in the process industry but also applications to the control of a diversity of processes ranging from robot manipulators [63] to clinical anaesthesia [67]. Applications in the cement industry, drying towers and in robot arms, are described in [31], whilst developments for distillation columns,

PVC plants, steam generators or servos are presented in [102], [105]. The good performance of these applications shows the capacity of the MPC to achieve highly efficient control systems able to operate during long periods of time with hardly any intervention.

MPC presents a series of advantages over other methods, amongst which stand out:

- it is particularly attractive to staff with only a limited knowledge of control because the concepts are very intuitive and at the same time the tuning is relatively easy.
- it can be used to control a great variety of processes, from those with relatively simple dynamics to other more complex ones, including systems with long delay times or of non-minimum phase or unstable ones.
- the multivariable case can easily be dealt with.
- it intrinsically has compensation for dead times.
- it introduces feed forward control in a natural way to compensate for measurable disturbances.
- the resulting controller is an easy to implement linear control law.
- its extension to the treatment of constraints is conceptually simple and these can be systematically included during the design process.
- it is very useful when future references (robotics or batch processes) are known.
- it is a totally open methodology based on certain basic principles which allow for future extensions.

As is logical, however, it also has its drawbacks. One of these is that although the resulting control law is easy to implement and requires little computation, its derivation is more complex than that of the classical PID controllers. If the process dynamic does not change, the derivation of the controller can be done beforehand, but in the adaptive control case all the computation has to be carried out at every sampling time. When constraints are considered, the amount of computation required is even higher. Although this, with the computing power available today, is not an essential problem one should bear in mind that many industrial process control computers are not at their best as regards their computing power and, above all, that most of the available time at the process computer has normally to be used for purposes other than the control algorithm itself (communications, dialogues with the operators, alarms, recording, etc). Even so, the greatest drawback is the need for an appropriate model of the process to be available. The design algorithm is based on a prior knowledge of the model and it is independent of it, but

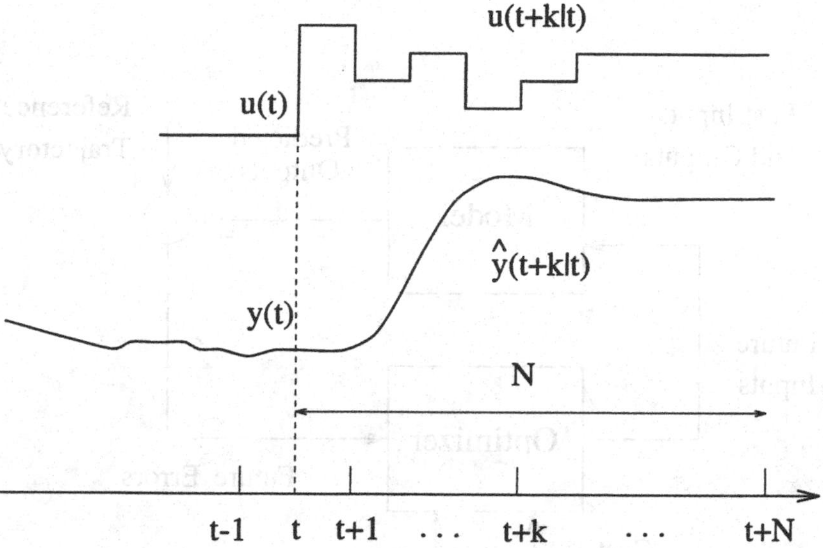

Figure 1.1: MPC Strategy

it is obvious that the benefits obtained will be affected by the discrepancies existing between the real process and the model used.

In practice, MPC has proved to be a reasonable strategy for industrial control, in spite of the original lack of theoretical results at some crucial points such as stability or robustness.

1.1 MPC Strategy

The methodology of all the controllers belonging to the MPC family is characterized by the following strategy, represented in figure 1.1:

1. The future outputs for a determined horizon N, called the prediction horizon, are predicted at each instant t using the process model. These predicted outputs $y(t+k \mid t)$ [1] for $k = 1 \dots N$ depend on the known values up to instant t (past inputs and outputs) and on the future control signals $u(t+k \mid t)$, $k = 0 \dots N-1$, which are those to be sent to the system and to be calculated.
2. The set of future control signals is calculated by optimizing a determined criterion in order to keep the process as close as possible to the reference trajectory $w(t+k)$ (which can be the setpoint itself or a close approximation of it). This criterion usually takes the form of a quadratic function

[1]The notation indicates the value of the variable at the instant $t+k$ calculated at instant t.

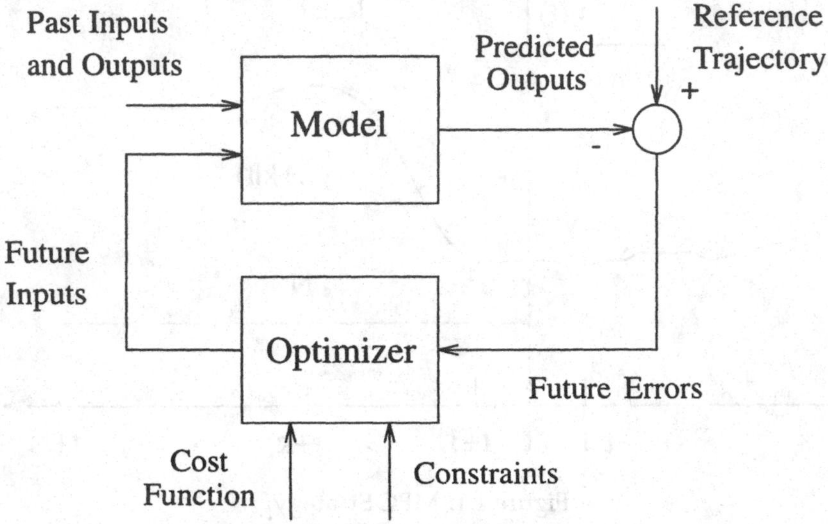

Figure 1.2: Basic structure of MPC

of the errors between the predicted output signal and the predicted reference trajectory. The control effort is included in the objective function in most cases. An explicit solution can be obtained if the criterion is quadratic, the model is linear and there are no constraints, otherwise an iterative optimization method has to be used. Some assumptions about the structure of the future control law are also made in some cases, such as that it will be constant from a given instant.

3. The control signal $u(t \mid t)$ is sent to the process whilst the next control signals calculated are rejected, because at the next sampling instant $y(t+1)$ is already known and step 1 is repeated with this new value and all the sequences are brought up to date. Thus the $u(t+1 \mid t+1)$ is calculated (which in principle will be different to the $u(t+1 \mid t)$ because of the new information available) using the receding horizon concept.

In order to implement this strategy, the basic structure shown in figure 1.2 is used. A model is used to predict the future plant outputs, based on past and current values and on the proposed optimal future control actions. These actions are calculated by the optimizer taking into account the cost function (where the future tracking error is considered) as well as the constraints.

The process model plays, in consequence, a decisive role in the controller. The chosen model must be capable of capturing the process dynamics so as to precisely predict the future outputs as well as being simple to implement and to understand. As MPC is not a unique technique but a set of different methodologies, there are many types of models used in various formulations.

One of the most popular in industry is the Truncated Impulse Response Model, which is very simple to obtain as it only needs the measurement of the output when the process is excited with an impulse input. It is widely accepted in industrial practice because it is very intuitive and can also be used for multivariable processes, although its main drawbacks are the large number of parameters needed and that only open-loop stable processes can be described this way. Closely related to this kind of model is the Step Response Model, obtained when the input is a step.

The Transfer Function Model is, perhaps, most widespread in the academic community and is used in most control design methods, as it is a representation that requires only a few parameters and is valid for all kind of processes. The State-Space Model is also used in some formulations, as it can easily describe multivariable processes.

The optimizer is another fundamental part of the strategy as it provides the control actions. If the cost function is quadratic, its minimum can be obtained as an explicit function (linear) of past inputs and outputs and the future reference trajectory. In the presence of inequality constraints the solution has to be obtained by more computationally taxing numerical algorithms. The size of the optimization problems depends on the number of variables and on the prediction horizons used and usually turn out to be relatively modest optimization problems which do not require sophisticated computer codes to be solved. However the amount of time needed for the constrained and robust cases can be various orders of magnitude higher than that needed for the unconstrained case and the bandwidth of the process to which constrained MPC can be applied is considerably reduced.

Notice that the MPC strategy is very similar to the control strategy used in driving a car. The driver knows the desired reference trajectory for a finite control horizon, and by taking into account the car characteristics (mental model of the car) decides which control actions (accelerator, brakes and steering) to take in order to follow the desired trajectory. Only the first control actions are taken at each instant, and the procedure is again repeated for the next control decisions in a receding horizon fashion. Notice that when using classical control schemes, such as PIDs, the control actions are taken based on past errors. If the car driving analogy is extended, as has been done by one of the commercial MPC vendors (SCAP) [71] in their publicity, the PID way of driving a car would be equivalent to driving the car just using the mirror as shown in figure 1.3. This analogy is not totally fair with PIDs, because more information (the reference trajectory) is used by MPC. Notice that if a future point in the desired reference trajectory is used as the setpoint for the PID, the differences between both control strategies would not seem so abysmal.

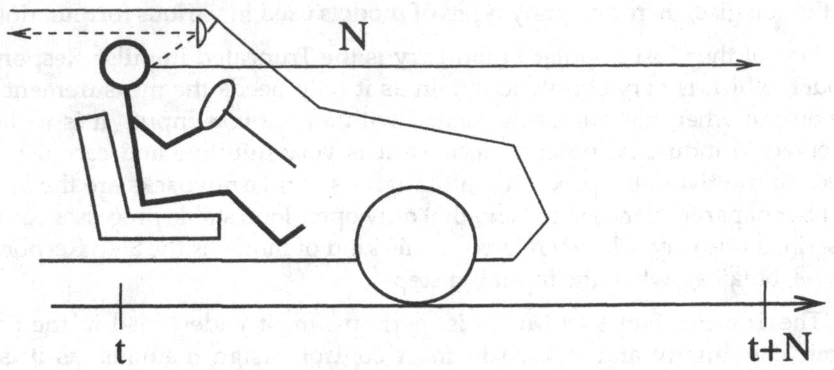

Figure 1.3: MPC analogy

1.2 Historical Perspective

From the end of the 1970's various articles appeared showing an incipient interest in MPC in the industry, principally the Richalet *et al.* publications [105][104] presenting Model Predictive Heuristic Control (MPHC) (later known as Model Algorithmic Control (MAC)) and those of Cutler and Ramakter [37] with Dynamic Matrix Control (DMC). A dynamic process model is explicitly used in both algorithms (impulse response in the first one and step response in the second) in order to predict the effect of the future control actions at the output; these are determined by minimizing the predicted error subject to operational restrictions. The optimization is repeated at each sampling period with up to date information about the process. These formulations were heuristic and algorithmic and took advantage of the increasing potential of digital computers at the time.

These controllers were closely related to the minimum time optimal control problem and to linear programming [129]. The receding horizon principle, one of the central ideas of MPC, was proposed by Propoi as long ago as 1963 [97], within the frame of "open-loop optimal feedback", which was extensively dealt with in the seventies.

MPC quickly became popular, particularly in chemical process industries, due to the simplicity of the algorithm and to the use of the impulse or step response model which, although possessing many more parameters than the formulations in the state space or input-output domain, is usually preferred as being more intuitive and requiring less *a priori* information for its identification. A complete report of its application in the petrochemical sector during the eighties can be found in [43]. The majority of these applications were carried out on multivariable systems including constraints. In spite of this success, these formulations lacked formal theories providing stability and robustness results; in fact, the finite-horizon case seemed too difficult to analyze apart from in very specific cases.

Another line of work arose independently around adaptive control ideas, developing strategies essentially for monovariable processes formulated with input/output models. Peterka's Predictor-Based Self-Tuning Control [94] can be included here. It was designed to minimize, for the most recent predicted values, the expected value of a quadratic criterion on a given control horizon (finite or asymptotically infinite) or Ydstie's Extended Horizon Adaptive Control (EHAC) [125]. This method tries to keep the future output (calculated by a Diophantine equation) close to the reference at a period of time after the process delay and permits different strategies. Extended Prediction Self Adaptive Control (EPSAC) by De Keyser *et al.* [55] proposes a constant control signal starting from the present moment[2] while using a sub-optimal predictor instead of solving a Diophantine equation. Generalized Predictive Control (GPC) developed by Clarke *et al.* in 1987 [34] also appears within this context. This uses ideas from Generalized Minimum Variance (GMV) [32] and is perhaps one of the most popular methods at the moment and will be the object of detailed study in following chapters. There are numerous predictive controller formulations based on the same common ideas, amongst which can be included: Multistep Multivariable Adaptive Control (MUSMAR) [47], Multipredictor Receding Horizon Adaptive Control (MURHAC) [66], Predictive Functional Control (PFC) [103] or Unified Predictive Control (UPC) [115].

MPC has also been formulated in the state space context [76]. This not only allows for the use of well known theorems of the state space theory, but also facilitates their generalization to more complex cases such as systems with stochastic disturbances and noise in the measured variables. By extending the step response model and using known state estimation techniques, processes with integrators can also be treated. The state estimation techniques arising from stochastic optimal control can be used for predictions without adding additional complications [62]. This perspective leads to simple tuning rules for stability and robustness: the MPC controller can be interpreted as being a compensator based on a state observer and its stability, performance and robustness are determined by the poles of the observer (which can be directly fixed by adjustable parameters) and the poles of the regulator (determined by the horizons, weightings, etc.). An analysis of the inherent characteristics of all the MPC algorithms (especially of the GPC) from the point of view of the gaussian quadratic optimal linear theory can be found in the book by Bitmead *et al.* [12].

Although the first works on GPC (Mohtadi [74]) proved some specific stability theorems using state-space relationships and studied the influence of filter polynomials on robustness improvement, the original lack of general

[2]Note that due to the receding horizon the real signal need not be kept constant.

stability results for finite horizon receding controllers was recognized as a drawback. Because of this, a fresh line of work on new predictive control methods with guaranteed stability appeared in the nineties. Two methods: CRHPC (Clarke and Scattolini [30]) and SIORHC (Mosca *et al.* [79]) were independently developed and were proved to be stable by imposing end-point equality constraints on the output after a finite horizon. Bearing in mind the same objective, Kouvaritakis *et al.* [58] presented *stable* GPC, a formulation which guarantees closed-loop stability by stabilizing the process prior to the minimization of the objective function.

Very impressive results have been obtained lately for what seemed to be a problem too difficult to tackle, that of the stability of constrained receding horizon controllers [100], [130], [108]. New results have also been obtained by using robust control design approaches by Campo and Morari [25] and Allwright [3]. The key idea is to take into account uncertainties about the process in an explicit manner and to design MPC in order to optimize the objective function for the worst situation of the uncertainties. These challenging results allow one to think that MPC will experience an even greater dissemination both in the academic world and within the control practitioner community. In this context, one of the leading manufacturers of distributed control equipment, Honeywell, has incorporated Robust Multivariable Predictive Control (RMPC^{TM}) into their TDC 3000 control system and announces it as containing several *break-thrus* in technology.

1.3 Industrial Technology

This section is focused on those predictive control technologies that have great impact on the industrial world and are commercially available, dealing with several topics such as a short application summary and the limitations of the existing technology.

Although there are companies that make use of technology developed in-house, that is not offered externally, the ones listed below can be considered representative of the current state of the art of Model Predictive Control technology. Their product names and acronyms are:

- DMC Corp. : Dynamic Matrix Control (DMC).
- Adersa: Identification and Command (IDCOM), Hierarchical Constraint Control (HIECON) and Predictive Functional Control (PFC).
- Honeywell Profimatics: Robust Model Predictive Control Technology (RMPCT) and Predictive Control Technology (PCT).
- Setpoint Inc. : Setpoint Multivariable Control Architecture (SMCA) and IDCOM-M (multivariable).

- Treiber Controls: Optimum Predictive Control (OPC).
- SCAP Europa: Adaptive Predictive Control System (APCS).

Some of these algorithms will be treated in more detail in following chapters. Notice that each product is not the algorithm alone, but it is accompanied by additional packages, usually identification or plant test packages.

Qin and Badgwell [98] present the results obtained from an industrial survey in 1997. The total number of applications reported in the paper is over 2200 and is quickly increasing . The majority of applications (67 %) are in the area of refining, one of the original application fields of MPC, where it has a solid background. An important number of applications can be found in petrochemicals and chemicals. Significant growth areas include pulp and paper, food processing, aerospace and automotive industries. Other areas such as gas, utility, furnaces or mining and metallurgy also appear in the report. Some applications in the cement industry or in pulp factories can be found in [71].

DMC corporation reports the largest total number of applications (26 %), while the four following vendors share the rest almost equally.

The existing industrial MPC technology has several limitations, as pointed out by Muske and Rawlings [82]. The most outstanding ones are:

- Over-parameterized models: most of the commercial products use the step or impulse response model of the plant, that are known to be over-parameterized. For instance, a first order process can be described by a transfer function model using only three parameters (gain, time constant and deadtime) whilst a step response model will require more than 30 coefficients to describe the same dynamics. Besides, these models are not valid for unstable processes. These problems can be overcome by using an auto-regressive parametric model.
- Tuning: the tuning procedure is not clearly defined since the trade-off between tuning parameters and closed loop behaviour is generally not very clear. Tuning in the presence of constraints may be even more difficult, and even for the nominal case, is not easy to guarantee closed loop stability; that is why so much effort must be spent on prior simulations. The feasibility of the problem is one of most challenging topics of MPC nowadays, and it will be treated in detail in chapter 7.
- Sub-optimality of the dynamic optimization: several packages provide sub-optimal solutions to the minimization of the cost function in order to speed up the solution time. It can be accepted in high speed applications (tracking systems) where solving the problem at every sampling time may not be feasible, but it is difficult to justify for process control applications unless it can be shown that the sub-optimal solution is always very nearly optimal.

- Model uncertainty: although model identification packages provide estimates of model uncertainty, only one product (RMPCT) uses this information in the control design. All other controllers can be detuned in order to improve robustness, although the relation between performance and robustness is not very clear.
- Constant disturbance assumption: although perhaps the most reasonable assumption is to consider that the output disturbance will remain constant in the future, better feedback would be possible if the distribution of the disturbance could be characterized more carefully.
- Analysis: a systematic analysis of stability and robustness properties of MPC is not possible in its original finite horizon formulation. The control law is in general time-varying and cannot be represented in the standard closed loop form, especially in the constrained case.

The technology is continually evolving and the next generation will have to face new challenges in open topics such as model identification, unmeasured disturbance estimation and prediction, systematic treatment of modelling error and uncertainty or such an open field as nonlinear model predictive control.

1.4 Outline of the Chapters

The book aims to study the most important issues of MPC with regards to its application to process control. In order to achieve this objective, it is organized as follows.

Chapter 2 describes the main elements that appear in any MPC formulation and reviews the best known methods. A brief reviewal of the the most outstanding methods is made. Chapter 3 focuses on commercial Model Predictive controllers. Because of its popularity, Generalized Predictive Control (GPC) is treated in greater detail in chapter 4. Two related methods which have shown good stability properties (CRHPC and SGPC) are also described.

Chapter 5 shows how GPC can easily be applied to a wide variety of plants in the process industry by using some Ziegler-Nichols types of tuning rules. By using these, the implementation of GPC is considerably simplified, and the computational burden and time that the implementation of GPC may bear, especially for the adaptive case, is avoided. The rules have been obtained for plants that can be modelled by the reaction curve method and plants having an integrating term, that is, most of the plants in the process industry. The robustness of the method is studied. In order to do this, both structured and unstructured uncertainties are considered. The closed loop is studied, defining the uncertainty limits that preserve stability of the real process when it is being controlled by a GPC designed for the nominal model.

The way of implementing GPC on multivariable processes, which can often be found in industry, is treated in chapter 6. Some examples dealing with implementation topics such as dead times are provided.

Although in practice all processes are subject to constraints, most of the existing controllers do not consider them explicitly. One of the advantages of MPC is that constraints can be dealt with explicitly. Chapter 7 is dedicated to showing how MPC can be implemented on processes subject to constraints. Although constraints play an important role in industrial practice, they are not considered in many formulations. The minimization of the objective function can no longer be done explicitly and a numerical solution is necessary. Existing numerical techniques are revised and some examples, as well as algorithms, are included.

Chapter 8 deals with a robust implementation of MPC. Although a robustness analysis was performed in chapter 5 for GPC of processes that can be described by the reaction curve method, this chapter indicates how MPC can be implemented by explicitly taking into account model inaccuracies or uncertainties. The controller is designed to minimize the objective function for the worst situation.

This book could not end without presenting some practical implementations of GPC. Chapter 9 presents results obtained in different plants. First, a real application of GPC to a Solar Power Plant is presented. This process presents changing perturbations that make it suitable for an adaptive control policy. The same controller is developed on a commercial distributed control system and applied to a pilot plant. Operating results show that a technique that has sometimes been rejected by practitioners because of its complexity can easily be programmed in any standard control system, obtaining better results and being as easy to use as traditional PID controllers. Finally the application of a GPC to a diffusion process of a sugar factory is presented.

Chapter 2

Model Based Predictive Controllers

This chapter describes the elements that are common to all Model-Based Predictive controllers, showing the various alternatives that are used in the different implementations. Some of the most popular methods will later be reviewed in order to demonstrate their most outstanding characteristics.

The last part of the chapter is dedicated to nonlinear predictive control, showing recent developments and new trends.

2.1 MPC Elements

All the MPC algorithms possess common elements and different options can be chosen for each one of these elements giving rise to different algorithms. These elements are:

- Prediction Model
- Objective Function
- Obtaining the control law

2.1.1 Prediction Model

The model is the corner-stone of MPC; a complete design should include the necessary mechanisms for obtaining the best possible model, which should be complete enough to fully capture the process dynamics and should also be capable of allowing the predictions to be calculated and at the same time, to be intuitive and to permit theoretic analysis. The use of the process model is determined by the necessity to calculate the predicted output at future instants $\hat{y}(t+k \mid t)$. The different strategies of MPC can use various models to represent the relationship between the outputs and the measurable inputs, some of which are manipulated variables and others can be considered to be measurable disturbances which can be compensated for by feedforward action. A disturbance model can also be taken into account in order to describe the behaviour which is not reflected by the process model, including the effect of non-measurable inputs, noise and model errors. The model can be separated into two parts: the actual process model and the disturbances model. Both parts are needed for the prediction.

Process Model

Practically every possible form of modelling a process appears in a given MPC formulation, the following being the most commonly used:

- Impulse response. Also known as weighting sequence or convolution model, it appears in MAC and as a special case in GPC and EPSAC. The output is related to the input by the equation

$$y(t) = \sum_{i=1}^{\infty} h_i u(t-i)$$

where h_i is the sampled output when the process is excited by a unitary impulse (see figure 2.1a). This sum is truncated and only N values are considered (thus only stable processes without integrators can be represented), having

$$y(t) = \sum_{i=1}^{N} h_i u(t-i) = H(z^{-1})u(t) \tag{2.1}$$

where $H(z^{-1}) = h_1 z^{-1} + h_2 z^{-2} + \cdots + h_N z^{-N}$, being z^{-1} the backward shift operator. Another inconvenience of this method is the large number of parameters necessary, as N is usually a high value (in the order of 40-50). The prediction will be given by:

$$\hat{y}(t+k \mid t) = \sum_{i=1}^{N} h_i u(t+k-i \mid t) = H(z^{-1})u(t+k \mid t)$$

This method is widely accepted in industrial practice because it is very intuitive and clearly reflects the influence of each manipulated variable on a determined output. Note that if the process is multivariable, the different outputs will reflect the effect of the m inputs in the following

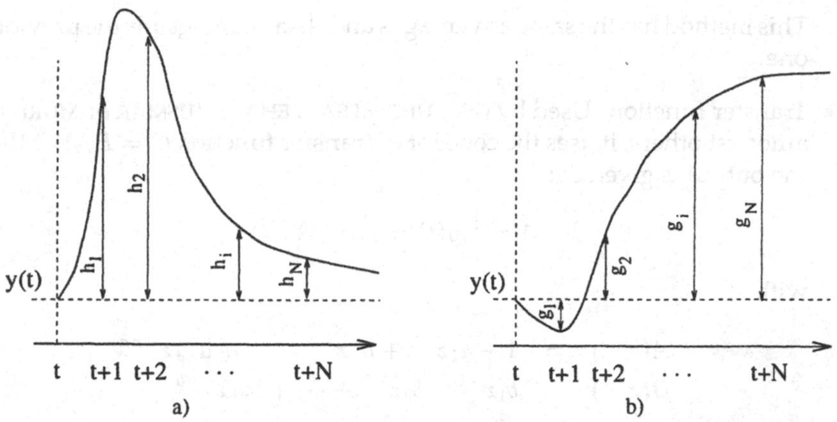

Figure 2.1: Impulse and step response

way:

$$y_j(t) = \sum_{k=1}^{m} \sum_{i=1}^{N} h_i^{kj} u^k(t-i)$$

One great advantage of this method is that no prior information about the process is needed, so that the identification process is simplified and at the same time it allows complex dynamics such as non-minimum phase or delays to be described with ease.

- Step response. Used by DMC and its variants, it is very similar to the previous one except that now the input signal is a step. For stable systems the truncated response is given by:

$$y(t) = y_0 + \sum_{i=1}^{N} g_i \, \Delta \, u(t-i) = y_0 + G(z^{-1})(1-z^{-1})u(t) \qquad (2.2)$$

where g_i are the sampled output values for the step input and $\Delta u(t) = u(t) - u(t-1)$ as shown in figure 2.1b. The value of y_0 can be taken to be 0 without loss of generality, so that the predictor will be:

$$\hat{y}(t+k \mid t) = \sum_{i=1}^{N} g_i \, \Delta \, u(t+k-i \mid t)$$

As an impulse can be considered as the difference between two steps with a lag of one sampling period, it can be written for a linear system that:

$$h_i = g_i - g_{i-1} \qquad\qquad g_i = \sum_{j=1}^{i} h_j$$

This method has the same advantages and disadvantages as the previous one.

- Transfer function. Used by GPC, UPC, EPSAC, EHAC, MUSMAR or MURHAC amongst others, it uses the concept of transfer function $G = B/A$ so that the output is given by:

$$A(z^{-1})y(t) = B(z^{-1})u(t)$$

with

$$\begin{aligned} A(z^{-1}) &= 1 + a_1 z^{-1} + a_2 z^{-2} + \cdots + a_{na} z^{-na} \\ B(z^{-1}) &= b_1 z^{-1} + b_2 z^{-2} + \cdots + b_{nb} z^{-nb} \end{aligned}$$

Thus the prediction is given by

$$\hat{y}(t+k \mid t) = \frac{B(z^{-1})}{A(z^{-1})} u(t+k \mid t)$$

This representation is also valid for unstable processes and has the advantage that it only needs a few parameters, although *a priori* knowledge of the process is fundamental, especially that of the order of the A and B polynomials.

- State space. Used in PFC for example, it has the following representation:

$$\begin{aligned} x(t) &= Mx(t-1) + Nu(t-1) \\ y(t) &= Qx(t) \end{aligned}$$

x being the state and M, N and Q the matrices of the system, of input and output respectively. The prediction for this model is given by [8]

$$\hat{y}(t+k \mid t) = Q\hat{x}(t+k \mid t) = Q[M^k x(t) + \sum_{i=1}^{k} M^{i-1} N u(t+k-i \mid t)]$$

It has the advantage that it can be used for multivariable processes in a straightforward manner. The control law is simply the feedback of a linear combination of the state vector, although sometimes the state basis chosen has no physical meaning. The calculations may be complicated with an additional necessity of including an observer if the states are not accessible.

- Others. Non-linear models can also be used to represent the process but the problem of their use springs from the fact that they cause the optimization problem to be more complicated. Neural nets [118] as well as fuzzy logic [113] are other forms of representation used in some applications.

Disturbances Model

The choice of the model used to represent the disturbances is as important as the choice of the process model. A model widely used is the Controlled Auto-Regressive and Integrated Moving Average (CARIMA) in which the disturbances, that is, the differences between the measured output and the one calculated by the model, are given by

$$n(t) = \frac{C(z^{-1})e(t)}{D(z^{-1})}$$

where the polynomial $D(z^{-1})$ explicitly includes the integrator $\Delta = 1 - z^{-1}$, $e(t)$ is a white noise of zero mean and the polynomial C is normally considered to equal one. This model is considered appropriate for two types of disturbances, random changes occurring at random instants (for example changes in the quality of the material) and "Brownian motion" and it is used directly in GPC, EPSAC, EHAC and UPC and with slight variations in other methods. Note that by including an integrator an offset-free steady state control is achieved.

Using the Diophantine equation

$$1 = E_k(z^{-1})D(z^{-1}) + z^{-k}F_k(z^{-1}) \tag{2.3}$$

one has

$$n(t) = E_k(z^{-1})e(t) + z^{-k}\frac{F_k(z^{-1})}{D(z^{-1})}e(t) \qquad n(t+k) = E_k(z^{-1})e(t+k) + F_k(z^{-1})n(t)$$

and the prediction will be

$$\hat{n}(t+k \mid t) = F_k(z^{-1})n(t) \tag{2.4}$$

If equation (2.4) is combined with a transfer function model (like the one used in GPC), making $D(z^{-1}) = A(z^{-1})(1 - z^{-1})$, the output prediction can be obtained:

$$\begin{aligned}
\hat{y}(t+k \mid t) &= \frac{B(z^{-1})}{A(z^{-1})}u(t+k \mid t) + F_k(z^{-1})\left(y(t) - \frac{B(z^{-1})}{A(z^{-1})}u(t)\right) \\
\hat{y}(t+k \mid t) &= F_k(z^{-1})y(t) + \frac{B(z^{-1})}{A(z^{-1})}(1 - z^{-k}F_k(z^{-1}))u(t+k \mid t)
\end{aligned}$$

and using (2.3) the following expression is obtained for the k-step ahead predictor

$$\hat{y}(t+k \mid t) = F_k(z^{-1})y(t) + E_k(z^{-1})B(z^{-1}) \,\Delta\, u(t+k \mid t)$$

In the particular case of ARIMA the constant disturbance

$$n(t) = \frac{e(t)}{1 - z^{-1}}$$

can be included whose best predictions will be $\hat{n}(t+k \mid t) = n(t)$. This disturbance model, together with the step response model is the one used on DMC and related methods.

An extension of the above are the drift disturbance used in PFC

$$n(t) = \frac{e(t)}{(1 - z^{-1})^2}$$

with $\hat{n}(t+k \mid t) = n(t) + (n(t) - n(t-1))k$ being the optimum prediction. Other polynomial models of high order can likewise be used.

Free and forced response

A typical characteristic of most MPC is the use of *free* and *forced* response concepts. The idea is to express the control sequence as the addition of the two signals:

$$u(t) = u_f(t) + u_c(t)$$

The signal $u_f(t)$ corresponds to the past inputs and is kept constant and equal to the last value of the manipulated variable in future time instants. That is:

$$\begin{aligned} u_f(t-j) &= u(t-j) \text{ for } j = 1, 2, \cdots \\ u_f(t+j) &= u(t-1) \text{ for } j = 0, 1, 2, \cdots \end{aligned}$$

The signal $u_c(t)$ is made equal to zero in the past and equal to the next control moves in the future. That is:

$$u_c(t-j) = 0 \text{ for } j = 1, 2, \cdots$$
$$u_c(t+j) = u(t+j) - u(t-1) \text{ for } j = 0, 1, 2, \cdots$$

The prediction of the output sequence is separated into two parts as can be seen in figure 2.2. One of them ($y_f(t)$), the *free* response, corresponds to the prediction of the output when the process manipulated variable is made equal to $u_f(t)$, and the other, the *forced* response ($y_c(t)$), corresponds to the prediction of the process output when the control sequence is made equal to $u_c(t)$. The *free* response corresponds to the evolution of the process due to its present state, while the forced response is due to the future control moves.

2.1.2 Objective Function

The various MPC algorithms propose different cost functions for obtaining the control law. The general aim is that the future output (y) on the considered horizon should follow a determined reference signal (w) and, at the same time, the control effort (Δu) necessary for doing so should be penalized. The general expression for such an objective function will be:

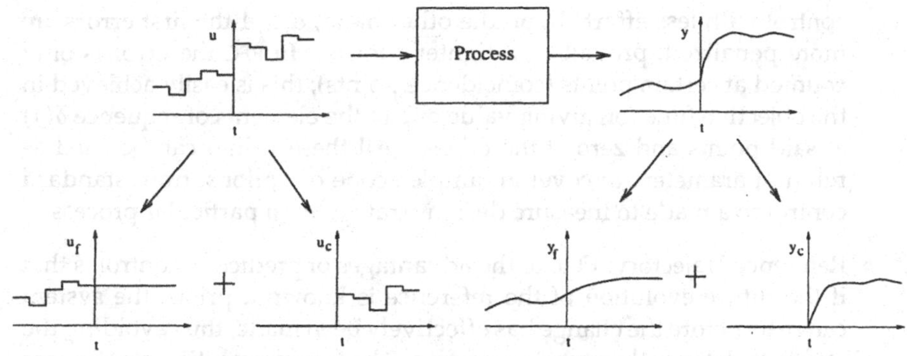

Figure 2.2: Free and forced responses

$$J(N_1, N_2, N_u) = \sum_{j=N_1}^{N_2} \delta(j)[\hat{y}(t+j \mid t)-w(t+j)]^2+\sum_{j=1}^{N_u} \lambda(j)[\Delta u(t+j-1)]^2 \quad (2.5)$$

In some methods the second term, that considers the control effort, is not taken into account, whilst in others (UPC) the values of the control signal (not its increments) also appear directly. In the cost function it is possible to consider:

- Parameters: N_1 and N_2 are the minimum and maximum cost horizons and N_u is the control horizon, which does not necessarily have to coincide with the maximum horizon, as will be seen later. The meaning of N_1 and N_2 is rather intuitive. They mark the limits of the instants in which it is desirable for the output to follow the reference. Thus, if a high value of N_1 is taken, it is because it is of no importance if there are errors in the first instants. This will originate a smooth response of the process. Note that in processes with dead time d there is no reason for N_1 to be less than d because the output will not begin to evolve until instant $t+d$. Also if the process is non-minimum phase, this parameter will allow the first instants of inverse response to be eliminated from the objective function. The coefficients $\delta(j)$ and $\lambda(j)$ are sequences that consider the future behaviour, usually constant values or exponential sequences are considered. For example it is possible to obtain an exponential weight of $\delta(j)$ along the horizon by using:

  $$\delta(j) = \alpha^{N_2-j}$$

  If α is given a value between 0 and 1, the errors farthest from instant t are penalized more than those nearest to it, giving rise to smoother control with less effort. If, on the other hand, $\alpha > 1$ the first errors are more penalized, provoking a tighter control. In PFC the error is only counted at certain points (coincidence points); this is easily achieved in the objective function giving value one to the elements of sequence $\delta(j)$ at said points and zero at the others. All these values can be used as tuning parameters to cover an ample scope of options, from standard control to a made to measure design strategy for a particular process.

- Reference Trajectory: One of the advantages of predictive control is that if the future evolution of the reference is known a priori, the system can react before the change has effectively been made, thus avoiding the effects of delay in the process response. The future evolution of reference $r(t+k)$ is known beforehand in many applications, such as robotics, servos or *batch* processes; in other applications a noticeable improvement in performance can be obtained even though the reference is constant by simply knowing the instant when the value changes and getting ahead of this circumstance. In the minimization (2.5), the majority of methods usually use a reference trajectory $w(t+k)$ which does not necessarily have to coincide with the real reference. It is normally a smooth approximation from the current value of the output $y(t)$ towards the known reference by means of the first order system:

$$w(t) = y(t) \qquad w(t+k) = \alpha w(t+k-1) + (1-\alpha)r(t+k) \qquad k = 1 \dots N \tag{2.6}$$

  α is a parameter between 0 and 1 (the closer to 1 the smoother the approximation) that constitutes an adjustable value that will influence the dynamic response of the system. In figure 2.3 the form of trajectory is shown from when the reference $r(t+k)$ is constant and for two different values of α; small values of this parameter provide fast tracking (w_1), if it is increased then the reference trajectory becomes w_2 giving rise to a smoother response.

  Another strategy is the one used in PFC, which is useful for variable setpoints:

$$w(t+k) = r(t+k) - \alpha^k(y(t) - r(t))$$

  The reference trajectory can be used to specify closed loop behaviour; this idea is used in GPC or EPSAC defining an auxiliary output

$$\psi(t) = P(z^{-1})y(t)$$

  the error in the objective function is given by $\psi(t+k) - w(t+k)$. The filter $P(z^{-1})$ has unit static gain and the generation of a reference trajectory with dynamics defined by $1/P(z^{-1})$ and an initial value of that of the measured output is achieved. In [33] it is demonstrated that if a *dead-beat* control in $\psi(t)$ is achieved, so that

$$\psi(t) = B(z^{-1})w(t)$$

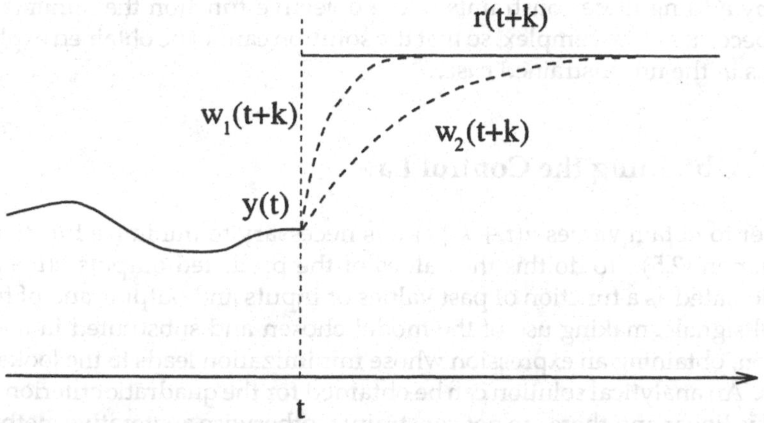

Figure 2.3: Reference Trajectory

$B(z^{-1})$ being a determined polynomial with unit gain, the closed loop response of the process will clearly be:

$$y(t) = \frac{B(z^{-1})}{P(z^{-1})} w(t)$$

that, in short, is equivalent to placing the closed loop poles at the zeros of design polynomial $P(z^{-1})$.

- Constraints: In practice all processes are subject to constraints. The actuators have a limited field of action as well as a determined slew rate, as is the case of the valves, limited by the positions of totally open or closed and by the response rate. Constructive reasons, safety or environmental ones or even the sensor scopes themselves can cause limits in the process variables such as levels in tanks, flows in piping or maximum temperatures and pressures; moreover, the operational conditions are normally defined by the intersection of certain constraints for basically economic reasons, so that the control system will operate close to the boundaries. All of this makes the introduction of constraints in the function to be minimized necessary. Many predictive algorithms intrinsically take into account constraints (MAC, DMC) and have therefore been very successful in industry, whilst others can incorporate them *a posteriori* (GPC)[16]. Normally bounds in the amplitude and in the slew rate of the control signal and limits in the output will be considered:

$$\begin{array}{rcll} u_{min} \leq & u(t) & \leq u_{max} & \forall t \\ du_{min} \leq & u(t) - u(t-1) & \leq du_{max} & \forall t \\ y_{min} \leq & y(t) & \leq y_{max} & \forall t \end{array}$$

by adding these constraints to the objective function the minimization becomes more complex, so that the solution cannot be obtained explicitly as in the unconstrained case.

2.1.3 Obtaining the Control Law

In order to obtain values $u(t+k \mid t)$ it is necessary to minimize functional J of equation (2.5). To do this the values of the predicted outputs $\hat{y}(t+k \mid t)$ are calculated as a function of past values of inputs and outputs and of future control signals, making use of the model chosen and substituted in the cost function, obtaining an expression whose minimization leads to the looked for values. An analytical solution can be obtained for the quadratic criterion if the model is linear and there are not constraints, otherwise an iterative method of optimization should be used. Whatever the method, obtaining the solution is not easy because there will be $N_2 - N_1 + 1$ independent variables, a value which can be high (in the order of 10 to 30). In order to reduce this degree of freedom a certain structure may be imposed on the control law. Furthermore, it has been found [54] that this structuralizing of the control law produces an improvement in robustness and in the general behaviour of the system, basically due to the fact that allowing the free evolution of the manipulated variables (without being structured) may lead to undesirable high frequency control signals and at the worst to instability. This control law structure is sometimes imposed by the use of the control horizon concept (N_u) used in DMC, GPC, EPSAC and EHAC, that consists of considering that after a certain interval $N_u < N_2$ there is no variation in the proposed control signals, that is:

$$\Delta u(t+j-1) = 0 \qquad j > N_u$$

which is equivalent to giving infinite weights to the changes in the control from a certain instant. The extreme case would be to consider N_u equal to 1 with which all future actions would be equal to $u(t)$[1]. Another way of structuring the control law is by using base functions, a procedure used in PFC and which consists of representing the control signal as a linear combination of certain predetermined base functions:

$$u(t+k) = \sum_{i=1}^{n} \mu_i(t) B_i(k) \tag{2.7}$$

the B_i are chosen according to the nature of the process and the reference, normally being polynomial type

$$B_0 = 1 \qquad B_1 = k \qquad B_2 = k^2 \ldots$$

[1]Remember that due to the receding horizon, the control signal is recalculated in the following sample.

As has previously been indicated, an explicit solution does not exist in the presence of constraints, so that quadratic programming methods have to be used (these methods will be studied in chapter 7). However, an explicit solution does exist for certain types of constraints, for example when the condition that the output attains the reference value at a determined instant is imposed, this method is used in Constrained Receding Horizon Predictive Control, (CRHPC) [30], which is very similar to GPC and which guarantees stability results.

2.2 Review of some MPC Algorithms

Some of the most popular methods will now be reviewed in order to demonstrate their most outstanding characteristics. Comparative studies can be found in [56], [43], [59] and [98]. The methods considered to be most representative, DMC, MAC, GPC, PFC, EPSAC and EHAC will briefly be dealt with. Some of them will be studied in greater detail in following chapters. Chapter 3 is devoted to DMC, MAC and PFC while GPC and its derivations are treated in chapter 4.

Dynamic Matrix Control

It uses the step response (2.2) to model the process, only taking into account the N first terms, therefore assuming the process to be stable and without integrators. As regards the disturbances, their value will be considered to be the same as at instant t along all the horizon, that is, to be equal to the measured value of the output (y_m) minus the one estimated by the model ($\hat{y}(t \mid t)$).

$$\hat{n}(t+k \mid t) = \hat{n}(t \mid t) = y_m(t) - \hat{y}(t \mid t)$$

and therefore the predicted value of the output will be

$$\hat{y}(t+k \mid t) = \sum_{i=1}^{k} g_i \, \Delta\, u(t+k-i) + \sum_{i=k+1}^{N} g_i \, \Delta\, u(t+k-i) + \hat{n}(t+k \mid t)$$

where the first term contains the future control actions to be calculated, the second one contains past values of the control actions and is therefore known, and the last one represents the disturbances. The cost function may consider future errors only, or it can include the control effort, in which case, it presents the generic form (2.5). One of the characteristics of this method making it very popular in the industry is the addition of constraints, in such a way that equations of the form:

$$\sum_{i=1}^{N} C_{yi}^{j} \hat{y}(t+k \mid t) + C_{ui}^{j} u(t+k-i) + c^{j} \leq 0 \qquad j = 1 \dots N_c$$

must be added to the minimization. Optimization (numerical because of the presence of constraints) is carried out at each sampling instant and the value of $u(t)$ is sent to the process as is normally done in all MPC methods. The inconveniences of this method are on one hand the size of the process model required and on the other hand the inability to work with unstable processes.

Model Algorithmic Control

Also known as Model Predictive Heuristic Control, it is marketed under the name of IDCOM (Identification-Command). It is very similar to the previous method with a few differences. Firstly, it uses an impulse response model (2.1) valid only for stable processes, in which the value of $u(t)$ appears instead of $\Delta u(t)$. Furthermore, it makes no use of the control horizon concept so that in the calculations as many control signals as future outputs appear. It introduces a reference trajectory as a first order system which evolves from the actual output to the setpoint according to a determined time constant, following expression (2.6). The variance of the error between this trajectory and the output is what one aims at minimizing in the objective function. The disturbances can be treated as in DMC or their estimations can be carried out by the following recursive expression:

$$\hat{n}(t+k \mid t) = \alpha \hat{n}(t+k-1 \mid t) + (1-\alpha)(y_m(t) - \hat{y}(t \mid t))$$

with $\hat{n}(t \mid t) = 0$. α is an adjustable parameter $(0 \leq \alpha < 1)$ closely related to the response time, the bandwidth and the robustness of the closed loop system [43]. It also takes into account constraints in the actuators as well as in the internal variables or secondary outputs. Various algorithms can be used for optimizing in presence of constraints, from the ones presented initially by Richalet *et al.* that can also be used for identifying the impulse response, to others that are shown in [110].

Predictive Functional Control

This controller was developed by Richalet [101] for the case of fast processes. It uses a state space model of the process and allows for nonlinear and unstable linear internal models. Nonlinear dynamics can be entered in the form of a nonlinear state space model. PFC has two distinctive characteristics: the use of *coincidence points* and *basis functions.*

The concept of coincidence points is used to simplify the calculation by considering only a subset of points in the prediction horizon. The desired and the predicted future outputs are required to coincide at these points, not in the whole prediction horizon.

The controller parameterizes the control signal using a set of polynomial basis functions, as given by equation (2.7). This allows a relative complex input profile to be specified over a large horizon using a small number of parameters. Choosing the family of basis functions establishes many of the features of the computed input profile. These functions can be selected in order to follow a polynomial setpoint with no lag, an important feature for mechanical servo control applications.

The cost function to be minimized is:

$$J = \sum_{j=1}^{n_H} [\hat{y}(t+h_j) - w(t+h_j)]^2$$

where $w(t+j)$ is usually a first order approach to the known reference.

The PFC algorithm can also accommodate maximum and minimum input acceleration constraints which are useful in mechanical servo control applications.

Extended Prediction Self Adaptive Control

The implementation of EPSAC is different to the previous methods. For predicting, the process is modelled by the transfer function

$$A(z^{-1})y(t) = B(z^{-1})u(t-d) + v(t)$$

where d is the delay and $v(t)$ the disturbance. The model can be extended by a term $D(z^{-1})d(t)$, with $d(t)$ being a measurable disturbance in order to include *feedforward* effect. Using this method the prediction is obtained as shown in [56]. One characteristic of the method is that the control law structure is very simple, as it is reduced to considering that the control signal is going to stay constant from instant t, that is $\Delta u(t+k) = 0$ for $k > 0$. In short: the control horizon is reduced to 1 and therefore the calculation is reduced to one single value: $u(t)$. To obtain this value a cost function is used of the form:

$$\sum_{k=d}^{N} \gamma(k)[w(t+k) - P(z^{-1})\hat{y}(t+k \mid t)]^2$$

$P(z^{-1})$ being a design polynomial with unit static gain and factor $\gamma(k)$ being a weighting sequence, similar to those appearing in (2.5). The control signal can be calculated analytically (which is an advantage over the previous methods) in the form:

$$u(t) = \frac{\sum_{k=d}^{N} h_k \gamma(k)[w(t+k) - P(z^{-1})\hat{y}(t+k \mid t)]}{\sum_{k=d}^{N} \gamma(k) h_k^2}$$

h_k being the discrete impulse response of the system.

Extended Horizon Adaptive Control

This formulation considers the process modelled by its transfer function without taking a model of the disturbances into account:

$$A(z^{-1})y(t) = B(z^{-1})u(t-d)$$

It aims at minimizing the discrepancy between the model and the reference at instant $t+N$: $\hat{y}(t+N \mid t) - w(t+N)$, with $N \geq d$. The solution to this problem is not unique (unless $N = d$)[125]; a possible strategy is to consider that the control horizon is 1, that is,

$$\Delta u(t+k-1) = 0 \qquad 1 < k \leq N-d$$

or to minimize the control effort:

$$J = \sum_{k=0}^{N-d} u^2(t+k)$$

There is an incremental version of EHAC that allows the disturbances in the load to be dealt with easily, it consists of considering

$$J = \sum_{k=0}^{N-d} \Delta u^2(t+k)$$

In this formulation a predictor of N steps in used as follows

$$\hat{y}(t+N \mid t) = y(t) + F(z^{-1}) \Delta y(t) + E(z^{-1})B(z^{-1}) \Delta u(t+N-d)$$

$E(z^{-1})$ and $F(z^{-1})$ being polynomials satisfying the equation

$$(1-z^{-1}) = A(z^{-1})E(z^{-1})(1-z^{-1}) + z^{-N}F(z^{-1})(1-z^{-1})$$

with the degree of E being equal to $N-1$. One advantage of this method is that a simple explicit solution can easily be obtained, resulting in

$$u(t) = u(t-1) + \frac{\alpha_0(w(t+N) - \hat{y}(t+N \mid t))}{\sum_{k=0}^{N-d} \alpha_i^2}$$

α_k being the coefficient corresponding to $\Delta u(t+k)$ in the prediction equation. Thus the control law only depends on the process parameters and can therefore easily be made self-tuning if it has an on-line identifier. As can be seen the only parameter of adjustment is the horizon of prediction N, which simplifies its use but provides little freedom for the design. One sees that the reference trajectory cannot be used because the error is only considered at one instant $(t+N)$, neither is it possible to ponder the control efforts at each point, so that certain frequencies in the performance cannot be eliminated.

Generalized Predictive Control

The output predictions of the Generalized Predictive Controller are based upon using a CARIMA model:

$$A(z^{-1})y(t) = B(z^{-1})z^{-d}\,u(t-1) + C(z^{-1})\frac{e(t)}{\Delta}$$

where the unmeasurable disturbance is given by a white noise coloured by $C(z^{-1})$. As its true value is difficult to know, this polynomial can be used for optimal disturbance rejection, although its role in robustness enhancement is more convincing.

The derivation of the optimal prediction is done by solving a Diophantine equation, whose solution can be found by an efficient recursive algorithm.

This algorithm, as with all algorithms using transfer function models, can easily be implemented in an adaptive mode by using an on-line estimation algorithm such as recursive least squares.

GPC uses a quadratic cost function of the form:

$$J(N_1, N_2, N_u) = \sum_{j=N_1}^{N_2} \delta(j)[\hat{y}(t+j \mid t) - w(t+j)]^2 + \sum_{j=1}^{N_u} \lambda(j)[\Delta u(t+j-1)]^2$$

where the weighting sequences $\delta(j)$ and $\lambda(j)$ are usually chosen constant or exponentially increasing and the reference trajectory $w(t+j)$ can be generated by a simple recursion which starts at the current output and tends exponentially to the setpoint.

The theoretical basis of the GPC algorithm has been widely studied, and it has been shown [33] that, for limiting cases of parameter choices, this algorithm is stable and also that well-known controllers such as mean level and deadbeat control are inherent in the GPC structure.

2.3 Nonlinear Predictive Control

Many processes are nonlinear to varying degrees of severity. Although in many situations the process will be operating in the neighborhood of a steady state, and therefore a linear representation will be adequate, there are some very important situations where this does not occur. On one hand, there are processes for which the nonlinearities are so severe (even in the vicinity of steady states) and so crucial to the closed loop stability, that a linear model is not sufficient. On the other hand, there are some processes that experience continuous transitions (start-ups, shutdowns, etc.) and spend a great deal of time away from a steady-state operating region, or even processes which are never in steady-state operation as is the case of batch processes where the whole operation is carried out in transient mode.

For these processes a linear control law will not be very effective, so nonlinear controllers will be essential for improved performance or simply for stable operation.

There is nothing in the basic concepts of MPC against the use of a nonlinear model. Therefore, the extension of MPC ideas to nonlinear processes is straightforward at least conceptually. However, this is not a trivial matter, and there are many open issues, such as:

- The availability of nonlinear models due to the lack of identification techniques for nonlinear processes.
- The computational complexities for solving the model predictive control of nonlinear processes.
- The lack of stability and robustness results for the case of nonlinear systems

Some of these problems are partially solved and MPC, with the use of nonlinear models, is becoming a field of intense research and will become more common as users demand higher performance.

2.3.1 Nonlinear Models

Developing adequate nonlinear empirical models may be very difficult and there is no model form that is clearly suitable to represent general nonlinear processes. Part of the success of standard MPC was due to the relative ease with which step and impulse responses or low order transfer functions could be obtained. Nonlinear models are much more difficult to construct, either from input/output data correlation or by the use of first principles from well known mass and energy conservation laws.

A major mathematical obstacle to a complete theory of nonlinear processes is the lack of a superposition principle for nonlinear systems. Because of this, the determination of models from process input/output data becomes a very difficult task. The amount of plant tests required to identify a nonlinear plant is much higher than that for a linear plant. If the plant is linear, in an ideal situation, only a step test has to be performed in order to know the step response of the plant. Because of the superposition principle, the response to a different size step can be obtained by multiplying the response to the step test by the ratio of both step sizes. This is not the case for nonlinear processes where tests with many different size steps must be performed in order to get the step response of the nonlinear plant. If the process is multivariable, the difference in the number of tests required is even higher. In general, if a linear system is tested with signals $u_1(t)$,$u_2(t)$,...,$u_n(t)$, and the corresponding responses are $y_1(t)$,$y_2(t)$,...,$y_n(t)$, the response to a signal which can be expressed as a linear combination of the tested input signals:

$$u(t) = \alpha_1 u_1(t) + \alpha_2 u_2(t) + \cdots + \alpha_n u_n(t)$$

is

$$y(t) = \alpha_1 y_1(t) + \alpha_2 y_2(t) + \cdots + \alpha_n y_n(t)$$

That is, a linear system does not need to be tested for any input signal sequence that is a linear combination of previously tested input sequences whilst this is not the case for a nonlinear system that must be analysed for all possible input signals.

If the deviation from linearity is not too large, some approximations can be made, which acknowledge that certain system characteristics change from operating point to operating point, but it assumes linearity in the neighborhood of a specific operating point. There are some approximations to the problem, such as a scheduled linearized MPC (see chapter 9), in which the model is linearized around several operating points and appropriately used with the linear MPC strategy as the process moves from one operating point to another. Also the extended linear MPC in which a basic linear model is used in combination with an explicit nonlinear model which captures the nonlinearities.

Different approaches exist that use Wiener models, others based on artificial neural networks, Volterra models, Hammerstein models, NARX models, fuzzy models, etc. which are more appropriate when the nonlinearities are more severe.

2.3.2 Techniques for Nonlinear Predictive Control

However, the choice of the appropriate model is not the only important issue. Using a nonlinear model changes the control problem from a convex quadratic program to a non-convex nonlinear problem, which is much more difficult to solve. Furthermore, in this situation there is no guarantee that the global optimum can be found, especially in real time control, when the optimum has to be obtained in a prescribed time.

Some solutions have been proposed to cope with this problem. In [20], the prediction of the process output is done by the addition of the *free* response (future response obtained if the system input is maintained at a constant value during the control and prediction horizons) obtained from a nonlinear model of the plant, and the *forced* response (the response obtained due to future control moves), computed from an incremental linear model of the plant. The predictions obtained this way are only an approximation because the superposition principle, which permits the mentioned division in *free* and *forced* response, only applies to linear systems. However, the approximation obtained this way is shown to be better than the ones obtained by using a linearized process model for computing both responses. If a quadratic cost function is used, the objective function is a quadratic function in the decision variables (future control moves) and the future control sequence can be computed in the unconstrained case, as the solution of a set of linear equations, leading to a simple control law. The only difference to standard linear MPC is that the *free* response is computed by a nonlinear model of the process. As the superposition principle does not hold for the nonlinear models, the approximation is only valid when the sequence of future control moves is small. Notice that this occurs when the process is operating in steady state with small perturbations. When the process is being changed from operating conditions or the external perturbations are high, the future control moves are usually high and the approximation is not very good.

A way to overcome this problem has been suggested in [53] for EPSAC. The key idea is that the manipulated variable sequence can be considered to be the addition of a base control sequence ($u_b(t+j)$) plus a sequence of increments of the manipulated variables ($u_i(t+j)$). That is:

$$u(t+j) = u_b(t+j) + u_i(t+j)$$

The process output j-step ahead prediction is computed as the sum of the response of the process ($y_b(t+j)$) due to the base input sequence plus the response of the process ($y_i(t+j)$) due to the future control increments on the base input sequence $u_i(t+j)$:

$$y(t+j) = y_b(t+j) + y_i(t+j)$$

As a nonlinear model is used to compute $y_b(t+j)$ while $y_i(t+j))$ is computed from a linear model of the plant, the cost function is quadratic in the decision variables ($u_i(t+j)$) and it can be solved by a QP algorithm as in standard MPC. The superposition principle does not hold for nonlinear processes and the process output generated this way and the process output generated by the nonlinear controller will only coincide in the case when the sequence of future control moves is zero.

If this is not the case, the base control sequence is made equal to the last base control sequence plus the optimal control increments found by the QP algorithm. The procedure is repeated until the sequence of future controls is driven close enough to zero.

The initial conditions for the base control sequence can first be made equal to the last control signal applied to the process. Notice that this is the case when computing the *free* response in standard MPC. A better initial guess can be made by making the base sequence equal to the optimal control sequence determined for the last sampling instant with the corresponding time shift.

The convergence conditions of the algorithm are very difficult to obtain as they depend on the severity of the nonlinear characteristics of the process, on past inputs and outputs, future reference sequence and perturbations.

In some cases, the nonlinear model can be transformed into a linear model by appropriate transformations. Consider for example the process described by the following state space model:

$$\begin{aligned} x(t+1) &= f(x(t), u(t)) \\ y(t) &= g(x(t)) \end{aligned}$$

The method consists of finding state and input transformation functions $z(t) = h(x(t))$ and $u(t) = p(x(t), v(t))$ such that:

$$\begin{aligned} z(t+1) &= \mathbf{A}z(t) + \mathbf{B}v(t)) \\ y(t) &= \mathbf{C}z(t) \end{aligned}$$

The method has two important drawbacks:

- The transformation functions $z(t) = h(x(t))$ and $u(t) = p(x(t), v(t))$ can be obtained for few cases.
- The constraints, which are usually linear, are transformed into a nonlinear set of constraints.

That is, even in the cases when the model can be linearized by suitable transformations, the problem is transformed from minimizing a nonlinear function (non quadratic) with linear constraints into minimizing a quadratic function with nonlinear constraints.

The general way of solving the problem is to use the complete nonlinear model of the plant for computing the output prediction. By doing this, a nonlinear constraint to the cost function minimization is added. Efficient QP algorithms cannot be used in these circumstances. However, the problem can be solved on line, in some cases, thanks to the rapid development of efficient nonlinear programming (NLP) algorithms capable of handling large numbers of variables and constraints. In spite of the significant advances which have been made in this field and in the computing power of modern control equipment, most of the issues related to nonlinear MPC are still unresolved and there is still much work to be done, from both theoretical analysis and practical implementation point of views.

Chapter 3

Commercial Model Predictive Control Schemes

As has been shown in previous chapters, there is a wide family of predictive controllers, each member of which being defined by the choice of the common elements such as the prediction model, the objective function and obtaining the control law.

This chapter is dedicated to overviewing some MPC algorithms widely used in industry. The first two belong to a major category of predictive control approaches, those which employ convolutional models, also called non-parametric methods. These approaches are based on step response or impulse response models and the most representative formulations are Dynamic Matrix Control (DMC) and Model Algorithmic Control (MAC). The third MPC algorithm presented in this chapter is Predictive Functional Control (PFC), which uses a set of basis functions to form the future control sequence.

It should be clear that the descriptions given here are necessarily incomplete, since only the general characteristics of each method are presented and each controller has proprietary features which are not known.

3.1 Dynamic Matrix Control

DMC was developed at the end of the seventies by Cutler and Ramaker [37] of Shell Oil Co. and has been widely accepted in the industrial world, mainly by petrochemical industries [98].

Nowadays DMC is something more than an algorithm and part of its success is due to the fact that the commercial product covers topics such as model identification or global plant optimization. In this section only the *standard* algorithm is analyzed, without addressing technical details such as software and hardware compatibilities, user interface requirements, personnel training or configuration and maintenance issues.

3.1.1 Prediction

The process model employed in this formulation is the step response of the plant, while the disturbance is considered to keep constant along the horizon. The procedure to obtain the predictions is as follows:

As a step response model is employed:

$$y(t) = \sum_{i=1}^{\infty} g_i \triangle u(t-i)$$

the predicted values along the horizon will be:

$$\hat{y}(t+k \mid t) = \sum_{i=1}^{\infty} g_i \triangle u(t+k-i) + \hat{n}(t+k \mid t) =$$

$$= \sum_{i=1}^{k} g_i \triangle u(t+k-i) + \sum_{i=k+1}^{\infty} g_i \triangle u(t+k-i) + \hat{n}(t+k \mid t)$$

Disturbances are considered to be constant, that is, $\hat{n}(t+k \mid t) = \hat{n}(t \mid t) = y_m(t) - \hat{y}(t \mid t)$. Then it can be written that:

$$\hat{y}(t+k \mid t) = \sum_{i=1}^{k} g_i \triangle u(t+k-i) + \sum_{i=k+1}^{\infty} g_i \triangle u(t+k-i) + y_m(t) -$$

$$-\sum_{i=1}^{\infty} g_i \triangle u(t-i) = \sum_{i=1}^{k} g_i \triangle u(t+k-i) + f(t+k)$$

where $f(t+k)$ is the free response of the system, that is, the part of the response that does not depend on the future control actions and is given by:

$$f(t+k) = y_m(t) + \sum_{i=1}^{\infty} (g_{k+i} - g_i) \triangle u(t-i) \tag{3.1}$$

If the process is asymptotically stable, the coefficients g_i of the step response tend to a constant value after N sampling periods, so it can be considered that

$$g_{k+i} - g_i \approx 0, \qquad i > N$$

and therefore the free response can be computed as:

$$f(t+k) = y_m(t) + \sum_{i=1}^{N} (g_{k+i} - g_i) \triangle u(t-i)$$

Notice that if the process is not asymptotically stable, then N does not exist and $f(t+k)$ cannot be computed (although a generalization exists in the case of the instability being produced by pure integrators)

Now the predictions can be computed along the prediction horizon ($k = 1, \ldots, p$), considering m control actions.

$$\begin{aligned}
\hat{y}(t+1 \mid t) &= g_1 \triangle u(t) + f(t+1) \\
\hat{y}(t+2 \mid t) &= g_2 \triangle u(t) + g_1 \triangle u(t+1) + f(t+2) \\
&\vdots \\
\hat{y}(t+p \mid t) &= \sum_{i=p-m+1}^{p} g_i \triangle u(t+p-i) + f(t+p)
\end{aligned}$$

Defining the system's *dynamic matrix* **G** as:

$$\mathbf{G} = \begin{bmatrix} g_1 & 0 & \cdots & 0 \\ g_2 & g_1 & \cdots & 0 \\ \vdots & \vdots & \ddots & \vdots \\ g_m & g_{m-1} & \cdots & g_1 \\ \vdots & \vdots & \ddots & \vdots \\ g_p & g_{p-1} & \cdots & g_{p-m+1} \end{bmatrix}$$

it can be written that:

$$\hat{\mathbf{y}} = \mathbf{G}\mathbf{u} + \mathbf{f} \tag{3.2}$$

Observe that **G** is made up of m (the control horizon) columns of the system's step response appropriately shifted down in order. $\hat{\mathbf{y}}$ is a p-dimensional vector containing the system predictions along the horizon, **u** represents the m-dimensional vector of control increments and **f** is the free response vector.

This is the expression that relates the future outputs with the control increments, so it will be used to calculate the necessary action to achieve a specific system behaviour.

3.1.2 Measurable Disturbances

Measurable disturbances can easily be added to the prediction equations, since they can be treated as system inputs. Expression (3.2) can be used to calculate the predicted disturbances:

$$\hat{\mathbf{y}}_d = \mathbf{D}\,\mathbf{d} + \mathbf{f}_d$$

where $\hat{\mathbf{y}}_d$ is the contribution of the measurable disturbance to the system output, $\mathbf{D}$ is a matrix similar to $\mathbf{G}$ containing the coefficients of the system response to a step in the disturbance, $\mathbf{d}$ is the vector of disturbance increment and $\mathbf{f}_d$ is the part of the response that does not depend on the disturbance.

In the most general case of measurable and non-measurable disturbances, the complete free response of the system (the fraction of the output that does not depend on the manipulated variable) can be considered as the sum of four effects: the response to the input $u(t)$, to the measurable disturbance $d(t)$, to the non-measurable disturbance and to the actual process state:

$$\mathbf{f} = \mathbf{f}_u + \mathbf{D}\,\mathbf{d} + \mathbf{f}_d + \mathbf{f}_n$$

Therefore the prediction can be computed by the general known expression:

$$\hat{\mathbf{y}} = \mathbf{G}\mathbf{u} + \mathbf{f}$$

3.1.3 Control Algorithm

The industrial success of DMC has mainly come from its application to high-dimension multivariable systems with the use of constraints. This section describes the control algorithm starting from the simpler case of a monovariable system without constraints and later it is extended to the general multivariable and constrained cases.

The objective of a DMC controller is to drive the output as close to the setpoint as possible in a least-squares sense with the possibility of the inclusion of a penalty term on the input moves. Therefore, the manipulated variables are selected to minimize a quadratic objective that can consider the minimization of future errors alone:

$$J = \sum_{j=1}^{p} [\hat{y}(t+j \mid t) - w(t+j)]^2$$

or it can include the control effort, in which case it presents the generic form

$$J = \sum_{j=1}^{p} [\hat{y}(t+j \mid t) - w(t+j)]^2 + \sum_{j=1}^{m} \lambda[\Delta u(t+j-1)]^2$$

If there are no constraints, the solution to the minimization of the cost function $J = \mathbf{e}\mathbf{e}^T + \lambda\mathbf{u}\mathbf{u}^T$, where $\mathbf{e}$ is the vector of future errors along the

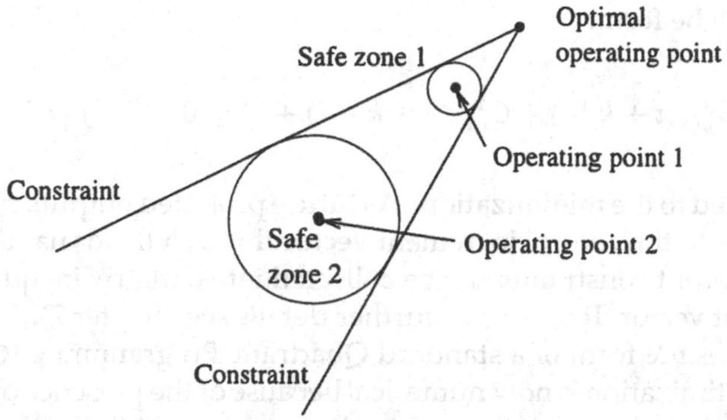

Figure 3.1: Economic operating point of a typical process

prediction horizon and **u** is the vector composed of the future control increments $\Delta u(t), \ldots, \Delta u(t+m)$, can be obtained analytically by computing the derivative of J and making it equal to 0, which provides the general result:

$$\mathbf{u} = (G^T G + \lambda I)^{-1} G^T (\mathbf{w} - \mathbf{f})$$

Remember that, as in all predictive strategies, only the first element of vector **u** ($\Delta u(t)$) is really sent to the plant. It is not advisable to implement the entire sequence over the next m intervals in automatic succession. This is because is impossible to perfectly estimate the disturbance vector and therefore it is also impossible to anticipate precisely the unavoidable disturbances that cause the actual output to differ from the predictions that are used to compute the sequence of control actions. Furthermore, the setpoint can also change over the next m intervals.

The constrained problem

Though computationally more involved than other simpler algorithms, the flexible constraints handling capabilities of the method (and MPC in general) are very attractive for practical applications, since the economic operating point of a typical process unit often lies at the intersection of constraints [96], as is shown in figure 3.1. It can be seen that, due to safety reasons, it is necessary to keep a safe zone around the operating point, since the effect of perturbations can make the process violate constraints. This zone can be reduced and therefore the economic profit improved if the controller is able to handle constraints (operating point 1).

Constraints both in inputs and outputs can be posed in such a way that equations of the form:

$$\sum_{i=1}^{N} C_{yi}^{j}\hat{y}(t+k \mid t) + C_{ui}^{j}u(t+k-i) + c^{j} \leq 0 \qquad j = 1 \ldots N_c$$

must be added to the minimization. As future projected outputs can be related directly back to the control increment vector through the dynamic matrix, all input and output constraints can be collected into a matrix inequality involving the input vector, $\mathbf{R}\mathbf{u} \leq \mathbf{c}$ (for further details see chapter 7). Therefore the problem takes the form of a standard Quadratic Programming (QP) formulation. The optimization is now numerical because of the presence of constraints and is carried out by means of standard commercial optimization QP code at each sampling instant and then the value of $u(t)$ is sent to the process, as is normally done in all MPC methods. In this case the method is known as QDMC, due to the Quadratic Programming algorithm employed.

Extension to the multivariable case

The basic scheme previously discussed extends to systems with multiple inputs and multiple outputs. The basic equations remain the same, except that the matrices and vectors become larger and appropriately partitioned. Since the multivariable case will be studied in more detail in chapter 6, only a few guidelines are provided here.

Based upon model linearity, the superposition principle can be used to obtain the predicted outputs provoked by the system inputs. The vector of predicted outputs is now defined as:

$$\hat{\mathbf{y}} = \left[y_1(t+1 \mid t), \ldots, y_1(t+p_1 \mid t), \ldots, y_{ny}(t+1 \mid t), \ldots, y_{ny}(t+p_{ny} \mid t)\right]^T$$

And the array of future control signals:

$$\mathbf{u} = [\Delta u_1(t), \ldots, \Delta u_1(t+m_1-1), \ldots, \Delta u_{nu}(t), \ldots, \Delta u_{nu}(t+m_{nu}-1)]^T$$

And the free response is defined as:

$$\mathbf{f} = \left[f_1(t+1 \mid t), \ldots, f_1(t+p_1 \mid t), \ldots, f_{ny}(t+1 \mid t), \ldots, f_{ny}(t+p_{ny} \mid t)\right]^T$$

taking into account that the free response of output i depends both on the past values of y_i and the past values of all control signals.

With the vector defined above, the prediction equations are the same as (3.2) simply considering matrix $\mathbf{G}$ to be:

$$\mathbf{G} = \begin{bmatrix} G_{11} & G_{12} & \cdots & G_{1nu} \\ G_{21} & G_{22} & \cdots & G_{2nu} \\ \vdots & \vdots & \ddots & \vdots \\ G_{ny1} & G_{ny2} & \cdots & G_{nynu} \end{bmatrix}$$

Each matrix G_{ij} contains the coefficients of the $i - th$ step response corresponding to the $j - th$ input.

3.2 Model Algorithmic Control

Maybe the most simple and intuitive formulation of Predictive Control is the one based on the key ideas of Richalet *et al.* [105], known as MAC and Model Predictive Heuristic Control (MPHC), whose software is called IDCOM (Identification-Command). This method is very similar to DMC with a few differences. It makes use of a truncated step response of the process and provides a simple explicit solution in the absence of constraints. This method has clearly been accepted by practitioners and is extensively used in many control applications [42] where most of its success is due to the process model used. It is known that transfer function models can give results with large errors when there is a mismatch in the model order. On the other hand, the impulse response representation is a good choice, since the identification of impulse responses is relatively simple.

3.2.1 Process Model and Prediction

The system output at instant t is related to the inputs by the coefficients of the truncated impulse response as follows:

$$y(t) = \sum_{j=1}^{N} h_j u(t-j) = H(z^{-1})u(t)$$

This model predicts that the output at a given time depends on a linear combination of past input values; the weights h_i are the impulse response coefficients. As the response is truncated to N elements, the system is assumed to be stable and causal. Using this internal model, a k-step ahead predictor can be written as

$$\hat{y}(t+k \mid t) = \sum_{j=1}^{N} h_j u(t+k-j) + \hat{n}(t+k \mid t)$$

where the sum can be divided into two terms:

$$f_r(t+k) = \sum_{j=k+1}^{N} h_j u(t+k-j) \qquad f_o(t+k) = \sum_{j=1}^{k} h_j u(t+k-j)$$

such that f_r represents the free response, being the expected value of $y(t+j)$ assuming zero future control actions, and f_o is the forced response, that is, the additional component of output response due to the proposed set of future control actions. It is now assumed that the disturbances will remain constant in the future with the same value as at instant t, that is, $\hat{n}(t+k \mid t) = \hat{n}(t \mid t)$ which is the measured output minus the output predicted by the nominal model:

$$\hat{n}(t+k \mid t) = \hat{n}(t \mid t) = y(t) - \sum_{j=1}^{N} h_j u(t-j)$$

Then the prediction is given by:

$$\hat{y}(t+k \mid t) = f_r + f_o + \hat{n}(t \mid t)$$

If M is the horizon and $\mathbf{u}_+$ the vector of proposed control actions (not increments), $\mathbf{u}_-$ of past control actions, $\mathbf{y}$ the predicted outputs, $\mathbf{n}$ the disturbances and the reference vector $\mathbf{w}$ being a smooth approach to the current setpoint:

$$\mathbf{u}_+ = \begin{bmatrix} u(t) \\ u(t+1) \\ \vdots \\ u(t+M-1) \end{bmatrix} \quad \mathbf{u}_- = \begin{bmatrix} u(t-N+1) \\ u(t-N+2) \\ \vdots \\ u(t-1) \end{bmatrix} \quad \mathbf{y} = \begin{bmatrix} \hat{y}(t+1) \\ \hat{y}(t+2) \\ \vdots \\ \hat{y}(t+M) \end{bmatrix}$$

$$\mathbf{n} = \begin{bmatrix} \hat{n}(t+1) \\ \hat{n}(t+2) \\ \vdots \\ \hat{n}(t+M) \end{bmatrix} \quad \mathbf{w} = \begin{bmatrix} w(t+1) \\ w(t+2) \\ \vdots \\ w(t+M) \end{bmatrix}$$

and defining the matrices

$$\mathbf{H_1} = \begin{bmatrix} h_1 & 0 & \cdots & 0 \\ h_2 & h_1 & \cdots & 0 \\ \cdots & \cdots & \ddots & \cdots \\ h_M & h_{M-1} & \cdots & h_1 \end{bmatrix} \mathbf{H_2} = \begin{bmatrix} h_N & \cdots & h_i & \cdots & h_2 \\ 0 & \cdots & h_j & \cdots & h_3 \\ \cdots & \ddots & \cdots & \cdots & \cdots \\ 0 & \cdots & h_N & \cdots & h_{M+1} \end{bmatrix}$$

the predictor can be written as:

$$\mathbf{y} = \mathbf{H_1}\,\mathbf{u}_+ + \mathbf{H_2}\,\mathbf{u}_- + \mathbf{n}$$

3.2.2 Control Law

The primary objective of the controller is to determine the sequence of control moves that will minimize the sum of the squared deviations of the predicted output from the reference trajectory.

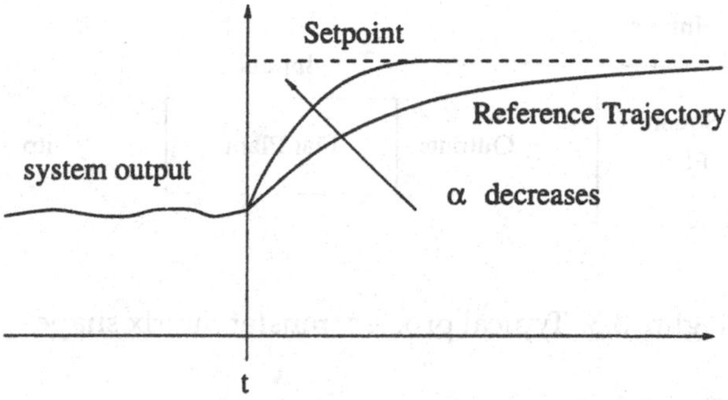

Figure 3.2: Influence of α on the reference tracking

The reference trajectory used in MAC is normally a smooth approximation from the current value of the system output towards the known reference by means of a first order system of the form:

$$w(t+k) = \alpha w(t+k-1) + (1-\alpha) r(t+k) \quad k = 1 \ldots N, \quad \text{with } w(t) = y(t)$$

It is important to note that the shape of the reference trajectory (that depends on the choice of α) determines the desired speed of approach to the set-point. This is of great interest in practice because it provides a natural way to control the aggressiveness of the algorithm: increasing the time constant leads to a slower but more robust controller (see figure 3.2). Therefore this is an important tuning parameter for this controller, and its choice is very closely linked to the robustness of the closed loop system [90]. Parameter α is a more direct and more intuitive tuning parameter than factors such as weighting sequences or horizon lengths employed by other formulations.

The objective function minimizes the error as well as the control effort. If the future errors are expressed as

$$\mathbf{e} = \mathbf{w} - \mathbf{y} = \mathbf{w} - \mathbf{H_2 u_-} - \mathbf{n} - \mathbf{H_1 u_+} = \mathbf{w} - \mathbf{f} - \mathbf{H_1 u_+}$$

where vector $\mathbf{f}$ contains the terms depending on known values (past inputs, current output and references). Then the cost function can be written as

$$J = \mathbf{e}^T\mathbf{e} + \lambda \mathbf{u_+}^T\mathbf{u_+}$$

being λ the penalization factor for the input variable variations. If no constraints are considered, the solution can be obtained explicitly giving:

$$\mathbf{u_+} = (\mathbf{H_1}^T\mathbf{H_1} + \lambda\mathbf{I})^{-1}\mathbf{H_1}^T(\mathbf{w} - \mathbf{f}) \tag{3.3}$$

As it is a receding-horizon strategy, only the first element of this vector $u(t)$ is used, rejecting the rest and repeating the calculations at the next sampling time.

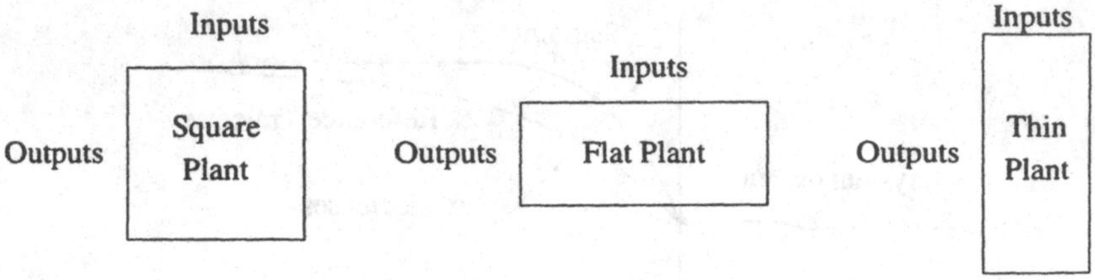

Figure 3.3: Typical process transfer matrix shapes

The calculation of the control law (3.3) is relatively simple compared to other formulations although it requires the inversion of an $M \times M$ matrix. Notice that if the number of future inputs to be calculated is chosen as a value $P < M$, then this matrix is of dimension $P \times P$, since $\mathbf{H_1}$ is of dimension $M \times P$, hence reducing the necessary calculations.

The simplicity of the algorithm. as well as the possibility of including constraints, has converted this formulation into one of the most frequently used in industry nowadays.

3.2.3 Multivariable Processes

The extension to the multivariable case is exactly the same as that of DMC, so no more attention will be paid to the equations. However, some implementation issues of the IDCOM-M (multivariable) will be addressed in this section.

The IDCOM-M algorithm [48] uses two separate objective functions, one for the outputs and then, if there are extra degrees of freedom, one for the inputs. The degree of freedom available for the control depends on the plant structure. Figure 3.3 shows the shape of the process transfer matrix for three general cases.

The square plant case, which is rare in real situations, occurs when the plant has as many inputs as outputs and leads to a control problem with a unique solution. The flat plant case is more common (more inputs than outputs) and the extra degrees of freedom available can be employed in different objectives, such as moving the plant closer to an optimal operating point. In the last situation (thin plant case, where there are more outputs than inputs) it is not possible to meet all of the control objectives, and some specifications must be relaxed.

Thus, for flat plants, IDCOM-M incorporates the concept of *Ideal Resting Values* (IRV) for the inputs. In this case, in addition to the primary objective (minimize the output errors), the controller also tries to minimize the sum of squared deviations of the inputs from their respective IRVs, which may come from a steady-state optimizer (by default the IRV for a given input is set to its current measured value). So the strategy involves a two-step optimization problem, that is solved using a quadratic programming approach: the primary problem involves the choice of the control sequence required to drive the controlled variables close to the set-points, and the second one involves optimizing the use of control effort in achieving the objective of the primary problem.

The input optimization makes the most effective use of available degrees of freedom without influencing the optimal output solution. Even when there are no excess inputs, the ideal resting values concept, that no other MPC strategy incorporates, is of great interest when, for operational or economic reasons, there is a benefit in maintaining a manipulated variable at a specific steady-state value.

3.3 Predictive Functional Control

The Predictive Functional Controller PFC was proposed by Richalet [101] for fast processes and is characterized by two distinctive features: the structuration of the control signal as a linear combination of certain predetermined *basis functions*, and the concept of *coincidence points* to evaluate the cost function along the horizon.

3.3.1 Formulation

Consider the following state space model

$$\begin{aligned} x_m(t) &= M x_m(t-1) + N u(t-1) \\ y_m(t) &= Q x_m(t) \end{aligned}$$

as representing the process behaviour. The prediction is obtained adding an autocompensation term calculated as a function of the observed differences between the model and the past outputs.

$$\hat{y}(t+k \mid t) = y_m(t+k) + e(t+k)$$

The future control signal is structured as a linear combination of the basis functions B_i, that are chosen according to the nature of the process and the reference:

$$u(t+k) = \sum_{i=1}^{n_B} \mu_i(t) B_i(k)$$

Normally these functions are polynomial type: steps ($B_1(k) = 1$), ramps ($B_2(k) = k$) or parabolas ($B_3(k) = k^2$), as the majority of references can be expressed as combinations of these functions. With this strategy a complex input profile can be specified using a small number of unknown parameters.

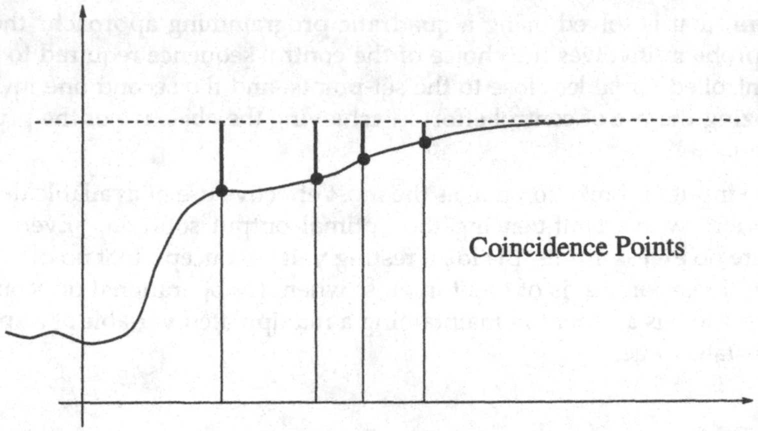

Figure 3.4: Coincidence points

The choice of the basis functions defines the input profile and can assure a predetermined behaviour (smooth signal, for instance). This can result in an advantage when controlling non linear systems. An important feature for mechanical servo applications is that if a polynomial basis is chosen, then the order can be selected to follow a polynomial setpoint.

The cost function to be minimized is:

$$J = \sum_{j=1}^{n_H} [\hat{y}(t+h_j) - w(t+h_j)]^2$$

where $w(t+j)$ is usually a first order approach to the known reference, as in (2.6) or:

$$w(t+k) = r(t+k) - \alpha^k (r(t) - y(t))$$

In order to smooth the control signal, a quadratic factor of the form $\lambda[\Delta u(k)]^2$ can be added to the cost function.

The predicted error is not considered along all the horizon, but only in certain instants $h_j, j = 1, \ldots, n_H$ called *coincidence points* (see figure 3.4). These points can be considered as tuning parameters and must be chosen taking into account their influence on the stability and robustness of the control system. Their number must be at least equal to the selected number of basis functions.

Calculation of the control law

In the case of SISO processes without constraints, the control law can be obtained as follows. First, the output is decomposed into free output and forced one, and the structurization of the control signal is employed to give:

$$y_m(t+k) = QM^k x_m(t) + \sum_{i=1}^{n_B} \mu_i(t) y_{B_i}(k)$$

being y_{B_i} the system response to the basis function B_i.

Now the cost function can be written as:

$$J = \sum_{j=1}^{n_H} [\hat{y}(t+h_j) - w(t+h_j)]^2 = \sum_{j=1}^{n_H} [\mu^T \mathbf{y}_B - d(t+h_j)]^2$$

where:

$$\mu = [\mu_1(t) \ldots \mu_{nB}(t)]^T$$
$$\mathbf{y}_B = [y_{B_1}(t+h_j) \ldots y_{B_{n_B}}(t+h_j)]^T$$
$$d(t+h_j) = \omega(t+h_j) - QM^j x_m(t) - e(t+h_j)$$

Minimizing J with respect to the coefficients μ:

$$\frac{\delta J}{\delta \mu} = 2((\mathbf{Y}_B \mathbf{Y}_B^T)\mu - \mathbf{Y}_B \mathbf{d}) = 0$$

being $\mathbf{Y}_B$ a matrix whose columns are the vectors $\mathbf{y}_B$ at the coincidence points h_j, $j = 1, \ldots n_h$ and $\mathbf{d} = [d(t+h_1) \ldots d(t+h_{n_H})]^T$.

Now the coefficients of the vector μ are computed and the control signal, taking into account the receding horizon strategy, is given by:

$$u(t) = \sum_{i=1}^{n_B} \mu_i(t) B_i(0)$$

The algorithm described above can only be used for stable models, since the pole cancellations can lead to stability problems when unstable or high-oscillatory modes appear. In this case, a procedure that decomposes the model into two stable ones can be employed [101]. The method can be used for non linear processes using non linear state space models.

3.4 Case Study: a Water Heater

This example shows the design of a DMC to control the outlet temperature of a water heater. Notice that a MAC can be designed following the same steps.

Consider a water heater where the cold water is heated by means of a gas burner. The outlet temperature depends on the energy added to the water through the gas burner (see figure 3.5). Therefore this temperature can be controlled by the valve which manipulates the gas flow to the heater.

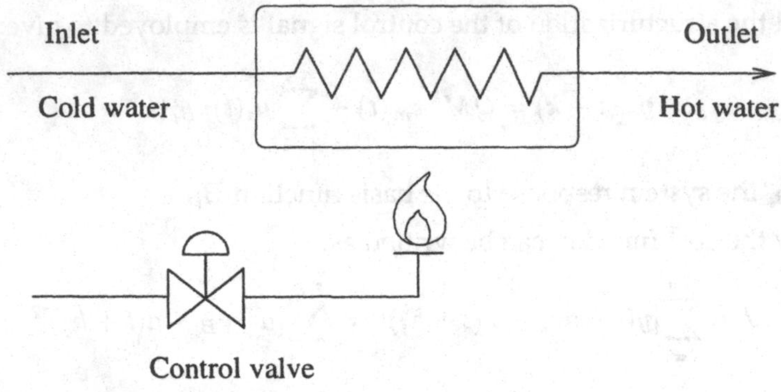

Figure 3.5: Water heater

The step response model of this process must be obtained in order to design the controller. The step response is obtained by applying a step in the control valve. Coefficients g_i can be obtained directly from the response shown in figure 3.6. It can be seen that the output stabilizes after 30 periods, so the model is given by:

$$y(t) = \sum_{i=1}^{30} g_i \, \Delta \, u(t-i),$$

where the coefficients g_i are shown in the following table:

g_1 0	g_2 0	g_3 0.271	g_4 0.498	g_5 0.687	g_6 0.845	g_7 0.977	g_8 1.087	g_9 1.179	g_{10} 1.256
g_{11} 1.320	g_{12} 1.374	g_{13} 1.419	g_{14} 1.456	g_{15} 1.487	g_{16} 1.513	g_{17} 1.535	g_{18} 1.553	g_{19} 1.565	g_{20} 1.581
g_{21} 1.592	g_{22} 1.600	g_{23} 1.608	g_{24} 1.614	g_{25} 1.619	g_{26} 1.623	g_{27} 1.627	g_{28} 1.630	g_{29} 1.633	g_{30} 1.635

The response shown in figure 3.6 corresponds to a system with a transfer function given by:

$$G(z) = \frac{0.2713z^{-3}}{1 - 0.8351z^{-1}}$$

Notice that although the g_i coefficients are obtained in practice from real plant tests, for this example where the response has been generated with a simple model, the step response coefficients can easily be obtained from the transfer function by the expression:

$$g_j = \sum_{i=1}^{j} a_i g_{j-i} + \sum_{i=0}^{j-1} b_i \qquad g_k = 0 \quad \text{for} \quad k \leq 0 \tag{3.4}$$

being a_i and b_i the coefficients of the denominator and numerator of the discrete transfer function respectively.

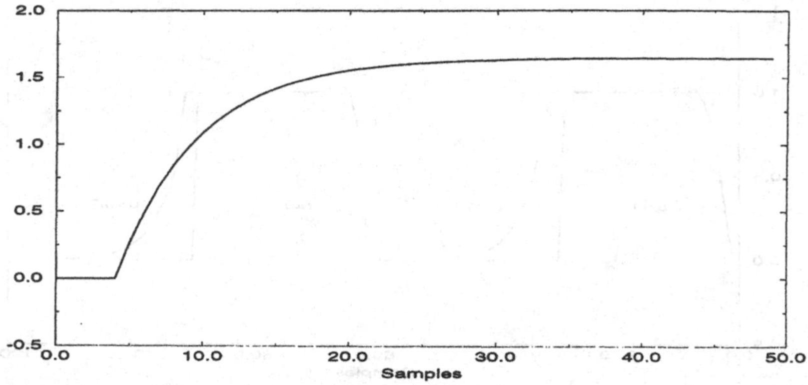

Figure 3.6: Step response

In this example the first two coefficients of the step model are zero, since the system has a dead time of two sampling periods.

Considering a prediction horizon of 10 and a control horizon of 5, the dynamic matrix is obtained from the coefficients of the step response and is given by:

$$\mathbf{G} = \begin{bmatrix} 0 & 0 & 0 & 0 & 0 \\ 0 & 0 & 0 & 0 & 0 \\ 0.271 & 0 & 0 & 0 & 0 \\ 0.498 & 0.271 & 0 & 0 & 0 \\ 0.687 & 0.498 & 0.271 & 0 & 0 \\ 0.845 & 0.687 & 0.498 & 0.271 & 0 \\ 0.977 & 0.845 & 0.687 & 0.498 & 0.271 \\ 1.087 & 0.977 & 0.845 & 0.687 & 0.498 \\ 1.179 & 1.087 & 0.977 & 0.845 & 0.687 \\ 1.256 & 1.179 & 1.087 & 0.977 & 0.845 \end{bmatrix}$$

Taking $\lambda = 1$, matrix $(G^TG + \lambda I)^{-1}G^T$ is calculated and therefore the control law is given by the product of the first row of this matrix (**K**) times the vector that contains the difference between the reference trajectory and the free response:

$$\Delta u(t) = \mathbf{K}(\mathbf{w} - \mathbf{f})$$

with

$$\mathbf{K} = [0 \;\; 0 \;\; 0.1465 \;\; 0.1836 \;\; 0.1640 \;\; 0.1224 \;\; 0.0780 \;\; 0.0410 \;\; 0.0101 \;\; -0.0157]$$

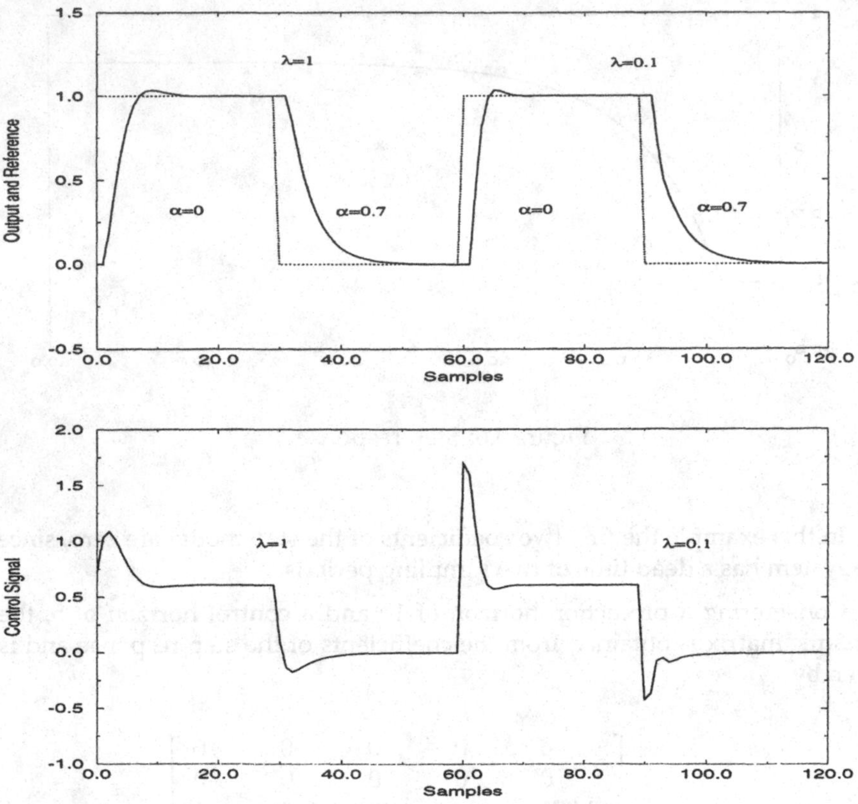

Figure 3.7: Controller behaviour

Where the free response is easily calculated using equation (3.1):

$$f(t+k) = y_m(t) + \sum_{i=1}^{30} (g_{k+i} - g_i) \bigtriangleup u(t-i)$$

Figure 3.7 shows the system response to a change in the outlet temperature setpoint for different shapes of the control weighting factor and the reference trajectory. The first setpoint change is made with a value of $\lambda = 1$ and $\alpha = 0$. In the second change α is changed to 0.7 and later the control weighting factor is changed to 0.1 for the same values of α. It can be seen that a small value of α makes the system response faster although with a slight oscillation, while a small value of λ gives bigger control actions. The combination $\lambda = 0.1$, $\alpha = 0$ provides the fastest response, but the control effort seems to be too vigorous.

The inlet temperature can become a disturbance, since any change in its value will disturb the process from its steady state operating point. This temperature can be measured and the controller can take into account its value in order to reject its effect before it appears in the system output. That is, it can be treated by DMC as a measurable disturbance and explicitly incorporated into the formulation. To do this, a model of the effect of the inlet temperature changes on the outlet temperature can easily be obtained by a step test.

In this example the disturbance is modelled by

$$y(t) = \sum_{i=1}^{30} d_i \, \triangle \, u(t-i),$$

with

d_1	d_2	d_3	d_4	d_5	d_6	d_7	d_8	d_9	d_{10}
0	0	0.050	0.095	0.135	0.172	0.205	0.234	0.261	0.285
d_{11}	d_{12}	d_{13}	d_{14}	d_{15}	d_{16}	d_{17}	d_{18}	d_{19}	d_{20}
0.306	0.326	0.343	0.359	0.373	0.386	0.397	0.407	0.417	0.425
d_{21}	d_{22}	d_{23}	d_{24}	d_{25}	d_{26}	d_{27}	d_{28}	d_{29}	d_{30}
0.433	0.439	0.445	0.451	0.456	0.460	0.464	0.468	0.471	0.474

which corresponds to the transfer function:

$$G(z) = \frac{0.05z^{-3}}{1 - 0.9z^{-1}}$$

Notice that the first 10 d_i coefficients are used to build matrix **D** in the same way as matrix **G**.

Figure 3.8 shows a simulation where a disturbance occurs from $t = 20$ to $t = 60$. In case the controller explicitly considers the measurable disturbances it is able to reject them, since the controller starts acting when the disturbance appears, not when its effect appears in the outlet temperature. On the other hand, if the controller does not take into account the measurable disturbance, it reacts later, when the effect on the output is considerable.

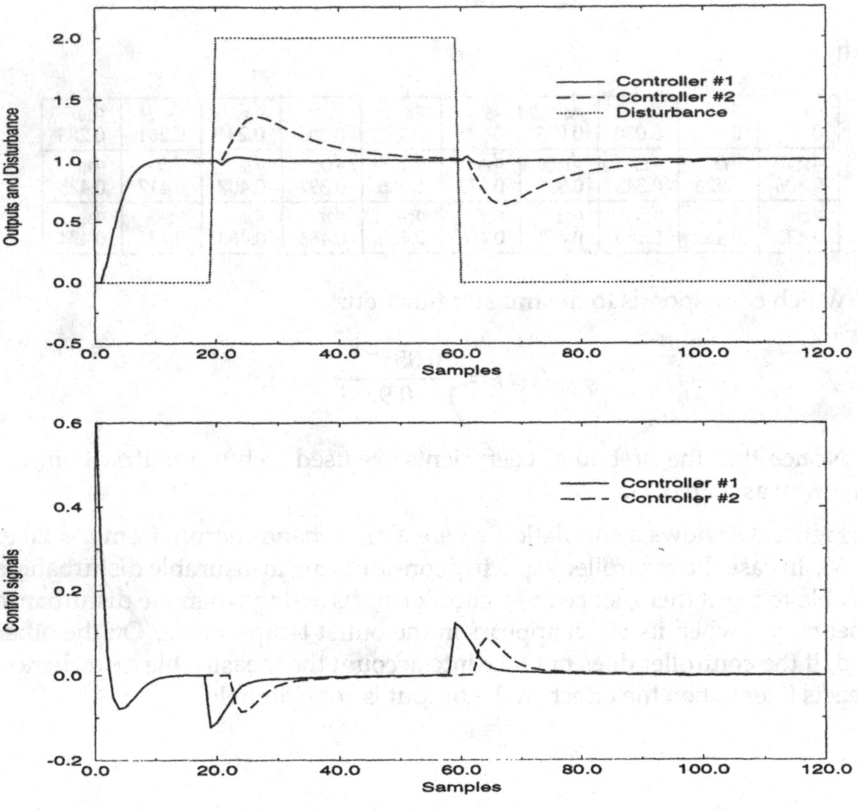

Figure 3.8: Disturbance rejection with (Controller #1) and without (Controller #2) considering measurable disturbances

Chapter 4

Generalized Predictive Control

This chapter describes one of the most popular predictive control algorithms: Generalized Predictive Control (GPC). The method is developed in detail, showing the general procedure to obtain the control law and its most outstanding characteristics. The original algorithm is extended to include the cases of measurable disturbances and change in the predictor. Close derivations of this controller as CRHPC and Stable GPC are also treated here, illustrating the way they can be implemented.

4.1 Introduction

The GPC method was proposed by Clarke *et al.* [34] and has become one of the most popular MPC methods both in industry and academia. It has been successfully implemented in many industrial applications [31], showing good performance and a certain degree of robustness. It can handle many different control problems for a wide range of plants with a reasonable number of design variables, which have to be specified by the user depending upon a prior knowledge of the plant and control objectives.

The basic idea of GPC is to calculate a sequence of future control signals in such a way that it minimizes a multistage cost function defined over a prediction horizon. The index to be optimized is the expectation of a quadratic function measuring the distance between the predicted system output and some predicted reference sequence over the horizon plus a quadratic function measuring the control effort. This approach has been used by Lelic and Zarrop [64] and Lelic and Wellstead [63] to obtain a generalized pole placement controller which is an extension of the well known pole placement controllers

[2], [123] and belongs to the class of extended horizon controllers.

Generalized Predictive Control has many ideas in common with the other predictive controllers previously mentioned since it is based upon the same concepts but it also has some differences. As will be seen later, it provides an analytical solution (in the absence of constraints), it can deal with unstable and non-minimum phase plants and incorporates the concept of control horizon as well as the consideration of weighting of control increments in the cost function. The general set of choices available for GPC leads to a greater variety of control objectives compared to other approaches, some of which can be considered as subsets or limiting cases of GPC.

4.2 Formulation of Generalized Predictive Control

Most single-input single-output (SISO) plants, when considering operation around a particular set-point and after linearization, can be described by

$$A(z^{-1})y(t) = z^{-d}B(z^{-1})u(t-1) + C(z^{-1})e(t)$$

where $u(t)$ and $y(t)$ are the control and output sequence of the plant and $e(t)$ is a zero mean white noise. A, B and C are the following polynomials in the backward shift operator z^{-1}:

$$\begin{aligned} A(z^{-1}) &= 1 + a_1 z^{-1} + a_2 z^{-2} + \ldots + a_{na} z^{-na} \\ B(z^{-1}) &= b_0 + b_1 z^{-1} + b_2 z^{-2} + \ldots + b_{nb} z^{-nb} \\ C(z^{-1}) &= 1 + c_1 z^{-1} + a_2 z^{-2} + \ldots + c_{nc} z^{-nc} \end{aligned}$$

where d is the dead time of the system. This model is known as a Controller Auto-Regressive Moving-Average (CARMA) model. It has been argued [34] that for many industrial applications in which disturbances are non-stationary an integrated CARMA (CARIMA) model is more appropriate. A CARIMA model is given by:

$$A(z^{-1})y(t) = B(z^{-1})z^{-d}\, u(t-1) + C(z^{-1})\frac{e(t)}{\Delta} \qquad (4.1)$$

with

$$\Delta = 1 - z^{-1}$$

For simplicity, in the following the C polynomial is chosen to be 1. Notice that if C^{-1} can be truncated it can be absorbed into A and B. The general case of a coloured noise will be treated later.

The Generalized Predictive Control (GPC) algorithm consists of applying a control sequence that minimizes a multistage cost function of the form

$$J(N_1, N_2, N_u) = \sum_{j=N_1}^{N_2} \delta(j)[\hat{y}(t+j \mid t) - w(t+j)]^2 + \sum_{j=1}^{N_u} \lambda(j)[\Delta u(t+j-1)]^2 \qquad (4.2)$$

where $\hat{y}(t+j \mid t)$ is an optimum j-step ahead prediction of the system output on data up to time t, N_1 and N_2 are the minimum and maximum costing horizons, N_u is the control horizon, $\delta(j)$ and $\lambda(j)$ are weighting sequences and $w(t+j)$ is the future reference trajectory, which can be calculated as shown in (2.6). In [34] $\delta(j)$ is considered to be 1 and $\lambda(j)$ is considered to be constant.

The objective of predictive control is to compute the future control sequence $u(t)$, $u(t+1)$,... in such a way that the future plant output $y(t+j)$ is driven close to $w(t+j)$. This is accomplished by minimizing $J(N_1, N_2, N_u)$.

In order to optimize the cost function the optimal prediction of $y(t+j)$ for $j \geq N_1$ and $j \leq N_2$ will be obtained. Consider the following Diophantine equation:

$$1 = E_j(z^{-1})\tilde{A}(z^{-1}) + z^{-j}F_j(z^{-1}) \text{ with } \tilde{A}(z^{-1}) = \Delta A(z^{-1}) \tag{4.3}$$

The polynomials E_j and F_j are uniquely defined with degrees $j-1$ and na respectively. They can be obtained dividing 1 by $\tilde{A}(z^{-1})$ until the remainder can be factorized as $z^{-j}F_j(z^{-1})$. The quotient of the division is the polynomial $E_j(z^{-1})$.

If equation (4.1) is multiplied by $\Delta E_j(z^{-1})z^j$

$$\begin{aligned}\tilde{A}(z^{-1})E_j(z^{-1})y(t+j) &= E_j(z^{-1})B(z^{-1})\Delta u(t+j-d-1) + \\ &+ E_j(z^{-1})e(t+j)\end{aligned} \tag{4.4}$$

Considering (4.3), equation (4.4) can be written as:

$$(1 - z^{-j}F_j(z^{-1}))y(t+j) = E_j(z^{-1})B(z^{-1})\Delta u(t+j-d-1) + E_j(z^{-1})e(t+j)$$

which can be rewritten as:

$$y(t+j) = F_j(z^{-1})y(t) + E_j(z^{-1})B(z^{-1})\Delta u(t+j-d-1) + E_j(z^{-1})e(t+j) \tag{4.5}$$

As the degree of polynomial $E_j(z^{-1}) = j-1$ the noise terms in equation (4.5) are all in the future. The best prediction of $y(t+j)$ is therefore:

$$\hat{y}(t+j \mid t) = G_j(z^{-1})\Delta u(t+j-d-1) + F_j(z^{-1})y(t)$$

where $G_j(z^{-1}) = E_j(z^{-1})B(z^{-1})$

It is very simple to show that the polynomials E_j and F_j can be obtained recursively. The recursion of the Diophantine equation has been demonstrated in [34]. A simpler demonstration is given in the following. There are other formulations of GPC not based on the recursion of the Diophantine equation [1].

Consider that polynomials E_j and F_j have been obtained by dividing 1 by $\tilde{A}(z^{-1})$ until the remainder of the division can be factorized as $z^{-j}F_j(z^{-1})$.

These polynomials can be expressed as:

$$\begin{aligned}F_j(z^{-1}) &= f_{j,0} + f_{j,1}z^{-1} + \cdots + f_{j,na}z^{-na} \\ E_j(z^{-1}) &= e_{j,0} + e_{j,1}z^{-1} + \cdots + e_{j,j-1}z^{-(j-1)}\end{aligned}$$

Suppose that the same procedure is used to obtain E_{j+1} and F_{j+1}, that is, dividing 1 by $\tilde{A}(z^{-1})$ until the remainder of the division can be factorized as $z^{-(j+1)}F_{j+1}(z^{-1})$ with

$$F_{j+1}(z^{-1}) = f_{j+1,0} + f_{j+1,1}z^{-1} + \cdots + f_{j+1,na}z^{-na}$$

It is clear that only another step of the division performed to obtain the polynomials E_j and F_j has to be taken in order to obtain the polynomials E_{j+1} and F_{j+1}. The polynomial E_{j+1} will be given by:

$$E_{j+1}(z^{-1}) = E_j(z^{-1}) + e_{j+1,j}z^{-j}$$

with $e_{j+1,j} = f_{j,0}$

The coefficients of polynomial F_{j+1} can then be expressed as:

$$f_{j+1,i} = f_{j,i+1} - f_{j,0}\,\tilde{a}_{i+1} \quad i = 0 \cdots na-1$$

The polynomial G_{j+1} can be obtained recursively as follows:

$$\begin{aligned}G_{j+1} &= E_{j+1}B = (E_j + f_{j,0}z^{-j})B \\ G_{j+1} &= G_j + f_{j,0}z^{-j}B\end{aligned}$$

That is, the first j coefficient of G_{j+1} will be identical to those of G_j and the remaining coefficients will be given by:

$$g_{j+1,j+i} = g_{j,j+i} + f_{j,0}\,b_i \qquad i = 0 \cdots nb$$

To solve the GPC problem the set of control signals $u(t)$, $u(t+1)$, ..., $u(t+N)$ has to be obtained in order to optimize expression (4.2). As the system considered has a dead time of d sampling periods, the output of the system will be influenced by signal $u(t)$ after sampling period $d+1$. The values N_1, N_2 and N_u defining the horizon can be defined by $N_1 = d+1$, $N_2 = d+N$ and $N_u = N$. Notice that there is no point in making $N_1 < d+1$ as terms added to expression (4.2) will only depend on the past control signals. On the other hand, if $N_1 > d+1$ the first points in the reference sequence, being the ones guessed with most certainty, will not be taken into account.

Now consider the following set of j ahead optimal predictions:

$$\begin{aligned}\hat{y}(t+d+1 \mid t) &= G_{d+1}\,\triangle\,u(t) + F_{d+1}y(t) \\ \hat{y}(t+d+2 \mid t) &= G_{d+2}\,\triangle\,u(t+1) + F_{d+2}y(t) \\ &\vdots \\ \hat{y}(t+d+N \mid t) &= G_{d+N}\,\triangle\,u(t+N-1) + F_{d+N}y(t)\end{aligned}$$

which can be written as:

$$\mathbf{y} = \mathbf{G}\mathbf{u} + \mathbf{F}(z^{-1})y(t) + \mathbf{G}'(z^{-1}) \bigtriangleup u(t-1) \tag{4.6}$$

where

$$\mathbf{y} = \begin{bmatrix} \hat{y}(t+d+1 \mid t) \\ \hat{y}(t+d+2 \mid t) \\ \vdots \\ \hat{y}(t+d+N \mid t) \end{bmatrix} \qquad \mathbf{u} = \begin{bmatrix} \Delta u(t) \\ \Delta u(t+1) \\ \vdots \\ \Delta u(t+N-1) \end{bmatrix}$$

$$\mathbf{G} = \begin{bmatrix} g_0 & 0 & \cdots & 0 \\ g_1 & g_0 & \cdots & 0 \\ \vdots & \vdots & \vdots & \vdots \\ g_{N-1} & g_{N-2} & \cdots & g_0 \end{bmatrix}$$

$$\mathbf{G}'(z^{-1}) = \begin{bmatrix} (G_{d+1}(z^{-1}) - g_0)z \\ (G_{d+2}(z^{-1}) - g_0 - g_1 z^{-1})z^2 \\ \vdots \\ (G_{d+N}(z^{-1}) - g_0 - g_1 z^{-1} - \cdots - g_{N-1}z^{-(N-1)})z^N \end{bmatrix}$$

$$\mathbf{F}(z^{-1}) = \begin{bmatrix} F_{d+1}(z^{-1}) \\ F_{d+2}(z^{-1}) \\ \vdots \\ F_{d+N}(z^{-1}) \end{bmatrix}$$

Notice that the last two terms in equation (4.6) only depend on the past and can be grouped into $\mathbf{f}$ leading to:

$$\mathbf{y} = \mathbf{G}\mathbf{u} + \mathbf{f}$$

Notice that if all initial conditions are zero, the free response $\mathbf{f}$ is also zero. If a unit step is applied to the input at time t; that is

$$\Delta u(t) = 1, \Delta u(t+1) = 0, \cdots, \Delta u(t+N-1) = 0$$

the expected output sequence $[\hat{y}(t+1), \hat{y}(t+2), \cdots, \hat{y}(t+N)]^T$ is equal to the first column of matrix $\mathbf{G}$. That is, the first column of matrix $\mathbf{G}$ can be calculated as the step response of the plant when a unit step is applied to the manipulated variable. The free response term can be calculated recursively by:

$$\mathbf{f}_{j+1} = z(1 - \tilde{A}(z^{-1}))\mathbf{f}_j + B(z^{-1}) \bigtriangleup u(t-d+j)$$

with $\mathbf{f}_0 = y(t)$ and $\Delta u(t+j) = 0$ for $j \geq 0$.

Expression (4.2) can be written as:

$$J = (\mathbf{G}\mathbf{u} + \mathbf{f} - \mathbf{w})^T(\mathbf{G}\mathbf{u} + \mathbf{f} - \mathbf{w}) + \lambda\mathbf{u}^T\mathbf{u} \tag{4.7}$$

where:

$$\mathbf{w} = [\ w(t+d+1) \quad w(t+d+2) \quad \cdots \quad w(t+d+N)\ ]^T$$

Equation (4.7) can be written as:

$$J = \frac{1}{2}\mathbf{u}^T\mathbf{H}\mathbf{u} + \mathbf{b}^T\mathbf{u} + \mathbf{f}_0 \tag{4.8}$$

where:

$$\begin{aligned} \mathbf{H} &= 2(\mathbf{G}^T\mathbf{G} + \lambda\mathbf{I}) \\ \mathbf{b}^T &= 2(\mathbf{f} - \mathbf{w})^T\mathbf{G} \\ \mathbf{f}_0 &= (\mathbf{f} - \mathbf{w})^T(\mathbf{f} - \mathbf{w}) \end{aligned}$$

The minimum of J, assuming there are no constraints on the control signals, can be found by making the gradient of J equal to zero, which leads to:

$$\mathbf{u} = -\mathbf{H}^{-1}\mathbf{b} = (\mathbf{G}^T\mathbf{G} + \lambda\mathbf{I})^{-1}\mathbf{G}^T(\mathbf{w} - \mathbf{f}) \tag{4.9}$$

Notice that the control signal that is actually sent to the process is the first element of vector $\mathbf{u}$, that is given by:

$$\triangle u(t) = \mathbf{K}(\mathbf{w} - \mathbf{f}) \tag{4.10}$$

where $\mathbf{K}$ is the first row of matrix $(\mathbf{G}^T\mathbf{G}+\lambda\mathbf{I})^{-1}\mathbf{G}^T$. This has a clear meaning, that can easily be derived from figure 4.1: if there are no future predicted errors, that is, if $\mathbf{w} - \mathbf{f} = \mathbf{0}$, then there is no control move, since the objective will be fulfilled with the free evolution of the process. However, in the other case, there will be an increment in the control action proportional (with a factor $\mathbf{K}$) to that future error. Notice that the action is taken with respect to *future* errors, not *past* errors, as is the case in conventional feedback controllers.

Notice that only the first element of $\mathbf{u}$ is applied and the procedure is repeated at the next sampling time. The solution to the GPC given involves the inversion (or triangularization) of an $N \times N$ matrix which requires a substantial amount of computation. In [34] the concept of control horizon is used to reduce the amount of computation needed, assuming that the projected control signals are going to be constant after $N_u < N$. This leads to the inversion of an $N_u \times N_u$ matrix which reduces the amount of computation (in particular, if $N_u = 1$ it is reduced to a scalar computation, as in EPSAC), but restricts the optimality of the GPC. A fast algorithm to implement self-tuning GPC for processes that can be modelled by the reaction curve method is presented in the next chapter. The use of Hopfield neural networks has also been proposed [99] to obtain fast GPCs.

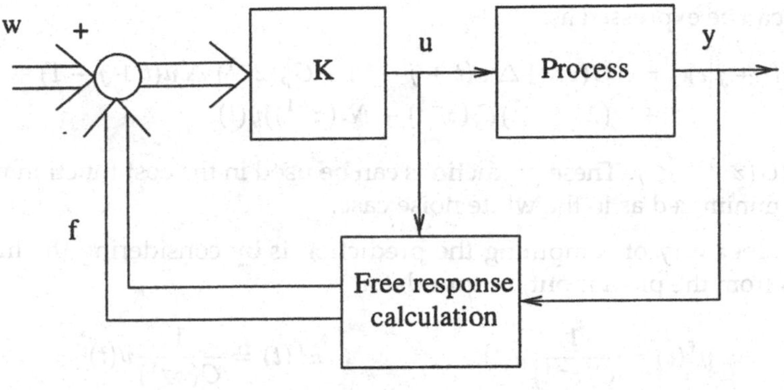

Figure 4.1: GPC control law

4.3 The Coloured Noise Case

When the noise polynomial $C(z^{-1})$ of equation (4.1) is not equal to 1 the prediction changes slightly. In order to calculate the predictor in this situation, the following Diophantine equation is solved:

$$C(z^{-1}) = E_j(z^{-1})\tilde{A}(z^{-1}) + z^{-j}F_j(z^{-1}) \tag{4.11}$$

with $\delta(E_j(z^{-1})) = j - 1$ and $\delta(F_j(z^{-1})) = \delta(\tilde{A}(z^{-1})) - 1$

Multiplying equation (4.1) by $\Delta E_j(z^{-1})z^j$ and using (4.11)

$$C(z^{-1})(y(t+j)-E_j(z^{-1})e(t+j)) = E_j(z^{-1})B(z^{-1})\Delta u(t+j-1)+F_j(z^{-1})y(t)$$

As the noise terms are all in the future, the expected value of the left hand side of the above equation is:

$$E[C(z^{-1})(y(t+j) - E_j(z^{-1})e(t+j))] = C(z^{-1})\hat{y}(t+j|t)$$

The expected value of the output can be generated by the equation:

$$C(z^{-1})\hat{y}(t+j|t) = E_j(z^{-1})B(z^{-1})\,\Delta\, u(t+j-1) + F_j(z^{-1})y(t) \tag{4.12}$$

Notice that this prediction equation could be used to generate the predictions in a recursive way. An explicit expression for the optimal j-step ahead prediction can be obtained by solving the Diophantine equation:

$$1 = C(z^{-1})M_j(z^{-1}) + z^{-k}N_j(z^{-1}) \tag{4.13}$$

with $\delta(M_j(z^{-1})) = j - 1$ and $\delta(N_j(z^{-1})) = \delta(C(z^{-1})) - 1$

Multiplying equation (4.12) by $M_j(z^{-1})$ and using (4.13),

$$\hat{y}(t+j|t) = M_jE_j(z^{-1})B(z^{-1})\Delta u(t+j-1)+M_j(z^{-1})F_j(z^{-1})y_(t)+N_j(z^{-1})y(t)$$

which can be expressed as:

$$\begin{aligned}\hat{y}(t+j|t) &= G(z^{-1})\,\Delta u(t+j-1) + G_p(z^{-1})\,\Delta u(t+j-1) + \\ &+ (M_j(z^{-1})F_j(z^{-1}) + N_j(z^{-1}))y(t)\end{aligned}$$

with $\delta(G(z^{-1})) < j$. These predictions can be used in the cost function which can be minimized as in the white noise case.

Another way of computing the prediction is by considering the filtered signals from the plant input/output data:

$$y^f(t) = \frac{1}{C(z^{-1})}y(t) \qquad u^f(t) = \frac{1}{C(z^{-1})}u(t)$$

so that the resulting overall model becomes

$$A(z^{-1})y^f(t) = B(z^{-1})u^f(t) + \frac{e(t)}{\Delta}$$

and the white noise procedure for computing the prediction can be used. The predicted signal $\hat{y}^f(t+j|t)$ obtained this way has to be filtered by $C(z^{-1})$ in order to get $\hat{y}(t+j|t)$.

4.4 An Example

In order to show the way a Generalized Predictive Controller can be implemented, a simple example is presented. The controller will be designed for a first order system for the sake of clarity.

The following discrete equivalence can be obtained when a first order continuous plant is discretized

$$(1 + az^{-1})y(t) = (b_0 + b_1 z^{-1})u(t-1) + \frac{e(t)}{\Delta}$$

In this example the delay d is equal to 0 and the noise polynomial $C(z^{-1})$ is considered to be equal to 1.

The algorithm to obtain the control law described in the previous section will be used on the above system, obtaining numerical results for the parameter values $a = -0.8$, $b_0 = 0.4$ and $b_1 = 0.6$, the horizons being $N_1 = 1$ and $N = N_u = 3$. As has been shown, predicted values of the process output over the horizon are first calculated and rewritten in the form of equation (4.6), and then the control law is computed using expression (4.9).

Predictor polynomials $E_j(z^{-1})$, $F_j(z^{-1})$ from $j = 1$ to $j = 3$ will be calculated solving the Diophantine equation (4.3), with

$$\tilde{A}(z^{-1}) = A(z^{-1})(1 - z^{-1}) = 1 - 1.8z^{-1} + 0.8z^{-2}$$

In this simple case where the horizon is not too long, the polynomials can be directly obtained dividing 1 by $\tilde{A}(z^{-1})$ with simple calculations. As has been explained above, they can also be computed recursively, starting with the values obtained at the first step of the division, that is:

$$E_1(z^{-1}) = 1 \qquad F_1(z^{-1}) = 1.8 - 0.8z^{-1}$$

Whatever the procedure employed, the values obtained are:

$$E_2 = 1 + 1.8z^{-1} \qquad F_2 = 2.44 - 1.44z^{-1}$$
$$E_3 = 1 + 1.8z^{-1} + 2.44z^{-2} \quad F_3 = 2.952 - 1.952z^{-1}$$

With these values and the polynomial $B(z^{-1}) = 0.4 + 0.6z^{-1}$, the values of $G_i(z^{-1})$ are:

$$\begin{aligned} G_1 &= 0.4 + 0.6z^{-1} \\ G_2 &= 0.4 + 1.32z^{-1} + 1.08z^{-2} \\ G_3 &= 0.4 + 1.32z^{-1} + 2.056z^{-2} + 1.464z^{-3} \end{aligned}$$

and so the predicted outputs can be written as:

$$\begin{aligned} \begin{bmatrix} \hat{y}(t+1 \mid t) \\ \hat{y}(t+2 \mid t) \\ \hat{y}(t+3 \mid t) \end{bmatrix} &= \begin{bmatrix} 0.4 & 0 & 0 \\ 1.32 & 0.4 & 0 \\ 2.056 & 1.32 & 0.4 \end{bmatrix} \begin{bmatrix} \Delta u(t) \\ \Delta u(t+1) \\ \Delta u(t+2) \end{bmatrix} + \\ &+ \underbrace{\begin{bmatrix} 0.6\,\Delta u(t-1) + 1.8y(t) - 0.8y(t-1) \\ 1.08\,\Delta u(t-1) + 2.44y(t) - 1.44y(t-1) \\ 1.464\,\Delta u(t-1) + 2.952y(t) - 1.952y(t-1) \end{bmatrix}}_{\mathbf{f}} \end{aligned}$$

The following step is to calculate $\mathbf{H}^{-1}\mathbf{b}$. If λ is taken as equal to 0.8

$$(\mathbf{G}^T\mathbf{G} + \lambda\mathbf{I})^{-1}\mathbf{G}^T = \begin{bmatrix} 0.133 & 0.286 & 0.147 \\ -0.154 & -0.165 & 0.286 \\ -0.029 & -0.154 & 0.1334 \end{bmatrix}$$

As only $\Delta u(t)$ is needed for the calculations, only the first row of the matrix is used, obtaining the following expression for the control law:

$$\begin{aligned} \Delta u(t) &= -0.604\,\Delta u(t-1) - 1.371y(t) + 0.805y(t-1) + \\ &+ \ 0.133w(t+1) + 0.286w(t+2) + 0.147w(t+3) \end{aligned}$$

where $w(t+i)$ is the reference trajectory which can be considered constant and equal to the current setpoint or a first order approach to the desired value. Then the control signal is a function of this desired reference and of past inputs and outputs and is given by:

$$\begin{aligned} u(t) &= 0.396u(t-1) + 0.604u(t-2) - 1.371y(t) + 0.805y(t-1) + \\ &+ \ 0.133w(t+1) + 0.286w(t+2) + 0.147w(t+3) \end{aligned} \tag{4.14}$$

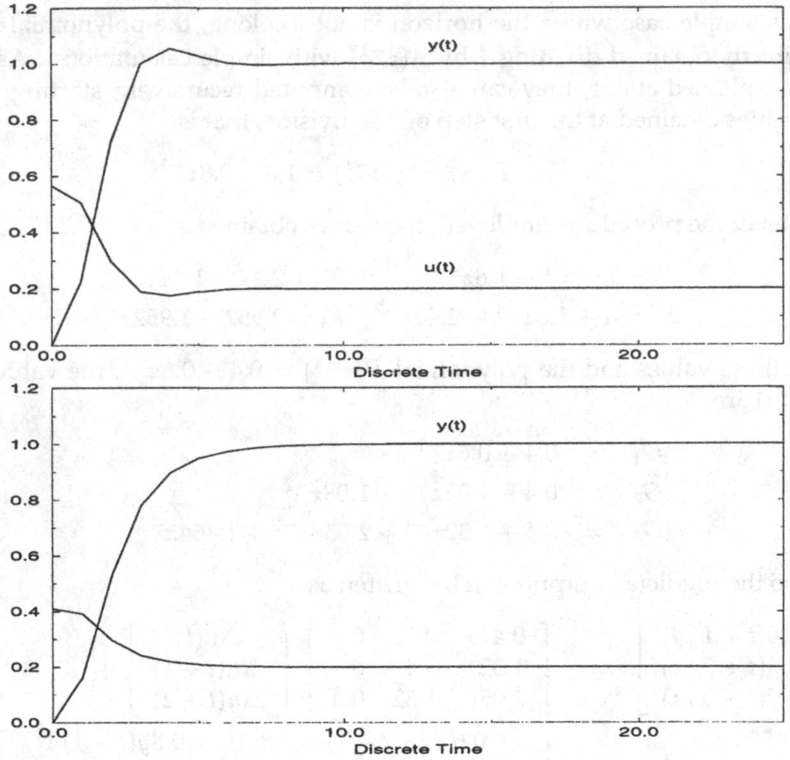

Figure 4.2: System response

Simulation results show the behaviour of the closed-loop system. In the first graph of figure 4.2 the reference is constant and equal to 1, meanwhile in the second one there is a smooth approach to the same value, obtaining a slightly different response, slower but without overshoot.

The GPC control law can also be calculated without the use of the Diophantine equation.

To obtain the control law it is necessary to know matrix **G** and the free response **f**, in order to compute $\mathbf{u} = (\mathbf{G}^T\mathbf{G} + \lambda\mathbf{I})^{-1}\mathbf{G}^T(\mathbf{w} - \mathbf{f})$. Matrix **G** is composed of the plant step response coefficients, so that the elements of the first column of this matrix are the first N coefficients, that can be computed as:

$$g_j = -\sum_{i=1}^{j} a_i g_{j-i} + \sum_{i=0}^{j-1} b_i \text{ with } g_k = 0 \quad \forall k < 0$$

being b_i and a_i the parameters of the numerator and denominator of the transfer function.

Therefore, as the prediction horizon is 3, $A = 1 - 0.8z^{-1}$ and $B = 0.4 +$

$0.6z^{-1}$:

$$\begin{aligned} g_0 &= b_0 = 0.4 \\ g_1 &= -a_1 g_0 + b_0 + b_1 = 1.32 \\ g_2 &= -a_1 g_1 - a_2 g_0 - a_3 g_0 + b_0 + b_1 = 2.056 \end{aligned}$$

and the matrix is given by:

$$\mathbf{G} = \begin{bmatrix} 0.4 & 0 & 0 \\ 1.32 & 0.4 & 0 \\ 2.056 & 1.32 & 0.4 \end{bmatrix}$$

which logically coincides with the one obtained by the previous method.

The free response can also be calculated without the use of the Diophantine equation, just noting that it is the response of the plant assuming that future controls equal the previous control $u(t-1)$ and that the disturbance is constant. Thus, using the transfer function:

$$\begin{aligned} y(t) &= 0.8y(t-1) + 0.4u(t-1) + 0.6u(t-2) \\ y(t+1) &= 0.8y(t) + 0.4u(t) + 0.6u(t-1) \end{aligned}$$

If both equations are added and $y(t+1)$ is extracted:

$$y(t+1) = 1.8y(t) - 0.8y(t-1) + 0.4 \triangle u(t) + 0.6 \triangle u(t-1)$$

Now, considering that in the free response only the control increments before instant t appear:

$$\begin{aligned} f(t+1) &= 1.8y(t) - 0.8y(t-1) + 0.6 \triangle u(t-1) \\ f(t+2) &= 1.8f(t+1) - 0.8y(t) = 2.44y(t) - 1.44y(t-1) + 1.08 \triangle u(t-1) \\ f(t+3) &= 1.8f(t+2) - 0.8f(t+1) = \\ &= 2.952y(t) - 1.952y(t-1) + 1.464 \triangle u(t-1) \end{aligned}$$

Vector **f** obtained this way is the same as the one previously obtained, so the control law is the one given by equation (4.14).

4.5 Closed Loop Relationships

Closed loop relations can be obtained for the unconstrained GPC. The closed loop system can be posed in the classical pole-placement structure of figure 4.3

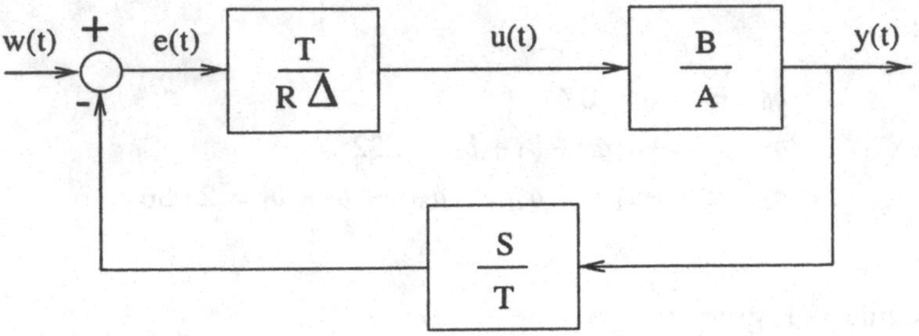

Figure 4.3: Classical pole-placement structure

The control law can be stated as:

$$R(z^{-1}) \triangle u(t) = T(z^{-1})w(t) - S(z^{-1})y(t) \tag{4.15}$$

being R, S and T polynomials in the backward shift operator. This control law can be considered as composed of a feedforward term (T/R) and a feedback part (S/R). In this situation it is possible to obtain the closed loop transfer function and derive some properties such as stability and robustness. First, the general GPC control scheme of figure 4.3 must be rearranged to take the form of equation (4.15).

The control law of equation (4.9) gives the future control sequence $\mathbf{u}$. As a receding strategy is being used, only the first element of that sequence $\Delta u(t \mid t)$ is actually sent to the process, therefore the control action is given by:

$$\triangle u(t) = \mathbf{K}(\mathbf{w} - \mathbf{f}) = \sum_{i=N_1}^{N_2} k_i[w(t+i) - f(t+i)] \tag{4.16}$$

being $\mathbf{K}$ the first row of matrix $(\mathbf{G}^T\mathbf{G} + \lambda\mathbf{I})^{-1}\mathbf{G}^T$.

The general case in which the $C(z^{-1})$ polynomial is not equal to zero will be considered to obtain the free response. In many situations this polynomial is not identified, since identification is not easy due to its time-varying characteristics and the difficulty of the CARIMA model to describe general deterministic disturbances. In these cases it is substituted by the so-called T polynomial that can be regarded as a fixed observer or a prefilter, as will be discussed later.

Then the plant model is given by:

$$A(z^{-1})y(t) = B(z^{-1})\, u(t-1) + T(z^{-1})\frac{e(t)}{\triangle}$$

The Diophantine equation that must be solved now includes the T polynomial:

$$T(z^{-1}) = E_j(z^{-1}) \triangle A(z^{-1}) + z^{-j}F_j(z^{-1}) \tag{4.17}$$

Using this equation and the plant model, the future output value is given by:

$$y(t+j) = \frac{B(z^{-1})}{A(z^{-1})}u(t+j-1) + E_j(z^{-1})e(t+j) + \frac{F_j(z^{-1})}{A(z^{-1})\Delta}e(t)$$

replacing the $e(t)$ from the plant model and using (4.17):

$$y(t+j) = \frac{F_j}{T}y(t) + \frac{E_j B}{T} \Delta u(t+j-1) + E_j e(t+j)$$

The best prediction is obtained replacing $e(t+j)$ by its expected value (zero):

$$\hat{y}(t+j \mid t) = \frac{F_j}{T}y(t) + \frac{E_j B}{T} \Delta u(t+j-1)$$

This expression is a function of known values and future control actions. The control actions can be separated into the past ones (those taken before instant t) and future ones (which must be calculated by the controller) using the Diophantine equation:[1]

$$E_j(z^{-1})B(z^{-1}) = H_j(z^{-1})T(z^{-1}) + z^{-j}I_j(z^{-1}) \tag{4.18}$$

that leads to the prediction equation:

$$\begin{aligned} \hat{y}(t+j \mid t) &= H_j \Delta u(t+j) + \frac{I_j}{T} \Delta u(t-1) + \frac{F_j}{T} y(t) = \\ &= H_j \Delta u(t+j) + I_j \Delta u^f(t-1) + F_j y^f(t) \end{aligned} \tag{4.19}$$

Using the filtered variables $y^f(t) = \frac{y(t)}{T}$ and $\Delta u^f(t-1) = \frac{\Delta u(t-1)}{T}$, this equation provides the same prediction along the horizon as given by (4.6) when $T(z^{-1}) = 1$, where the coefficients of H_j are the elements of matrix $\mathbf{G}$ and I_j are the rows of vector $\mathbf{G}'$.

Now the free response of the system (the one that is needed for the control law) is given by:

$$\mathbf{f} = \mathbf{I}(z^{-1}) \Delta u^f(t-1) + \mathbf{F}(z^{-1})y^f(t) = \mathbf{I}(z^{-1})\frac{\Delta u(t-1)}{T(z^{-1})} + \mathbf{F}(z^{-1})\frac{y(t)}{T(z^{-1})}$$

Once the free response has been obtained when the T polynomial is considered, it can be included in the expression of the control law given by (4.16):

$$\begin{aligned} \Delta u(t) &= \mathbf{K}(\mathbf{w} - \mathbf{f}) = \sum_{i=N_1}^{N_2} k_i[w(t+i) - f(t+i)] = \\ &= \sum_{i=N_1}^{N_2} k_i w(t+i) - \sum_{i=N_1}^{N_2} k_i \frac{I_i(z^{-1})}{T(z^{-1})} \Delta u(t-1) - \sum_{i=N_1}^{N_2} k_i \frac{F_i(z^{-1})}{T(z^{-1})} y(t) \end{aligned}$$

[1] notice that this equation with $T(z^{-1}) = 1$ is implicitly used to derive $\mathbf{G}$ and $\mathbf{G}'$ in (4.6).

Omitting the term z^{-1} and reordering the last equation:

$$\left[T + z^{-1}\sum_{i=N_1}^{N_2} k_i I_i\right] \triangle u(t) = T\sum_{i=N_1}^{N_2} k_i w(t) - \sum_{i=N_1}^{N_2} k_i F_i y(t)$$

where it has been considered that the future reference trajectory keeps constant along the horizon or its evolution is unknown and therefore $w(t+i)$ is taken as equal to $w(t)$. In the other case, the first term of the right hand side should be expressed as $T\sum_{i=N_1}^{N_2} k_i z^i w(t)$ and therefore the following relations could change slightly.

The values of polynomials R and S can be obtained comparing the last equation with (4.15), and are given by:

$$R(z^{-1}) = \frac{T(z^{-1}) + z^{-1}\sum_{i=N_1}^{N_2} k_i I_i}{\sum_{i=N_1}^{N_2} k_i}$$

$$S(z^{-1}) = \frac{\sum_{i=N_1}^{N_2} k_i F_i}{\sum_{i=N_1}^{N_2} k_i}$$

The closed loop characteristic equation comes from the inclusion of the control action given by (4.15) in the plant model expressed as:

$$A \triangle y(t) = B \triangle u(t-1) + Te(t)$$

Therefore, if the control action:

$$\Delta u(t) = \frac{T}{R} w(t) - \frac{S}{R} y(t)$$

is replaced in the plant model, the following expression is obtained:

$$A \triangle y(t) = Bz^{-1}(\frac{T}{R} w(t) - \frac{S}{R} y(t)) + Te(t)$$

Extracting $y(t)$ from this equation provides the closed loop relation that gives the output as a function of the reference and the disturbance:

$$y(t) = \frac{BTz^{-1}}{RA\triangle + BSz^{-1}} w(t) + \frac{TR}{RA\triangle + BSz^{-1}} e(t) \tag{4.20}$$

and consequently the characteristic equation is given by:

$$RA\triangle + BSz^{-1} = 0$$

With a few manipulations and using (4.18), the characteristic polynomial can be decomposed as:

$$RA\triangle + BSz^{-1} = \frac{1}{\sum_{i=N_1}^{N_2} k_i}(T\tilde{A} + T\sum_{i=N_1}^{N_2} k_i z^{i-1}(B - \tilde{A}H_i)) = TP_c$$

And therefore the equation (4.20) turns to:

$$y(t) = \frac{Bz^{-1}}{P_c}w(t) + \frac{R}{P_c}e(t)$$

where it is shown that the T polynomial is cancelled in the closed loop transfer function between output and reference, as is the case in any observer, and that stability and performance are driven by the roots of polynomial P_c. However, it is difficult to establish clear dependencies of these roots on the tuning parameters N_1, N_2, N_U and λ. It is interesting to note that $P_c(1) = B(1)$, which guarantees offset-free response since the static gain of the transfer function between output and reference is always one.

When the GPC is written in the general pole-placement structure, some stability properties can be derived from the transfer function. A paper by Clarke and Mohtadi [33] presents some properties related to stability. It is proven that stability can be guaranteed if the tuning parameters (horizons and control-weighting factor) are correctly chosen. In following sections, two formulations related to GPC with guaranteed stability will be treated in more detail.

4.6 The Role of the T polynomial

Although the T polynomial does not appear in the transfer function between the output and the reference, this is not the case for the transfer function between the output and the disturbance. From equation (4.19) it can be seen that both the output $y(t)$ and the control move $\Delta u(t)$ appear in the prediction, and therefore in the control law, filtered by $1/T$. Thus, from a practical point of view, it means that the T polynomial can be treated as a filter. By ensuring that the degree of T is big enough, the roll-off of the filter attenuates the component of prediction error caused by model mismatch, which is particularly important at high frequencies. Notice that low frequency disturbances can be removed by the Δ term that appears in the prediction.

The high frequency disturbances are mainly due to the presence of high-frequency unmodelled dynamics and unmeasurable load disturbances. If there is no unmodelled dynamics, the effect of T is the rejection of disturbances, with no influence on reference tracking. In this case T can be used to *detune* the response to unmeasurable high-frequency load disturbances, preventing excessive control actions.

On the other hand, T is used as a design parameter that can influence robust stability. In this case the predictions will not be optimal but robustness in the face of uncertainties can be achieved, in a similar interpretation to that used by Ljung [68]. Then this polynomial can be considered as a prefilter as well as an observer. The effective use of observers is known to play an essential role in the robust realization of predictive controllers (see [33] for the effect of prefiltering on robustness).

4.6.1 Selection of the T Polynomial

The selection of the filter polynomial T is not a trivial matter. Although some guidelines are given in [106] for mean-level and dead-beat GPC, a systematic design strategy for the T filter has not been completely established. Usually it is assumed that the stronger filtering (considered as stronger that filtering with smaller bandwidth or with bigger slope if the bandwidth is the same) has better robustness properties against high frequency uncertainties. But this fact is not always true, as is shown with some counter-examples in [128]. In this paper and in [127] Yoon and Clarke present the guidelines for the selection of T . These guidelines are based upon the robustness margin improvement at high frequencies and conclude stating that, for open-loop stable processes, the best choice is

$$T(z^{-1}) = A(z^{-1})(1 - \beta z^{-1})^{N_1 - \delta(P)}$$

Where β is close to the dominant root of A, N_1 is the minimum prediction horizon and $\delta(P)$ is the degree of polynomial P (the filter used to generate a reference trajectory with specified dynamics, see section 2.1.2).

Notice that this idea of filtering for improving robustness also lies in Internal Model Control (IMC)[77] where, once the controller that provides the desired performance is obtained, it is detuned with a filter in order to improve robustness.

4.6.2 Relationships with other Formulations

The prefiltering with T can be compared to $\mathcal{H}_\infty$ optimization based on the Q parametrization [77], obtaining equivalent robustness results [126]. It implies that prefiltering with polynomial T is an alternative to the optimal Q whose computation is demanding, especially in the adaptive case.

Robustness is improved by the introduction of polynomial T but, on the other hand, this fact implies that the prediction is no longer optimal. In a certain way, this idea is similar to the one used in the Linear Quadratic Gaussian (LQG) regulator to recover the good robustness properties that the Linear Quadratic Regulator (LQR) looses with the inclusion of the observer. This recovery is achieved by means of the LQG/LTR method, that consists of the Loop Transfer Recovery, (LTR), in such a way that it approaches the open-loop transfer function of the LQR method. This can be done by acting on the Kalman filter parameters, working with fictitious covariances (see [45]). In this way robustness is gained although prediction deteriorates. In both cases, the loss of optimality in the prediction or estimation is not considered a problem, since the controller works and is robust.

4.7 The P Polynomial

Clarke in the presentation of Generalized Predictive Control [34] points out the possibility of the use of an additional polynomial $P(z^{-1})$ as a design element in a similar way as that employed in the Minimum Variance Controller [29] as a weighting polynomial, that would be used for model following.

The $P(z^{-1})$ polynomial can appear when defining an auxiliary output as happens for instance in EPSAC (see chapter 2):

$$\psi(t) = P(z^{-1})y(t)$$

in such a way that it affects the output. This polynomial allows the control objectives to expand by using its roots as design parameters. In this way *dead-beat*, pole-placement or LQ control can be achieved.

This can easily be incorporated into the standard GPC formulation by considering the augmented plant

$$A \triangle P(z^{-1})y(t) = BA \triangle P(z^{-1})u(t-1)$$

with P(1)=1 to guarantee $\psi(t) = y(t)$ in steady state.

This is equivalent to defining filtered auxiliary signals $\nu(t) = P(z^{-1})u(t)$ and $\psi(t) = P(z^{-1})y(t)$. Thus the plant is given by:

$$A \triangle \psi(t) = BA \triangle \nu(t-1)$$

Now the error that appears in the cost function is defined by $w(t+j) - \psi(t+j)$, which is equivalent to considering a reference trajectory generated by $1/P(z^{-1})$. A *dead-beat* control of $\psi(t)$ can be achieved acting on $\nu(t)$, whose closed loop transfer function is given by:

$$\psi(t) = \frac{B(z^{-1})}{B(1)} w(t)$$

This means that

$$y(t) = \frac{B(z^{-1})}{B(1)P(z^{-1})} w(t) \tag{4.21}$$

and the GPC algorithm is solved to provide the auxiliary control increment $\Delta\nu(t)$, from which the system input is calculated as:

$$u(t) = u(t-1) + \frac{\Delta\nu(t)}{P(z^{-1})}$$

As can be observed (4.21) is the typical response of a pole-placement method, with poles placed at zeros of the chosen $P(z^{-1})$. That is, the output is made to track the dynamics specified by $P(z^{-1})$.

4.8 Consideration of Measurable Disturbances

Many processes are affected by external disturbances caused by the variation of variables that can be measured. This situation is typical in processes whose outputs are affected by variations of the load regime. Consider for instance a cooled jacket continuous reactor where the temperature is controlled by manipulating the water flow entering the cooling jacket. Any variation of the reactive flows will influence the reactor temperature. These types of perturbations, also known as load disturbances, can easily be handled by the use of feedforward controllers. Known disturbances can be taken explicitly into account in MPC, as will be seen in the following.

Consider a process described by the following In this case the CARIMA model must be changed to include the disturbances:

$$A(z^{-1})y(t) = B(z^{-1})u(t-1) + D(z^{-1})v(t) + \frac{1}{\Delta}C(z^{-1})e(t) \tag{4.22}$$

where the variable $v(t)$ is the measured disturbance at time t and $D(z^{-1})$ is a polynomial defined as:

$$D(z^{-1}) = d_0 + d_1 z^{-1} + d_2 z^{-2} + \cdots + d_{n_d} z^{-n_d}$$

Multiplying equation (4.22) by $\Delta E_j(z^{-1})z^j$:

$$\begin{aligned} E_j(z^{-1})\tilde{A}(z^{-1})y(t+j) &= E_j(z^{-1})B(z^{-1}) \Delta u(t+j-1) + \\ &+ E_j(z^{-1})D(z^{-1}) \Delta v(t+j) + E_j(z^{-1})e(t+j) \end{aligned}$$

By using (4.3) and after some manipulation we get:

$$\begin{aligned} y(t+j) &= F_j(z^{-1})y(t) + E_j(z^{-1})B(z^{-1}) \Delta u(t+j-1) + \\ &+ E_j(z^{-1})D(z^{-1}) \Delta v(t+j) + E_j(z^{-1})e(t+j) \end{aligned}$$

Notice that because the degree of $E_j(z^{-1})$ is $j-1$, the noise terms are all in the future. By taking the expectation operator and considering that $E[e(t)] = 0$, the expected value for $y(t+j)$ is given by:

$$\begin{aligned} \hat{y}(t+j|t) = E[y(t+j)] &= F_j(z^{-1})y(t) + E_j(z^{-1})B(z^{-1}) \Delta u(t+j-1) + \\ &+ E_j(z^{-1})D(z^{-1}) \Delta v(t+j) \end{aligned}$$

By making the polynomial $E_j(z^{-1})D(z^{-1}) = H_j(z^{-1}) + z^{-j}H'_j(z^{-1})$, with $\delta(H_j(z^{-1})) = j-1$, the prediction equation can now be written as:

$$\begin{aligned} \hat{y}(t+j|t) = G_j(z^{-1}) \Delta u(t+j-1) + H_j(z^{-1}) \Delta v(t+j) + \\ + G'_j(z^{-1}) \Delta u(t-1) + H'_j(z^{-1}) \Delta v(t) + F_j(z^{-1})y(t) \end{aligned} \tag{4.23}$$

Notice that the last three terms of the right hand side of this equation depend on past values of the process output, measured disturbances and input variables and correspond to the free response of the process considered if the control signals and measured disturbances are kept constant; while the first term only depends on future values of the control signal and can be interpreted as the forced response. That is, the response obtained when the initial conditions are zero $y(t-j)=0$, $\Delta u(t-j-1)=0$, $\Delta v(t-j)$ for $j>0$.

The second term of equation (4.23) depends on the future deterministic disturbances. In some cases, when they are related to the process load, future disturbances are known. In other cases, they can be predicted using trends or by other means. If this is the case, the term corresponding to future deterministic disturbances can be computed. If the future load disturbances are supposed to be constant and equal to the last measured value (i.e. $v(t+j)=v(t)$), then $\Delta v(t+j)=0$ and the second term of this equation vanishes.

Equation (4.23) can be rewritten as:

$$\hat{y}(t+j|t) = G_j(z^{-1})\,\Delta\, u(t+j-1) + H_j(z^{-1})\,\Delta\, v(t+j) + f_j$$

with $f_j = G'_j(z^{-1})\,\Delta\, u(t-1) + H'_j(z^{-1})\,\Delta\, v(t) + F_j(z^{-1})y(t)$.

Let us now consider a set of N j-ahead predictions:

$$\begin{aligned}
\hat{y}(t+1|t) &= G_1(z^{-1})\,\Delta\, u(t) + H_j(z^{-1})\,\Delta\, v(t+1) + f_1\\
\hat{y}(t+2|t) &= G_2(z^{-1})\,\Delta\, u(t+1) + H_j(z^{-1})\,\Delta\, v(t+2) + f_2\\
&\;\vdots\\
\hat{y}(t+N|t) &= G_N(z^{-1})\,\Delta\, u(t+N-1) + H_j(z^{-1})\,\Delta\, v(t+N) + f_N
\end{aligned}$$

Because of the recursive properties of the E_j polynomial, these expressions can be rewritten as:

$$\begin{bmatrix} \hat{y}(t+1|t)\\ \hat{y}(t+2|t)\\ \vdots\\ \hat{y}(t+j|t)\\ \vdots\\ \hat{y}(t+N|t) \end{bmatrix} = \begin{bmatrix} g_0 & 0 & \cdots & 0 & \cdots & 0\\ g_1 & g_0 & \cdots & 0 & \cdots & 0\\ \vdots & \vdots & \ddots & \vdots & \vdots & \vdots\\ g_{j-1} & g_{j-2} & \cdots & g_0 & \vdots & 0\\ \vdots & \vdots & \vdots & \vdots & \ddots & \vdots\\ g_{N-1} & g_{N-2} & \cdots & \cdots & \cdots & g_0 \end{bmatrix} \begin{bmatrix} \Delta u(t)\\ \Delta u(t+1)\\ \vdots\\ \Delta u(t+j-1)\\ \vdots\\ \Delta u(t+N-1) \end{bmatrix}$$

$$+ \begin{bmatrix} h_0 & 0 & \cdots & 0 & \cdots & 0\\ h_1 & h_0 & \cdots & 0 & \cdots & 0\\ \vdots & \vdots & \ddots & \vdots & \vdots & \vdots\\ h_{j-1} & \cdots & h_1 & h_0 & \vdots & 0\\ \vdots & \vdots & \vdots & \ddots & \ddots & \vdots\\ h_{N-1} & \cdots & \cdots & \cdots & h_1 & h_0 \end{bmatrix} \begin{bmatrix} \Delta v(t+1)\\ \Delta v(t+2)\\ \vdots\\ \Delta v(t+j-1)\\ \vdots\\ \Delta v(t+N) \end{bmatrix} + \begin{bmatrix} f_1\\ f_2\\ \vdots\\ f_j\\ \vdots\\ f_N \end{bmatrix}$$

where $H_j(z^{-1}) = \sum_{i=1}^{j} h_i z^{-i}$, being h_i the coefficients of the system step response to the disturbance.

By making $\mathbf{f}' = \mathbf{Hv} + \mathbf{f}$, the prediction equation is now:

$$\mathbf{y} = \mathbf{Gu} + \mathbf{f}'$$

which has the same shape as the general prediction equation used in the case of zero measured disturbances. The future control signal can be found in the same way, simply using as free response the process response due to initial conditions (including external disturbances) and future "known" disturbances.

4.9 Use of a Different Predictor in GPC

In this section it is shown that a GPC is equivalent to a structure based on an optimal predictor plus a classical two degree of freedom controller. If the optimal predictor is replaced by a Smith predictor [114], a new controller with similar nominal performance and better robust properties is obtained. This has great interest in the case of time-delay systems. The controller uses the same procedure to compute the control signal although the future outputs are calculated using the Smith predictor instead of the optimal predictor.

4.9.1 Equivalent Structure

In order to show the equivalence between the GPC and a structure composed of an optimal predictor and a classical controller, a CARIMA model with white integrated noise is used to compute the prediction. Let us consider a process with a dead time d, $T(z^{-1}) = 1$, $N_1 = d+1$, $N_2 = d+N$, $N_u = N$, and the weighting sequences $\delta(j) = 1$, $\lambda(j) = \lambda$. Thus it is possible to write:

$$\begin{aligned}\hat{y}(t+d+j \mid t) &= (1-a_1)\hat{y}(t+d+j-1 \mid t) + \\ &+(a_1-a_2)\hat{y}(t+d+j-2 \mid t) + \ldots + a_{na}\hat{y}(t+d+j-n_a-1 \mid t) + \\ &+b_0 \Delta u(t+j-1) + \ldots + b_{nb} \Delta u(t+j-1-nb)\end{aligned} \quad (4.24)$$

If this equation is applied recursively for $j = 1, 2, \cdots, N$ we get

$$\begin{bmatrix} \hat{y}(t+d+1 \mid t) \\ \hat{y}(t+d+2 \mid t) \\ \vdots \\ \hat{y}(t+d+N \mid t) \end{bmatrix} = \mathbf{G} \begin{bmatrix} \Delta u(t) \\ \Delta u(t+1) \\ \vdots \\ \Delta u(t+N-1) \end{bmatrix} + \mathbf{H} \begin{bmatrix} \Delta u(t-1) \\ \Delta u(t-2) \\ \vdots \\ \Delta u(t-nb) \end{bmatrix} +$$

$$+ \mathbf{S} \begin{bmatrix} \hat{y}(t+d \mid t) \\ \hat{y}(t+d-1 \mid t) \\ \vdots \\ \hat{y}(t+d-na \mid t) \end{bmatrix}$$

where **G**, **H** and **S** are constant matrices of dimension $N \times N$, $N \times n_b$ and $N \times n_a + 1$ respectively. This equation can be written in a vector form as follows:

$$\hat{\mathbf{y}} = \mathbf{G}\mathbf{u} + \mathbf{H}\mathbf{u}' + \mathbf{S}\mathbf{y}' = \mathbf{G}\mathbf{u} + \mathbf{f}$$

where it is clear that $\mathbf{f} = \mathbf{H}\mathbf{u}' + \mathbf{S}\mathbf{y}'$ is composed of the terms in the past and correspond to the free response of the system.

If $\hat{\mathbf{y}}$ is introduced in the cost function, $J(N)$ is a function of $\mathbf{y}'$, $\mathbf{u}$, $\mathbf{u}'$ and the reference sequence. Minimizing $J(N)$ with respect to **u**, that is, $\Delta u(t)$, $\Delta u(t+1)$ ... $\Delta u(t+N-1)$) leads to

$$\mathbf{M} \begin{bmatrix} \Delta u(t) \\ \Delta u(t+1) \\ \vdots \\ \Delta u(t+N-1) \end{bmatrix} = \mathbf{P}_0 \begin{bmatrix} \hat{y}(t+d \mid t) \\ \hat{y}(t+d-1 \mid t) \\ \vdots \\ \hat{y}(t+d-na \mid t) \end{bmatrix} + \mathbf{P}_1 \begin{bmatrix} \Delta u(t-1) \\ \Delta u(t-2) \\ \vdots \\ \Delta u(t-nb) \end{bmatrix} + $$

$$+ \mathbf{P}_2 \begin{bmatrix} w(t+d+1) \\ w(t+d+2) \\ \vdots \\ w(t+d+N) \end{bmatrix}$$

Where $\mathbf{M} = \mathbf{G}^T\mathbf{G} + \lambda\mathbf{I}$ and $\mathbf{R} = \mathbf{G}^T$ are of dimension $N \times N$, $\mathbf{P}_0 = -\mathbf{G}^T S$ of dimension $N \times n_a + 1$ and $\mathbf{P}_1 = -\mathbf{G}^T\mathbf{H}$ of dimension $N \times n_b$. As in a receding horizon algorithm only the value of $\Delta u(t)$ is computed, if **q** is the first row of matrix $\mathbf{M}^{-1}$, $\Delta u(t)$ is given by:

$$\Delta u(t) = \mathbf{q}\mathbf{P}_0 \begin{bmatrix} \hat{y}(t+d \mid t) \\ \hat{y}(t+d-1 \mid t) \\ \vdots \\ \hat{y}(t+d-na \mid t) \end{bmatrix} + \mathbf{q}\mathbf{P}_1 \begin{bmatrix} \Delta u(t-1) \\ \Delta u(t-2) \\ \vdots \\ \Delta u(t-nb) \end{bmatrix} +$$

$$+ \mathbf{q}\mathbf{P}_2 \begin{bmatrix} w(t+d+1) \\ w(t+d+2) \\ \vdots \\ w(t+d+N) \end{bmatrix}$$

Therefore the control increment $\Delta u(t)$ can be written as:

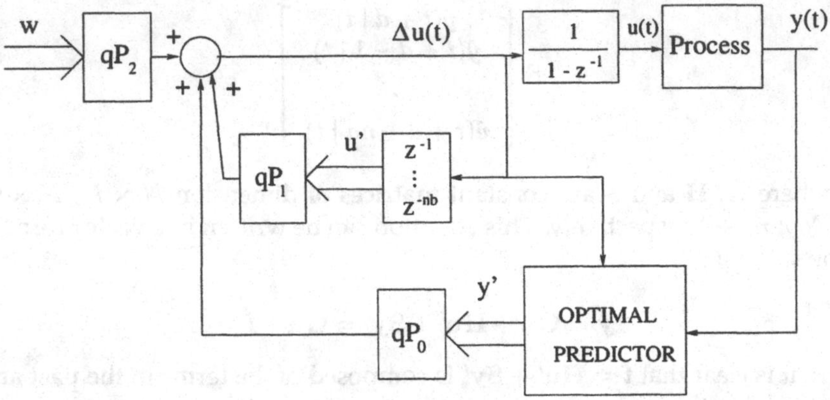

Figure 4.4: Control Scheme

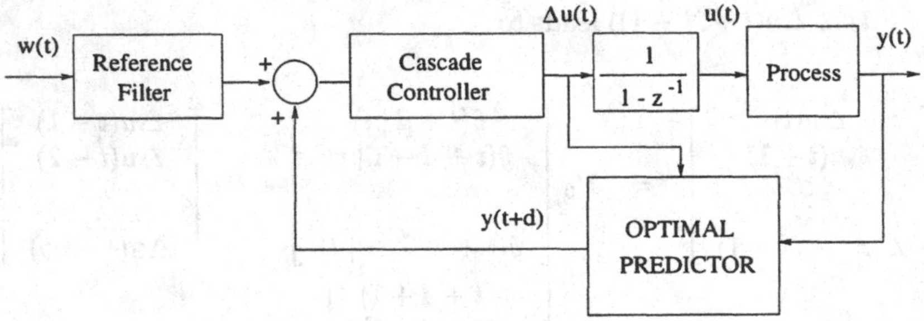

Figure 4.5: Equivalent control structure of the GPC

$$\Delta u(t) = \mathbf{qP_0 y'} + \mathbf{qP_1 u'} + \mathbf{qP_2 w}$$

The resulting control scheme (see figure 4.4) is a linear feedback of predictions $\hat{y}(t+d \mid t)$, ..., $\hat{y}(t+d-na \mid t)$ generated by an optimal predictor. That is, prediction over a time horizon equal to the process dead time. The controller coefficients are computed for each choice of N and λ.

The block diagram presented in figure 4.4 can easily be transformed into the one shown in figure 4.5, which shows that a GPC is equivalent to a classical controller plus an optimal predictor. This classical controller is composed of a reference filter and a cascade block. If the plant can be modelled by a first order system with a dead time then the classical controller results in a simple PI [85]. In general, the primary controller is of the same order as the model of the plant.

This relation can be used to study how to improve the robustness of the GPC using a different predictor structure. Note that the computation of

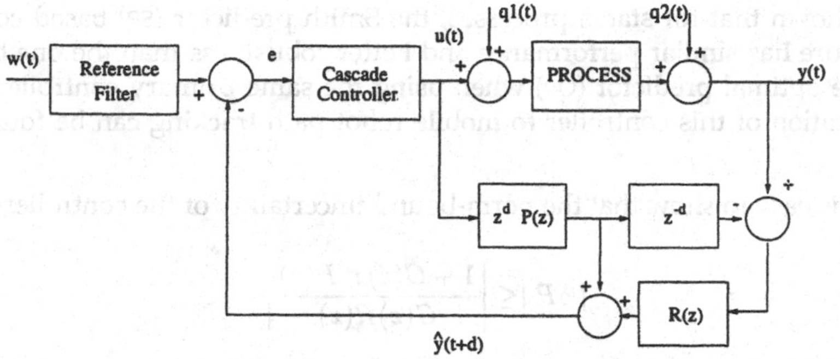

Figure 4.6: Equivalent structure for the OP in the GPC

the classical controller is done independently to the predictor structure, so different predictors can be compared using the same controller parameters of the GPC.

The prediction can be obtained directly from the polynomials A and B and the delay d considering the equation (4.24) for $j = 1$:

$$\hat{y}(t+1 \mid t) = (1-\tilde{A}(z^{-1}))zy(t)+\tilde{B}(z^{-1})u(t-d) \quad \text{with } \tilde{B}(z^{-1}) = (1-z^{-1})B(z^{-1})$$

Using the same procedure for $j = 2....$, $\hat{y}(t+j \mid t)$ is computed as:

$$\hat{y}(t+j \mid t) = ((1-\tilde{A}(z^{-1}))z)^j y(t) + \sum_{i=1}^{j}(1-\tilde{A}(z^{-1}))^{i-1}\tilde{B}(z^{-1})u(t-d+j-1)$$

The predicted output at instant $t+d$ is therefore:

$$\hat{y}(t+d \mid t) = ((1-\tilde{A}(z^{-1}))z)^d y(t) + (1-(1-\tilde{A}(z^{-1}))^d)z^d Pu(t)$$

being $P(z) = \frac{B(z^{-1})z^{-1}}{A(z^{-1})}z^{-d}$ the plant model.

Defining $R(z) = (1-\tilde{A}(z^{-1}))^d z^d$, the prediction can be written as:

$$\hat{y}(t+d \mid t) = R(z)y(t) + (z^d - R(z))P(z)u(t)$$

The closed-loop block diagram of the whole control system is shown in figure 4.6. Now it is clear that the GPC has a structure similar to the well known dead-time compensators like the Smith predictor (obtained when $R(z) = 1$) and that in the absence of a dead time the final control law is a classical two-degree of freedom controller.

The theoretical comparative results about the robustness and performance of GPC and the GPC which uses a Smith predictor can be found in [87], where it is shown that for stable processes, the Smith predictor (SP) based control structure has similar performance and better robustness than the one based on the optimal predictor (OP) when using the same primary controller. An application of this controller to mobile robot path tracking can be found in [89].

It is easy to show that the norm-bound uncertainty of the controller is:

$$|\ \delta P\ | \leq \left| \frac{1 + C(z) z^d P(z)}{C(z) R(z)} \right|$$

Notice that the norm-bound uncertainty is inversely proportional to $|\ R(z)\ |$ and that for the GPC the block $R(z)$ has high pass characteristics, so the controller has low values of the norm-bound uncertainty at high frequencies. On the other hand, the Smith predictor based controller has $R(z) = 1$ and consequently a higher robustness index.

As has been mentioned in this chapter, the robustness of the GPC can be improved by the use of a prefilter in the prediction equations. The effect of this filter (the T polynomial) can also be analysed in figure 4.6 because, when T is included in the predictor $R(z)$ is a function of $T(z)$. If T is chosen appropriately then $R(z)$ could have low pass characteristic and the robustness of the controller could be increased. On the other hand, the disturbance rejection response deteriorates when the robustness increases [6]. Note that for the Smith predictor based controller a filter $F(z)$ could also be included, but in this case the tuning of F is simpler than the tuning of T in the GPC case as for the new structure $R(z) = F(z)$. Note that to improve the robustness F must be chosen as a low pass filter [88].

For the computation of the controller the following steps must be taken:

- compute the prediction of the output (from t to $t + d$) using the open loop model of the plant without considering disturbances

- correct each open loop prediction adding the mismatch between the output and the prediction :

$$\hat{y}(t + d - i \mid t) = \hat{y}(t + d - i \mid t) + y(t - i) - \hat{y}(t - i)$$

- compute the control law as in the normal GPC using the coefficients of $\mathbf{q}$, $\mathbf{P}_0$, $\mathbf{P}_1$ and $\mathbf{P}_2$.

Note that to compute the control law it is also possible to use the forced and free response concepts used in standard GPC. In this case the forced response can be computed as done in GPC but the free response must be computed using the Smith predictor from t to $t + d$ and the optimal predictor from $t + d + 1$ to $t + N$.

4.9.2 A Comparative Example

In order to illustrate the robustness properties of the proposed GPC an example comparing the SPGPC and the GPC, is presented. It corresponds to a temperature control of a heat-exchanger, where the model used in the predictor is a first order system with a dead time, such as the one presented in [86]. This system represents a typical industrial process and because of this it is normally used to evaluate the performance of industrial controllers. In the example ARIMA disturbances are considered.

The relation between the output temperature and the input flow in the heat-exchanger was obtained using the reaction curve method and is given by:

$$P(s) = \frac{0.12e^{-3s}}{1+6s}$$

Using a sample time $T_s = 0.6s$ the obtained discrete plant is

$$P(z) = \frac{bz^{-1}}{1-az^{-1}}z^{-d}$$

where the nominal values of the parameters are: $d_n = 5, a_n = 0.905$ and $b_n = 0.0114$. In practice there are errors in the estimation of all parameters but for this example only dead time uncertainty is considered, with a maximum value of $\delta d = 2$.

The GPC is computed in order to obtain (for the nominal case) a step response faster than the open loop one. Thus, the GPC parameters were chosen as $N = 15, \lambda = 0.8$. The SPGPC has the same parameters.

The closed-loop behaviour of the GPC and the SPGPC for the nominal case is shown in figure 4.7.a. At $t = 0$ a step change in the reference is performed and at $t = 100$ samples a 10% step load disturbance is applied to the system. The noise is generated with an ARIMA model with uniform distribution in ± 0.005. As can be observed, both systems have similar setpoint tracking and disturbance rejection behaviour for the nominal case.

In the next simulation the delay of the plant is set to $d = 7$ and again a step change in the reference is considered at $t = 0$. Figure 4.7.b shows that the SPGPC based control system is stable and the GPC one is unstable.

It must be stated that the comparison has been made with a GPC without T polynomial and the introduction of this in the formulation could improve its response when the dead time uncertainty appears, but at the same time the nominal disturbance rejection response will be deteriorated.

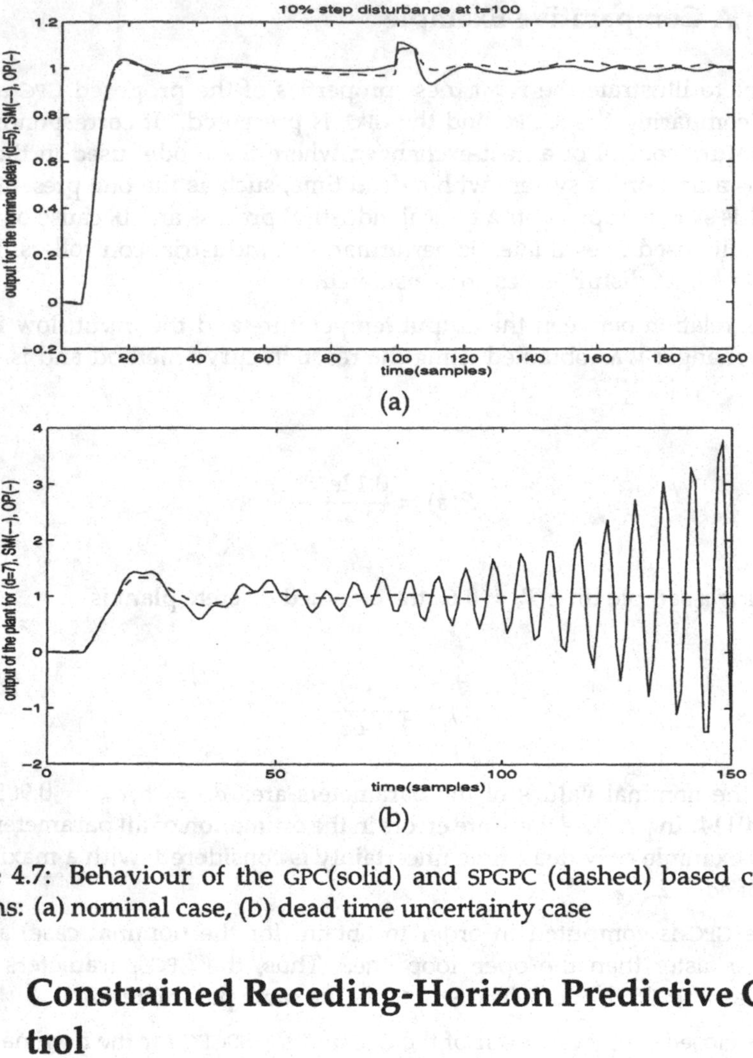

Figure 4.7: Behaviour of the GPC(solid) and SPGPC (dashed) based control systems: (a) nominal case, (b) dead time uncertainty case

4.10 Constrained Receding-Horizon Predictive Control

In spite of the great success of GPC in industry, there was an original lack of theoretic results about the properties of predictive control and an initial gap in important questions such as stability and robustness. In fact, the majority of stability results are limited to the infinite horizon case and there is a lack of a clear theory relating the closed loop behaviour to design parameters, such as horizons and weighting sequences.

Bearing in mind the need to solve some of these drawbacks, a variation of the standard formulation of GPC appears developed by Clarke and Scattolini called Constrained Receding-Horizon Predictive Control (CRHPC) [30], [84], [80], which allows stability and robustness results to be obtained for small horizons. The idea basically consists of deriving a future control sequence so that the predicted output over some future time range is constrained to be at the reference value exactly, as shown in figure 4.8. Some degrees of freedom of the future control signals are employed to force the output, whilst the rest is available to minimize the cost function over a certain interval.

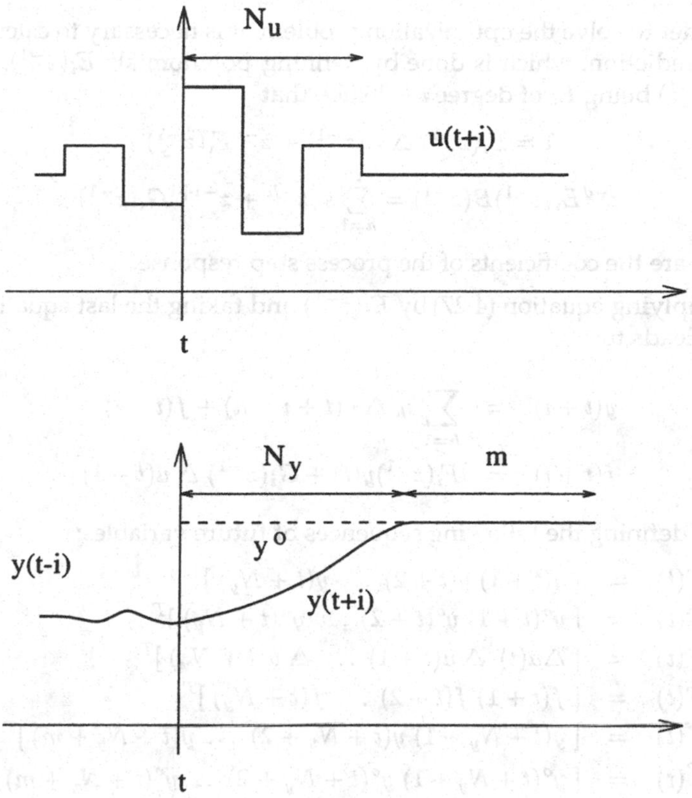

Figure 4.8: Constrained Receding Horizon Predictive Control

4.10.1 Computation of the Control Law

The computation of the control law is performed in a similar way to GPC, although the calculations become a little more complicated. Control signals must minimize the objective function:

$$J(N_y, N_u) = \sum_{i=1}^{N_y} \mu(i)[\hat{y}(t+i \mid t) - y^o(t+i)]^2 + \sum_{i=0}^{N_u} \lambda(i)[\Delta u(t+j)]^2 \quad (4.25)$$

where $y^o(t+i)$ is the reference, subject to constraints

$$\begin{array}{rcll} y(t+N_y+i) & = & y^o(t+N_y+1) & i=1\ldots m \\ \Delta u(t+N_u+j) & = & 0 & j>0 \end{array} \quad (4.26)$$

The design parameters are the values of the horizons, the weighting sequences $\mu(i)$ and $\lambda(i)$ and the value of m, which defines the instants of coincidence between output and reference. The system is modelled by:

$$\Delta A(z^{-1})y(t) = B(z^{-1}) \Delta u(t-d) \tag{4.27}$$

In order to solve the optimization problem, it is necessary to calculate the output prediction, which is done by defining polynomials $E_i(z^{-1})$, $F_i(z^{-1})$ and $G_i(z^{-1})$ being E_i of degree $i-1$ such that

$$1 = E_i(z^{-1}) \Delta A(z^{-1}) + z^{-i}F_i(z^{-1})$$

$$z^{-d}E_i(z^{-1})B(z^{-1}) = \sum_{h=1}^{i} s_h z^{-h} + z^{-i+1}G_i(z^{-1})$$

where s_h are the coefficients of the process step response.

Multiplying equation (4.27) by $E_i(z^{-1})$ and taking the last equations into account, leads to:

$$\begin{aligned} y(t+i) &= \sum_{h=1}^{i} s_h \Delta u(t+i-h) + f(t+i) \\ f(t+i) &= F_i(z^{-1})y(t) + G_i(z^{-1}) \Delta u(t-1) \end{aligned}$$

Now defining the following sequences of future variables:

$$\begin{aligned} Y(t) &= [\, y(t+1)\; y(t+2)\; \ldots\; y(t+N_y)\,]^T \\ Y^o(t) &= [\, y^o(t+1)\; y^o(t+2)\; \ldots\; y^o(t+N_y)\,]^T \\ \Delta U(t) &= [\, \Delta u(t)\; \Delta u(t+1)\; \ldots\; \Delta u(t+N_u)\,]^T \\ F(t) &= [\, f(t+1)\; f(t+2)\; \ldots\; f(t+N_y)\,]^T \\ \overline{Y}(t) &= [\, y(t+N_y+1)\; y(t+N_y+2)\; \ldots\; y(t+N_y+m)\,]^T \\ \overline{Y}^o(t) &= [\, y^o(t+N_y+1)\; y^o(t+N_y+2)\; \ldots y^o(t+N_y+m)\,]^T \\ \overline{F}^o(t) &= [\, f(t+N_y+1)\; f(t+N_y+2)\; \ldots\; f(t+N_y+m)\,]^T \end{aligned}$$

the following matrices, of dimension $N_y \times (N_u+1)$ and $m \times (N_u+1)$ respectively:

$$G = \begin{bmatrix} s_1 & 0 & 0 & \cdots & 0 \\ s_2 & s_1 & 0 & \cdots & 0 \\ \vdots & \vdots & \vdots & \vdots & \vdots \\ s_{Ny} & s_{Ny-1} & s_{Ny-2} & \cdots & s_{Ny-Nu} \end{bmatrix}$$

$$\overline{G} = \begin{bmatrix} s_{Ny+1} & s_{Ny} & \cdots & s_{Ny-Nu+1} \\ s_{Ny+2} & s_{Ny+1} & \cdots & s_{Ny-Nu+2} \\ \vdots & \vdots & \vdots & \vdots \\ s_{Ny+m} & s_{Ny+m-1} & \cdots & s_{Ny-Nu+m} \end{bmatrix}$$

and the weighting sequences

$$\begin{aligned} M(t) &= \text{diag}\,\{\mu(1), \mu(2), \ldots \mu(N_y)\} \\ \Lambda(t) &= \text{diag}\,\{\lambda(0), \lambda(1), \ldots \lambda(N_u)\} \end{aligned}$$

the following relations hold:

$$\begin{aligned} Y(t) &= G\,\Delta U(t) + F(t) \\ \overline{Y}(t) &= \overline{G}\,\Delta U(t) + \overline{F}(t) \end{aligned}$$

Then the cost function (equation 4.25) and the constraints (4.26) can be written as:

$$J = [Y(t) - Y^o(t)]^T\, M(t)\, [Y(t) - Y^o(t)] + \Delta U^T(t)\, \Lambda(t)\, \Delta U(t)$$
$$\overline{G}\,\Delta U(t) + \overline{F}(t) = \overline{Y}^o(t)$$

The solution can be obtained by the use of Lagrange multipliers. If the common case of constant weighting sequence is considered, that is, $M(t) = \mu I$, $\Lambda(t) = \lambda I$, the solution can be written as:

$$\begin{aligned} \Delta U(t) = (\mu G^T G + \lambda I)^{-1}[\mu G^T (Y^o(t) - F(t)) + \overline{G}^T (\overline{G}(\mu G^T G + \lambda I)^{-1}\overline{G}^T)^{-1} \\ \times(\overline{Y}^o(t) - \overline{F}(t) - \mu\overline{G}(\mu G^T G + \lambda I)^{-1} G^T (Y^o(t) - F(t)))] \end{aligned} \quad (4.28)$$

As it is a receding horizon strategy, only the first element of vector $\Delta U(t)$ is used, repeating the calculation at the next sampling time. This method provides an analytical solution that, as can be observed in (4.28) proves to be more complex than the standard GPC solution. Computational burden can be considerable since various matrix operations, including inversion, must be made, although some calculations can be optimized knowing that G is triangular and the matrices to be inverted are symmetrical. This factor can be decisive in the adaptive case, where all vectors and matrices can change at every sampling time.

Obtaining the control signal requires the inversion of matrix $\overline{G}(\mu G^T G + \lambda I)^{-1}\overline{G}^T)$. From the definition of matrix $\overline{G}$, it can be derived that condition $m \leq N_u + 1$ must hold, which can be interpreted as that the number m of output constraints cannot be bigger than the number of control signal variations N_u+1. Another condition for invertibility is that $m \leq n+1$, since the coefficient s_i of the step response is a linear combination of the previous $n+1$ values (being n the system order). Notice that this last condition constrains the order of the matrix to invert ($m \times m$) to relative small values, as in the majority of situations the value of m will not be bigger than two or three.

4.10.2 Properties

As has been stated, one of the advantages of this method is the availability of stability results for finite horizons, compensating in a certain way the computational burden it carries. The following results present the outstanding properties of CRHPC, whose demonstration can be found in [84], based upon a state-space formulation of the control law (4.28). The fact that the output follows the reference over some range guarantees the monotonicity of the associated Riccati equation and in consequence stability, based upon the results by Kwon and Pearson [61].

Property 1. If $N_y = N_u > n + d + 1$ and $m = n + 1$ then the closed loop system is asymptotically stable.

Property 2. If $N_u = n + d$ and $m = n + 1$ the control law results in a stable dead-beat control.

Property 3. If the system is asymptotically stable, $\mu = 0, m = 1$ and there exists ν such that either

$$s_\nu \leq s_{\nu+1} \leq \ldots \leq s_\infty, \;\; s_\nu > s_\infty/2, \;\; s_\infty > 0$$

or

$$s_\nu \geq s_{\nu+1} \geq \ldots \geq s_\infty, \;\; s_\nu < s_\infty/2, \;\; s_\infty < 0$$

Then, for any $N_y = N_u \geq \nu - 1$ the closed loop system is asymptotically stable.

Property 4. Under the latter conditions, the closed loop system is asymptotically stable for $N_u = 0$ and $N_y \geq \nu - 1$.

Property 5. If the open loop system is asymptotically stable, $\mu = 0, m = 1$, $N_u = 0$ and constants $K > 0$ and $0 < \eta < 1$ exist such that

$$\mid s_i - s_\infty \mid \leq K\eta^i, \;\; i \geq 0$$

Then, for any N_y such that

$$\mid s_{Ny+1} \mid > K\frac{1+\eta}{1-\eta}\eta^{Ny+1}$$

the closed loop system is asymptotically stable.

So, it is seen that the method is able to stabilize any kind of processes, such as unstable or non-minimum phase ones. As in GPC, the use of filtered inputs and outputs can result in a pole-placement control. Using the $P(z^{-1})$ polynomial introduced in [35] shows the close relationship between predictive control and pole-placement [30]. This polynomial appears when defining an auxiliary output $\psi(t) = P(z^{-1})y(t)$, which substitutes $y(t)$ in (4.25) and (4.26) in such a way that the desired closed-loop poles are the roots of $z^{n+1}P(z^{-1})$.

4.11 Stable GPC

To overcome the lack of stability results for GPC, a new formulation has been developed by Rossiter and Kouvaritakis which ensures that the associated cost function is monotonically decreasing, guaranteeing closed-loop stability. For this reason, this algorithm is called Stable Generalized Predictive Control SGPC [58], [109] and it is based on stabilizing the loop before the application of the control strategy. Now the control actions that must be calculated are the future values of the reference that are sent to the closed loop, instead of the system inputs, that are functions of these. The stabilizing inner loop controller is designed to obtain a finite closed-loop impulse response; this fact simplifies the implementation of the algorithm at the same time as ensuring the monotonicity of the cost function.

4.11.1 Formulation of the Control Law

The model considered for obtaining the control law is:

$$G(z) = \frac{z^{-1}b(z)}{a(z)} = \frac{b_1 z^{-1} + b_2 z^{-2} + \cdots + b_n z^{-n}}{1 + a_1 z^{-1} + a_2 z^{-2} + \cdots + a_n z^{-n}} \tag{4.29}$$

where, for the sake of simplicity and without any loss of generality, the order of numerator and denominator is considered to be the same; if the delay is bigger than one, it is enough to set the first terms of $b(z)$ to zero. As has been stated, before optimizing the cost function, the loop is stabilized by means of polynomials $X(z)$ and $Y(z)$ as shown in figure 4.9. Signal c is the closed-loop reference and is the value that will appear in the cost function.

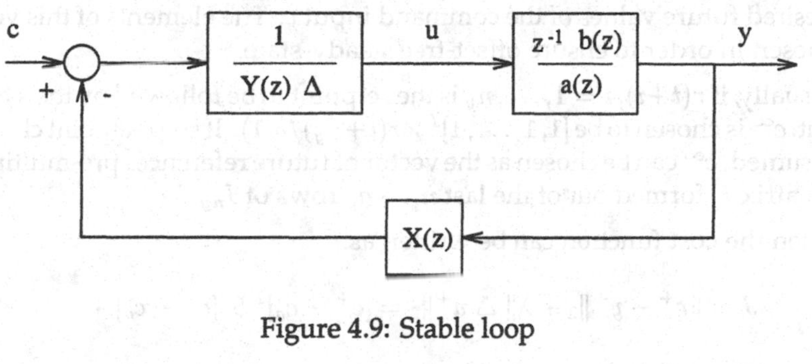

Figure 4.9: Stable loop

Polynomial $X(z)$ and $Y(z)$ satisfy the following relations:

$$a(z)Y(z)\,\Delta\,(z) + z^{-1}b(z)X(z) = 1$$
$$K(z) = \frac{X(z)}{Y(z)\,\Delta\,(z)}$$

with $\Delta(z) = 1 - z^{-1}$ and $K(z)$ the overall feedback controller. With these relations, it can be deduced that:

$$\begin{aligned} y(z) &= z^{-1}b(z)c(z) \\ \Delta u(z) &= A(z)c(z) \end{aligned}$$

being $A(z) = a(z)\Delta$.

The cost function that, like standard GPC, measures the discrepancies between future outputs and reference as well as the necessary control effort can be obtained making use of these relations. The cost must be expressed as a function of the future values of c in such a way that the reference signals for the stabilized system can be obtained from its minimization. In order to achieve this objective, the following vectors are considered:

$$\begin{aligned} y^+ &= [\, y(t+1)\; y(t+2)\; \dots\; y(t+n_y)\,]^T \\ c^+ &= [\, c(t+1)\; c(t+2)\; \dots\; c(t+n_c)\,]^T \\ \Delta u^+ &= [\, \Delta u(t)\; \Delta\, u(t+1)\; \dots\; \Delta\, u(t+n_y-1)\,]^T \\ c^- &= [\, c(t)\; c(t-1)\; \dots\; c(t-n)\,]^T \end{aligned}$$

where n_y and n_c are the output and reference horizons.

Simulating equation (4.29) forward in time it can be written:

$$\begin{aligned} y^+ &= \Gamma_b c^+ + H_b c^- + M_b c^\infty \\ \Delta u^+ &= \Gamma_a c^+ + H_A c^- + M_A c^\infty \end{aligned}$$

where the last terms of each equation represent the free response y_f and Δu_f of y and Δu respectively. Matrices Γ_A, Γ_b, H_A, H_b, M_A and M_b can easily be derived as shown in [58] while c^∞ is the vector of $n_y - n_c$ rows that contains the desired future values of the command input c. The elements of this vector are chosen in order to ensure offset-free steady-state.

Usually, if $r(t+i), i = 1, \dots, n_y$ is the setpoint to be followed by the system output, c^∞ is chosen to be $[1, 1, \dots, 1]^T \times r(t+n_y)/b(1)$. If step setpoint changes are assumed, c^∞ can be chosen as the vector of future references pre-multiplied by a matrix E formed out of the last $n_y - n_c$ rows of I_{ny}.

Then the cost function can be written as:

$$J = \|r^+ - y^+\|_2 + \lambda \|\, \Delta\, u^+\|_2 = [c^+ - c_o]^T S^2 [c^+ - c_o] + \gamma$$

where

$$\begin{aligned} S^2 &= \Gamma_b^T \Gamma_b + \lambda \Gamma_A^T \Gamma_A \\ c_o &= S^{-2}[\Gamma_b^T (r^+ - y_f) - \lambda \Gamma_A^T \,\Delta\, u_f] \\ \gamma &= \|r^+ - y_f\|_2 + \lambda \|\,\Delta\, u_f\|_2 - \|c_o\|_2 \end{aligned}$$

As γ is a known constant, the control law of SGPC comes from the minimization of cost $J = \|S(c^+ - c_o)\|$, and is defined by

$$\Delta u(t) = A(z)c(t) \qquad c(t) = \frac{p_r(z)}{p_c(z)} r(t + ny)$$

where $p_r(z)$ and $p_c(z)$ are polynomials defined as:

$$p_r = e^T T[\Gamma_b^T (I - M_b E) - \Gamma_b^T \Gamma_A E]$$
$$p_c = e^T T[\Gamma_b^T H_b + \lambda \Gamma_b H_A]$$

and $T = (\Gamma_b^T \Gamma_b + \lambda \Gamma_A \Gamma_A)^{-1}$. e is the first standard basis vector and the polynomials are in descending order of z.

The stability of the algorithm is a consequence of the fact that y and Δu are related to c by means of Finite Impulse Response operators ($z^{-1}b(z)$ and $A(z)$), which can be used to establish that the cost is monotonically decreasing [109] and can therefore be interpreted as a Lyapunov function guaranteeing stability.

Chapter 5

Simple Implementation of GPC for Industrial Processes

One of the reasons for the success of the traditional PID controllers in industry is that PID are very easy to implement and tune by using heuristic tuning rules such as the Ziegler-Nichols rules frequently used in practice. A Generalized Predictive Controller, as shown in the previous chapter, results in a linear control law which is very easy to implement once the controller parameters are known. The derivation of the GPC parameters requires, however, some mathematical complexities such as solving recursively the Diophantine equation, forming the matrices G, G′, f and then solving a set of linear equations. Although this is not a problem for people in the research control community where mathematical packages are normally available, it may be discouraging for those practitioners used to much simpler ways of implementing and tuning controllers.

The previously mentioned computation has to be carried out only once when dealing with processes with fixed parameters , but if the process parameters change, the GPC's parameters have to be derived again, perhaps in real time, at every sampling time if a self-tuning control is used. This again may be a difficulty because on one hand, some distributed control equipment has only limited mathematical computation capabilities for the controllers, and on the other hand, the computation time required for the derivation of the GPC parameters may be excessive for the sampling time required by the process and the number of loops implemented.

The goal of this chapter is to show how a GPC can very easily be implemented and tuned for a wide range of processes in industry. It will be shown that a GPC can be implemented with a limited set of instructions, available in most distributed control systems, and that the computation time required, even for tuning, is very short. The method to implement the GPC is based on the fact that a wide range of processes in industry can be described by a few parameters and that a set of simple Ziegler-Nichols type of functions relating GPC parameters to process parameters can be obtained. By using these functions the implementation and tuning of a GPC results almost as simple as the implementation and tuning of a PID.

The influence of modelling errors is also analyzed in this chapter, with a section dedicated to perform a robustness analysis of the method when modelling errors are taken into account.

5.1 Plant Model

Most processes in industry when considering small changes around an operating point can be described by a linear model of, normally, very high order. The reason for this is that most industrial processes are composed of many dynamic elements, usually first order, so the full model is of an order equal to the number of elements. In fact, each mass or energy storage element in the process provides a first order element in the model. Consider for instance a long pipe used for heat exchanging purposes, as is the case of a cooler or a steam generator. The pipe can be modelled by breaking it into a set of small pieces, each of which can be considered as a first order system. The resulting model will have an order equal to the number of pieces used to model the pipe, that is, a very high order model. These very high order models would be difficult to use for control purposes but, fortunately, it is possible to approximate the behaviour of such high order processes by a system with one time constant and a dead time.

As is shown in [38], we may consider a process having N first order elements in series, each having a time constant τ/N. That is, the resulting transfer function will be

$$G(s) = \frac{1}{(1 + \frac{\tau}{N}s)^N}$$

Changing the value of N from 1 to ∞ the response shifts from exact first order to pure dead time (equal to τ). When a time constant is much larger than the others (as is usual in many processes) the smaller time constants work together to produce a lag that acts as a pure dead time. In this situation the dynamical effects are mainly due to this larger time constant, as can be seen in figure 5.1. It is therefore possible to approximate the model of a very high order, complex, dynamic process with a simplified model consisting of a first order process combined with a dead time element. This type of system can then be described by the following transfer function:

$$G(s) = \frac{K}{1 + \tau s} e^{-s\tau_d} \tag{5.1}$$

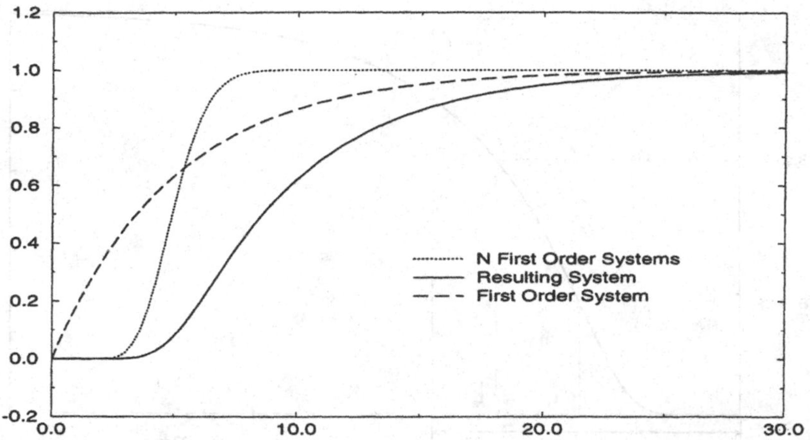

Figure 5.1: System Response

where K is the process static gain, τ is the time constant or process lag, and τ_d is the dead time or delay.

5.1.1 Plant Identification: The Reaction Curve Method

Once the model structure is defined, the next step is to choose the correct value for the parameters. In order to identify these parameters, a suitable stimulus must be applied to the process input. In the Reaction Curve Method a step perturbation, that is an input with a wide frequency content, is applied to the process and the output is recorded in order to fit the model to the data.

The step response or reaction curve of the process looks like figure 5.2. Here, a step of magnitude Δu is produced in the manipulated variable and the time response of the process variable $y(t)$ is shown. The process parameters of equation (5.1) can be obtained by measuring two times, t_1 being the time when the output reaches 28.3 per cent of its steady state value Δy, and t_2 when the output reaches 63.2 per cent. Using these values, the process parameters are given by:

$$\begin{aligned} K &= \frac{\Delta y}{\Delta u} \\ \tau &= 1.5(t_2 - t_1) \\ \tau_d &= 1.5(t_1 - \tfrac{1}{3}t_2) \end{aligned}$$

A similar and perhaps more intuitive way of obtaining the process parameters consists of finding the inflection point of the response and drawing the line that represents the slope at that point [49]. The static gain is obtained by the former expression and the times τ and τ_d come directly from the response,

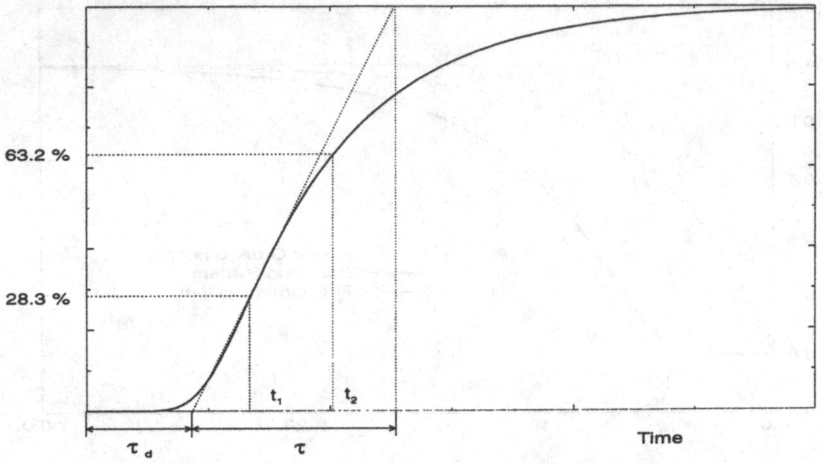

Figure 5.2: Reaction Curve

as can be seen in the figure. The values obtained are very similar in both approaches.

The Reaction Curve method is probably one of the most popular methods used in industry for tuning regulators, as is the case of the Ziegler-Nichols method for tuning PIDs, and is used in the pretune stage of some commercial adaptive and auto-tuning regulators.

5.2 The Dead Time Multiple of Sampling Time Case

5.2.1 Discrete Plant Model

When the dead time τ_d is an integer multiple of the sampling time T ($\tau_d = dT$), the corresponding discrete transfer function of equation (5.1) has the form

$$G(z^{-1}) = \frac{bz^{-1}}{1 - az^{-1}} z^{-d} \tag{5.2}$$

where discrete parameters a, b and d can easily be derived from the continuous parameters by discretization of the continuous transfer function, resulting in the following expressions:

$$a = e^{-\frac{T}{\tau}} \qquad b = K(1-a) \qquad d = \frac{\tau_d}{T}$$

If a CARIMA (Controlled Auto-Regressive and Integrated Moving-Average) model is used to model the random disturbances in the system and the noise polynomial is chosen to be 1 , the following equation is obtained

$$(1 - az^{-1})y(t) = bz^{-d}u(t-1) + \frac{\varepsilon(t)}{\Delta}$$

where u(t) and y(t) are the control and output sequences of the plant and $\varepsilon(t)$ is a zero mean white noise and $\Delta = 1 - z^{-1}$. This equation can be transformed into:

$$y(t+1) = (1+a)y(t) - ay(t-1) + b\,\Delta\, u(t-d) + \varepsilon(t+1) \qquad (5.3)$$

5.2.2 Problem Formulation

As has been shown in the previous chapter, the Generalized Predictive Control (GPC) algorithm consists of applying a control sequence that minimizes a multistage cost function of the form

$$J(N_1, N_2) = \sum_{j=N_1}^{N_2} \delta(j)[\hat{y}(t+j \mid t) - w(t+j)]^2 + \sum_{j=1}^{N_2-d} \lambda(j)[\Delta u(t+j-1)]^2 \qquad (5.4)$$

Notice that the minimum output horizon N_1 should be set to a value greater than the dead-time d as the output for smaller time horizons cannot be affected by the first action $u(t)$. In the following N_1 and N_2 will be considered to be: $N_1 = d+1$ and $N_2 = d+N$, N being the control horizon.

If $\hat{y}(t+d+j-1 \mid t)$ and $\hat{y}(t+d+j-2 \mid t)$ are known, it is clear, from equation (5.3), that the best expected value for $\hat{y}(t+d+j \mid t)$ is given by:

$$\begin{aligned}\hat{y}(t+d+j \mid t) &= (1+a)\hat{y}(t+d+j-1 \mid t) - a\hat{y}(t+d+j-2 \mid t) + \\ &+ b\Delta\, u(t+j-1)\end{aligned} \qquad (5.5)$$

If equation (5.5) is applied recursively for $j = 1, 2, \cdots, i$ we get

$$\hat{y}(t+d+i \mid t) = G_i(z^{-1})\hat{y}(t+d \mid t) + D_i(z^{-1})\,\Delta\, u(t+i-1) \qquad (5.6)$$

where $G_i(z^{-1})$ is of degree 1 and $D_i(z^{-1})$ is of degree $i-1$. Notice that when $\delta(i) = 1$ and $\lambda(i) = \lambda$, the polynomials $D_i(z^{-1})$ are equal to the polynomials $G_i(z^{-1})$ given in [34] for the case of $d = 0$ and that the terms $f(t+i)$ given in that reference are equal to $G_i(z^{-1})y(t)$ of equation (5.6).

If $\hat{y}(t+d+i \mid t)$ is introduced in equation (5.4), $J(N)$ is a function of $\hat{y}(t+d \mid t)$, $\hat{y}(t+d-1 \mid t)$, $\Delta u(t+N_2-d-1)$, $\Delta u(t+N_2-d-2)$ ... $\Delta u(t)$ and the reference sequence.

Minimizing $J(N)$ with respect to $\Delta u(t)$, $\Delta u(t+1)$ ... $\Delta u(t+N-1)$ leads to

$$\mathbf{M\, u} = \mathbf{P\, y} + \mathbf{R\, w} \qquad (5.7)$$

where

$$\begin{aligned}\mathbf{u} &= [\Delta u(t)\ \Delta u(t+1)\ \cdots\ \Delta u(t+N-1)]^T \\ \mathbf{y} &= [\hat{y}(t+d \mid t)\ \hat{y}(t+d-1 \mid t)]^T \\ \mathbf{w} &= [w(t+d+1)\ w(t+d+2)\ \cdots\ w(t+d+N)]^T\end{aligned}$$

$\mathbf{M}$ and $\mathbf{R}$ are matrices of dimension $N \times N$ and $\mathbf{P}$ of dimension $N \times 2$. Let us call $\mathbf{q}$ the first row of matrix $\mathbf{M}^{-1}$. Then $\Delta u(t)$ is given by

$$\Delta u(t) = \mathbf{q}\,\mathbf{P}\,\mathbf{y} + \mathbf{q}\,\mathbf{R}\,\mathbf{w} \tag{5.8}$$

When the future setpoints are unknown, $w(t+d+i)$ is supposed to be equal to the current reference $r(t)$. The reference sequence can be written as:

$$\mathbf{w} = [1 \cdots 1] r(t)$$

The control increment $\Delta u(t)$ can be written as:

$$\Delta u(t) = l_{y1}\hat{y}(t+d \mid t) + l_{y2}\hat{y}(t+d-1 \mid t) + l_{r1}r(t) \tag{5.9}$$

Where $\mathbf{q}\,\mathbf{P} = [l_{y1} l_{y2}]$ and $l_{r1} = \sum_{i=1}^{N} q_i \sum_{j=1}^{N} r_{ij}$. The coefficients l_{y1}, l_{y2}, l_{r1}, are functions of a, b, $\delta(i)$ and $\lambda(i)$. If the GPC is designed considering the plant to have a unit static gain, the coefficients in (5.9) will only depend on $\delta(i)$ and $\lambda(i)$ (which are supposed to be fixed) and on the pole of the plant which will change for the adaptive control case. Notice that by doing this, a normalized weighting factor λ is used and that it should be corrected accordingly for systems with different static gains.

The resulting control scheme is shown in figure 5.3. The estimated plant parameters are used to compute the controller coefficients (l_{y1}, l_{y2}, l_{r1}). The values $\hat{y}(t+d \mid t)$, $\hat{y}(t+d-1 \mid t)$ are obtained by the use of the predictor given by equation (5.5). The control signal is divided by the process static gain in order to get a system with a unitary static gain.

Notice that the controller coefficients do not depend on the dead time d and for fixed values of $\delta(i)$ and $\lambda(i)$, they will be a function of the estimated pole ($\hat{a}$). The standard way of computing the controller coefficients would be by computing the matrices $\mathbf{M}$, $\mathbf{P}$ and $\mathbf{R}$ and solving equation (5.7) followed by the generation of the control law of equation (5.9). This involves the triangularization of a $N \times N$ matrix which could be prohibitive for some real time applications.

As suggested in [21], the controller coefficients can be obtained by interpolating in a set of previously computed values as shown in figure 5.4. Notice

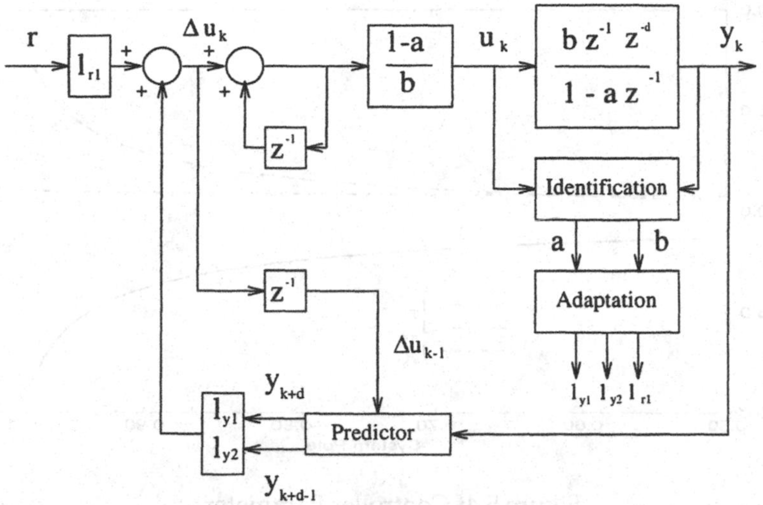

Figure 5.3: Control Scheme

that this can be accomplished in this case because the controller coefficients only depend on one parameter. The number of points of the set used depends on the variability of the process parameters and on the accuracy needed. The set does not need to be uniform and more points can be computed in regions where the controller parameters vary substantially in order to obtain a better approximation or to reduce the computer memory needed.

The predictor needed in the algorithm to calculate $\hat{y}(t+d \mid t), \hat{y}(t+d-1 \mid t)$ is obtained by applying equation (5.5) sequentially for $j = 1 - d \cdots 0$. Notice that it basically consists of a model of the plant which is projected towards the future with the values of past inputs and outputs and it only requires straightforward computation.

5.2.3 Computation of the Controller Parameters

The algorithm described above can be used to compute controller parameters of GPC for plants which can be described by equation (5.2) (most industrial plants can be described this way) over a set covering the region of interest.

Notice that the sampling time of a digital controller is chosen in practice according to the plant time response. Sampling time between 1/15 and 1/4 of T_{95} (the time needed by the system to reach 95 percent of the final output value) is recommended in [49]. The pole of the discrete form of the plant transfer function is therefore going to vary between 0.5 and 0.95 for most industrial processes when sampled at appropriate rates.

The curves shown in figure 5.4 correspond to the controller parameters

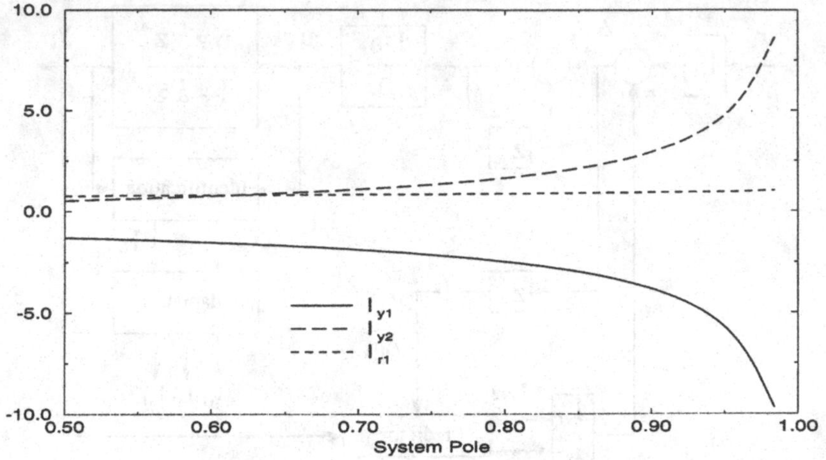

Figure 5.4: Controller Parameters

(l_{y1}, l_{y2}, l_{r1}) obtained for $\delta(i) = \delta^i$ and $\lambda(i) = \lambda^i$ with $\delta = 1$, $\lambda = 0.8$ and $N = 15$. The pole of the system has been changed with a 0.0056 step from 0.5 to 0.99. Notice that due to the fact that the closed-loop static gain must be equal to 1, the sum of the three parameters equals zero. This result implies that only two of the three parameters need to be known.

By looking at figure 5.4 it can be seen that the functions relating the controller parameters to the process pole can be approximated by functions of the form:

$$l_{yi} = k_{1i} + k_{2i}\frac{a}{k_{3i} - a} \qquad i = 1, 2 \tag{5.10}$$

The coefficients k_{ji} can be calculated by a least squares adjustment using the set of known values of l_{yi} for different values of a. Equation (5.10) can be written as:

$$al_{yi} = l_{yi}k_{3i} - k_{1i}k_{3i} + a(k_{i1}a - k_{2i})$$

Repeating this equation for the N_p points used to obtain the approximation we get:

$$\begin{bmatrix} a^1 l_{yi}^1 \\ a^2 l_{yi}^2 \\ \vdots \\ a^{N_p} l_{yi}^{N_p} \end{bmatrix} = \begin{bmatrix} l_{yi}^1 & 1 & a^1 \\ l_{yi}^2 & 1 & a^2 \\ \vdots & \vdots & \vdots \\ l_{yi}^{N_p} & 1 & a^{N_p} \end{bmatrix} \begin{bmatrix} x_1 \\ x_2 \\ x_3 \end{bmatrix} + \begin{bmatrix} e^1 \\ e^2 \\ \vdots \\ e^{N_p} \end{bmatrix} \tag{5.11}$$

where l_i^j, a^j, e^j, $j = 1...N_p$ are the N_p values of the system pole, the pre-calculated parameters, and the approximation errors, $x_1 = k_{3i}$, $x_2 = -k_{1i}k_{3i}$ and $x_3 = k_{1i} - k_{2i}$.

Equation (5.11) can be written in matrix form

$$\mathbf{Y} = \mathbf{M}\,\mathbf{X} + \mathbf{E}$$

In order to calculate the optimum values of X we minimize $E^T E$ obtaining

$$\mathbf{X} = (\mathbf{M}^T\,\mathbf{M})^{-1}\,\mathbf{M}^T\,\mathbf{Y}$$

The desired coefficients can now be evaluated as:

$$\begin{aligned} k_{3i} &= x_1 \\ k_{1i} &= -x_2/k_{3i} \\ k_{2i} &= k_{1i} - x_3 \end{aligned}$$

In the case of $\lambda = 0.8$ and for a control horizon of 15, the controller coefficients are given by:

$$\begin{aligned} l_{y1} &= -0.845 - 0.564\frac{a}{1.05 - a} \\ l_{y2} &= 0.128 + 0.459\frac{a}{1.045 - a} \\ l_{r1} &= -l_{y1} - l_{y2} \end{aligned}$$

These expressions give a very good approximation to the true controller parameters and fit the set of computed data with a maximum error of less than one percent of the nominal value for the range of interest of the open loop pole.

5.2.4 Role of the Control-Weighting Factor

The control-weighting factor λ affects the control signal in equation (5.9). The bigger this value is, the smaller the control effort is allowed to be. If it is given a small value, the system response will be fast since the controller tends to minimize the error between the output and the reference, forgetting about control effort. The controller parameters l_{y1}, l_{y2} and l_{r1} and therefore the closed-loop poles depend on the values of λ.

Figure 5.5 shows the value of the modulus of the biggest closed-loop pole when changing λ from 0 to 0.8 by increments of 0.2. As can be seen, the modulus of the biggest closed-loop pole decreases with λ, indicating faster systems. For a value of λ equal 0, the closed-loop poles are zero, indicating dead-beat behaviour.

A set of functions was obtained by making λ change from 0.3 to 1.1 by increments of 0.1. It was found that the values of the parameters $k_{ij}(\lambda)$ of

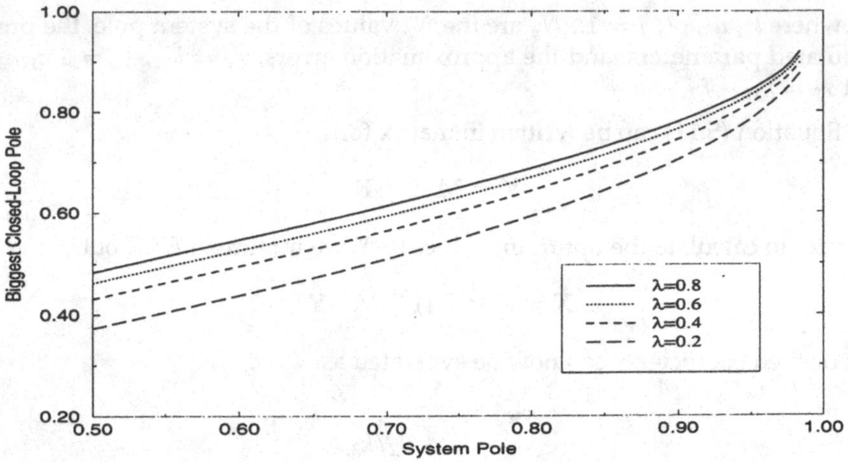

Figure 5.5: Influence of the control-weighting factor

equation (5.10) obtained could be approximated by functions with the form: $\mathrm{sgn}(k_{ij})e^{c_0+c_1\lambda+c_2\lambda^2}$.

By applying logarithm the coefficients c_1,c_2 and c_3 can be adjusted by using a polynomial fitting procedure. The following expressions were obtained for the grid of interest:

$$\begin{array}{rcl} k_{11} & = & -e^{0.3598-0.9127\lambda+0.3165\lambda^2} \\ k_{21} & = & -e^{0.0875-1.2309\lambda+0.5086\lambda^2} \\ k_{31} & = & 1.05 \\ k_{12} & = & e^{-1.7383-0.40403\lambda} \\ k_{22} & = & e^{-0.32157-0.81926\lambda+0.3109\lambda^2} \\ k_{32} & = & 1.045 \end{array} \qquad (5.12)$$

The values of the control parameters l_{y1} and l_{y2} obtained when introducing the k_{ij} given in equation (5.10) are a very good approximation to the real ones. The maximum relative error for $0.55 < a < 0.95$ and $0.3 \le \lambda \le 1.1$ is less than 3 per cent.

5.2.5 Implementation Algorithm

Once the λ factor has been decided, the values k_{ij} can very easily be computed by expressions (5.12) and the approximate adaptation laws given by equation (5.10) can easily be employed. The proposed algorithm in the adaptive case can be seen below.

0. Compute k_{ij} with expressions (5.12)

1. Perform an identification step

2. Make $l_i = k_{1i} + k_{2i}\frac{\hat{a}}{k_{3i}-\hat{a}}$ for $i = 1, 2$ and $l_{r1} = -l_{y1} - l_{y2}$

3. Compute $\hat{y}(t+d \mid t)$ and $\hat{y}(t+d-1 \mid t)$ using equation (5.5) recursively

4. Compute control signal $u(t)$ with:
$\Delta u(t) = l_{y1}\hat{y}(t+d \mid t) + l_{y2}\hat{y}(t+d-1 \mid t) + l_{r1}r(t)$

5. Divide the control signal by the static gain

6. Go to step 1

Notice that in a fixed-parameter case the algorithm is simplified since the controller parameters need to be computed only once (unless the control weighting factor λ is changed) and only steps 3 and 4 have to be carried out at every sampling time.

5.2.6 An Implementation Example

In order to show the straightforwardness of this method, an application to a typical process such as a simple furnace is presented. First, identification by means of the Reaction Curve method is performed and then the precalculated GPC is applied. The process basically consists of a water flow being heated by fuel which can be manipulated by a control valve. The output variable is the coil outlet temperature whereas the manipulated variable is the fuel flow.

The stationary values of the variables are: inlet temperature 20 $^\circ C$, outlet temperature 50 $^\circ C$ and fuel valve at 18.21 percent. Under these conditions, the fuel rate is changed to a value of 30 percent and the outlet temperature response is shown in figure 5.6, reaching a final value of $y = 69.41\ ^\circ C$. The plant parameters can be obtained directly from this response as was explained in section 3.1.1. First, the times t_1 (when the response reaches 28.3 % of its final value) and t_2 (63.2 %) are obtained, resulting in $t_1 = 5.9$ and $t_2 = 9.6$ seconds. Then the plant parameters are:

$$
\begin{aligned}
K &= \frac{\Delta y}{\Delta u} &= \frac{69.411 - 50}{30 - 18.21} &= 1.646 \\
\tau &= 1.5(t_2 - t_1) &= 1.5(9.6 - 5.9) &= 5.55 \\
\tau_d &= 1.5(t_1 - \tfrac{1}{3}t_2) &= 1.5(5.9 - \tfrac{9.6}{3}) &= 4.05
\end{aligned}
$$

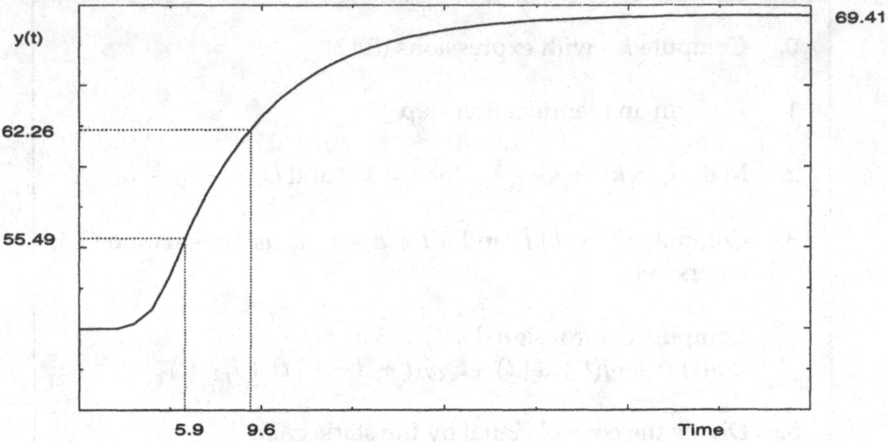

Figure 5.6: Outlet Temperature Response

with these values, and a sampling time of one second, the equivalent discrete transfer function results in

$$G(z^{-1}) = \frac{0.2713z^{-1}}{1 - 0.8351z^{-1}} z^{-4}$$

As the process is considered to have fixed parameters, the controller coefficients l_{y1}, l_{y2} and l_{r1} can be calculated off-line. In this case, choosing $\lambda = 0.8$, the coefficients are obtained from equation (5.12).

$$\begin{aligned} k_{11} &= -0.845 \\ k_{21} &= -0.564 \\ k_{31} &= 1.05 \\ k_{12} &= 0.128 \\ k_{22} &= 0.459 \\ k_{32} &= 1.045 \end{aligned}$$

In consequence the controller coefficients are:

$$l_{y1} = -3.0367 \qquad l_{y2} = 1.9541 \qquad l_{r1} = 1.0826$$

Therefore, at every sampling time it will only be necessary to compute the predicted outputs and the control law. The predictions needed are $\hat{y}(t+4 \mid t)$ and $\hat{y}(t+3 \mid t)$, that will be calculated from the next equation with $i = 1 \cdots 4$

$$\hat{y}(t+i \mid t) = (1+a)\hat{y}(t+i-1 \mid t) - a\hat{y}(t+i-2 \mid t) + b \triangle u(t+i-5)$$

and the control law, G being the static gain

$$u(t) = u(t-1) + (l_{y1}\,\hat{y}(t+4 \mid t) + l_{y2}\,\hat{y}(t+3 \mid t) + l_{r1}\,r)/G$$

The implementation of the controller in a digital computer will result in a simple program, a part of whose code written in language C is shown in figure 5.7. Two arrays u, and y are used. The first is used to store the past values of the control signal and the latter to store the values of the outputs. In this process with a dead time of four, the arrays are:

$$\mathbf{y} = [\,y(t-1), y(t), \hat{y}(t+1), \hat{y}(t+2), \hat{y}(t+3), \hat{y}(t+4)\,]$$
$$\mathbf{u} = [\,u(t), u(t-1), u(t-2), u(t-3), u(t-4), u(t-5)\,]$$

Notice that **y** contains the predicted outputs and also the outputs in t and $t-1$, because these last two values are needed in the first two predictions $\hat{y}(t+1), \hat{y}(t+2)$.

```
/* Predictor */
for (i=2; i<=5; i++)
y[i]=1.8351*y[i-1]-0.8351*y[i-2]+0.2713*(u[5-i]-u[6-i]);

/* Control Law */
u[0]=u[1]+(-3.0367*y[5]+1.9541*y[4]+1.08254*r)/1.646;

/* Updating */
for (i=5; i>0; i--)   u[i] = u[i-1];
y[0] = y[1];
```

Figure 5.7: C code of implementation algorithm

The closed loop response to a setpoint change of $+10\,^0C$ is plotted in figure 5.8, where the evolution of the control signal can also be seen . Additionally, the control weighting factor λ was changed at $t = 60$ from the original value of 0.8 to a smaller one of 0.3; notice that the control effort increases and the output tends to be faster. On the other hand, if λ takes a bigger value such as 1.3 the behaviour tends to be slower; this change was performed at $t = 100$.

5.3 The Dead Time non Multiple of the Sampling Time Case

5.3.1 Discrete Model of the Plant

When the dead time τ_d of the process is not an integer multiple of the sampling time T ($dT \leq \tau_d \leq (d+1)T$), equation (5.2) cannot be employed. In this case

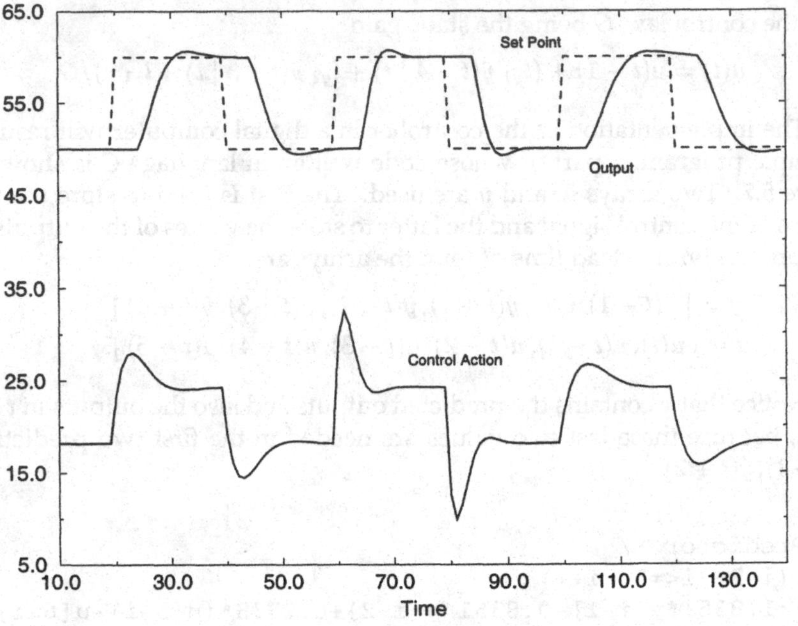

Figure 5.8: System Response

the fractional delay time can be approximated [40] by the first two terms of the Padé expansion and the plant discrete transfer function can be written as:

$$G(z^{-1}) = \frac{b_0 z^{-1} + b_1 z^{-2}}{1 - az^{-1}} z^{-d} \tag{5.13}$$

As can be seen this transfer function is slightly different from the previous model (equation(5.2)), presenting an additional zero; a new parameter appears in the numerator. Using the same procedure as in the previous case, a similar implementation of GPC can be obtained for this family of processes.

To obtain the discrete parameters a, b_0 and b_1, the following relations can be used [40]: first, the dead time is decomposed as $\tau_d = dT + \epsilon T$ with $0 < \epsilon < 1$. Then the parameters are:

$$a = e^{-\frac{T}{\tau}} \qquad b_0 = K(1-a)(1-\alpha) \qquad b_1 = K(1-a)\alpha \qquad \alpha = \frac{a(a^{-\epsilon} - 1)}{1 - a}$$

Since the derivation of the control law is very similar in this case to the previous section, some steps will be omitted for simplicity.

The function J to be minimized is also that of equation (5.4). Using the CARIMA model with the noise polynomial equal to 1, the system can be written as

$$(1 - az^{-1})y(t) = (b_0 + b_1 z^{-1})z^{-d}u(t-1) + \frac{\varepsilon(t)}{\Delta}$$

which can be transformed into:

$$y(t+1) = (1+a)y(t) - ay(t-1) + b_0\Delta u(t-d) + b_1\Delta u(t-d-1) + \varepsilon(t+1) \quad (5.14)$$

If $\hat{y}(t+d+i-1 \mid t)$ and $\hat{y}(t+d+i-2 \mid t)$ are known, it is clear, from equation (5.14), that the best expected value for $\hat{y}(t+d+i \mid t)$ is given by:

$$\begin{aligned}\hat{y}(t+d+i \mid t) &= (1+a)\hat{y}(t+d+i-1 \mid t) - a\hat{y}(t+d+i-2 \mid t) + \\ &+ b_0 \,\Delta\, u(t+i-1) + b_1 \,\Delta\, u(t+i-2)\end{aligned}$$

If $\hat{y}(t+d+i \mid t)$ is introduced in the function to be minimized, $J(N)$ is a function of $\hat{y}(t+d \mid t)$, $\hat{y}(t+d-1 \mid t)$, $\Delta u(t+N_2-d-1)$, $\Delta u(t+N_2-d-2)$ ... $\Delta u(t)$, $\Delta u(t-1)$ and the reference sequence.

Minimizing $J(N)$ with respect to $\Delta u(t)$, $\Delta u(t+1)$ ... $\Delta u(t+N-1)$ leads to

$$\mathbf{M\,u} = \mathbf{P\,y} + \mathbf{R\,w} + \mathbf{Q}\,\Delta\, u(t-1) \quad (5.15)$$

where

$$\begin{aligned}\mathbf{u} &= [\,\Delta u(t)\ \Delta\, u(t+1)\ \cdots\ \Delta\, u(t+N-1)\,]^T \\ \mathbf{y} &= [\,\hat{y}(t+d \mid t)\ \hat{y}(t+d-1 \mid t)\,]^T \\ \mathbf{w} &= [\,w(t+d+1)\ w(t+d+2)\ \cdots\ w(t+d+N)\,]^T\end{aligned}$$

M and **R** are matrices of dimension $N \times N$, **P** of dimension $N \times 2$ and **Q** of $N \times 1$.

Notice that the term **Q** $\Delta\, u(t-1)$ did not appear in the simpler plant because of the different plant parameters, hence the control law will not be the same. Let us call **q** the first row of matrix $\mathbf{M}^{-1}$. Then $\Delta u(t)$ is given by

$$\Delta\, u(t) = \mathbf{q\,P\,y} + \mathbf{q\,R\,w} + \mathbf{q\,Q}\ \Delta\, u(t-1) \quad (5.16)$$

If the reference sequence is considered to be $\mathbf{w} = [1 \cdots 1]r(t)$, the control increment $\Delta u(t)$ can be written as:

$$\Delta\, u(t) = l_{y1}\,\hat{y}(t+d \mid t) + l_{y2}\,\hat{y}(t+d-1 \mid t) + l_{r1}\,r(t) + l_{u1}\ \Delta u(t-1) \quad (5.17)$$

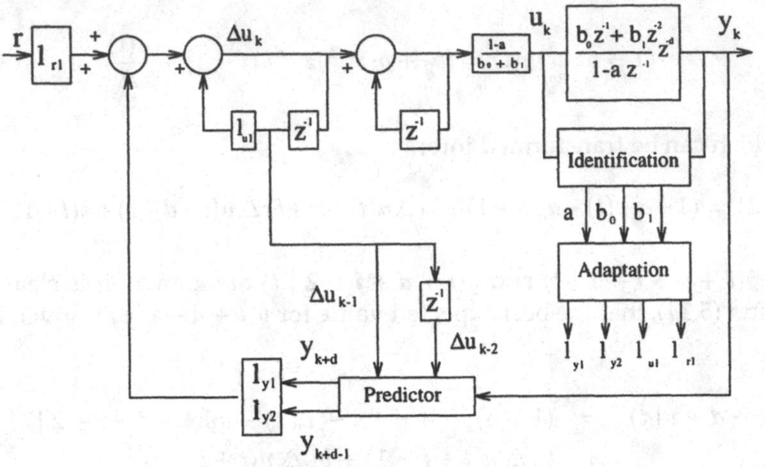

Figure 5.9: Control Scheme

Where $\mathbf{q}\ \mathbf{P} = [l_{y1}\ l_{y2}]$, $l_{r1} = \sum_{i=1}^{N} q_i \sum_{j=1}^{N} r_{ij}$ and $l_{u1} = \mathbf{q}\ \mathbf{Q}$. The resulting control scheme is shown in figure 5.9, where the values $\hat{y}(t+d \mid t), \hat{y}(t+d-1 \mid t)$ are obtained by the use of the predictor previously described. Notice that the predictor basically consists of a model of the plant which is projected towards the future with the values of past inputs and outputs.

5.3.2 Controller Parameters.

The plant estimated parameters can be used to compute the controller coefficients (l_{y1}, l_{y2}, l_{r1} and l_{u1}). These coefficients are functions of the plant parameters a, b_0, b_1, $\delta(i)$ and $\lambda(i)$. If the GPC is designed considering the plant to have a unit static gain, there exists a relationship between the plant parameters:

$$1 - a = b_0 + b_1$$

so only two of the three parameters will be needed to calculate the coefficients in (5.17). One parameter will be the system pole a and the other will be:

$$m = \frac{b_0}{b_0 + b_1}$$

Parameter m indicates how close the true dead time is to parameter d in the model used in equation (5.13). That is, if $m = 1$ the plant dead time is d and if $m = 0$ it is $d + 1$; so a range of m between 1 and 0 covers the fractional dead times between d and $d + 1$.

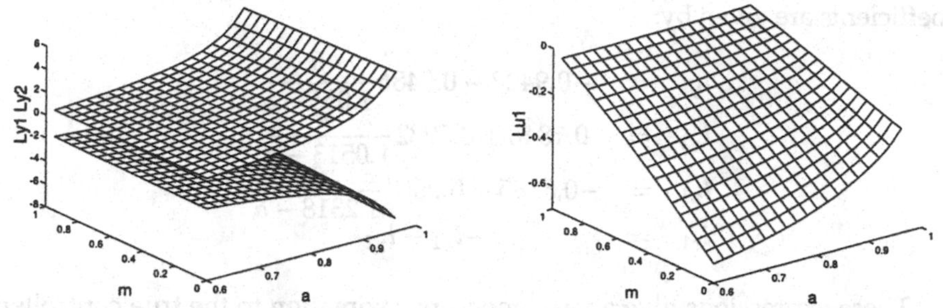

Figure 5.10: Controller Parameters l_{y1}, l_{y2} and l_{u1} as functions of a and m

Once the values of $\delta(i)$ and $\lambda(i)$ have been chosen and the plant parameters are known, the controller coefficients can easily be derived. The control signal is divided by the process static gain in order to get a system with a unitary static gain and reduce the number of parameters.

In order to avoid the heavy computational requirements needed to compute matrices **M**, **P**, **R** and **Q** and solve equation (5.15), the coefficients can be obtained by interpolating in a set of previously computed values as shown in figure 5.10. Notice that this can be accomplished in this case because the controller coefficients only depend on two parameters. As they have been obtained considering a unitary static gain, they must be corrected dividing the coefficients l_{y1}, l_{y2} and l_{r1} by this value.

The algorithm described above can be used to compute controller parameters of GPC for plants which can be described by equation (5.13) over a set covering the region of interest. This region is defined by values of the pole in the interval $[0.5, 0.95]$ and the other plant parameter m that will vary between 0 and 1.

The curves shown in figure 5.10 correspond to the controller parameters l_{y1}, l_{y2} and l_{u1} for $\delta(i) = \delta^i$ and $\lambda(i) = \lambda^i$ with $\delta = 1$, $\lambda = 0.8$ and $N = 15$. Notice that due to the fact that the closed-loop static gain must equal the value 1, the sum of parameters l_{y1}, l_{y2} and l_{r1} equals zero. This result implies that only three of the four parameters need to be known.

The expressions relating the controller parameters to the process parameters can be approximated by functions of the form:

$$k_{1i}(m) + k_{2i}(m)\frac{a}{k_{3i}(m) - a} \tag{5.18}$$

The coefficients $k_{ji}(m)$ depend on the value of m and can be calculated by a least squares fitting using the set of known values of l_{yi} for different values of a and m. Low order polynomials that give a good approximation for $k_{ji}(m)$ have been obtained by Bordons [13].

In the case of $m = 0.5, \lambda = 0.8$ and for a control horizon of 15, the controller coefficients are given by:

$$\begin{aligned} l_{y1} &= -0.9427 - 0.5486\frac{a}{1.055 - a} \\ l_{y2} &= 0.1846 + 0.5082\frac{a}{1.0513 - a} \\ l_{u1} &= -0.3385 + 0.0602\frac{a}{1.2318 - a} \\ l_{r1} &= -l_{y1} - l_{y2} \end{aligned}$$

These expressions give a very good approximation to the true controller parameters and fit the set of computed data with a maximum error of less than two percent of the nominal values for the range of interest of plant parameters.

The influence of the control weighting factor λ on the controller parameters can also be taken into account. For small values of λ, the parameters are bigger so as to produce a bigger control effort, thus this factor has to be considered in the approximative functions. With a procedure similar to that of previous sections, the values of $k_{ij}(m)$ in expressions (5.18) can be approximated as functions of λ, obtaining a maximum error of around three percent (with the worst cases at the limits of the region).

The algorithm in the adaptive case will consider the plant parameters and the control law and can be seen below.

1. Perform an identification step.
2. Compute $k_{ij}(m, \lambda)$.
3. Calculate l_{y1}, l_{y2} and l_{u1}. Make $l_{r1} = -l_{y1} - l_{y2}$
4. Compute $\hat{y}(t+d \mid t)$ and $\hat{y}(t+d-1 \mid t)$ using equation (5.5) recursively.
5. Compute $u(t)$ with:
   $\Delta u(t) = (l_{y1}\, \hat{y}(t+d \mid t) + l_{y2}\, \hat{y}(t+d-1 \mid t) +$
   $+ l_{r1}\, r(t))/G + l_{u1}\, \Delta\, u(t-1)$
6. Go to step 1.

5.3.3 Example

This example is taken from [116] and corresponds to the distillate composition loop of a binary distillation column. The manipulated variable is the reflux flowrate and the controlled variable is the distillate composition. Although the process is clearly nonlinear, it can be modelled by a first order model plus a dead time of the form $G(s) = Ke^{-s\tau_d}/(1+\tau s)$ at different operating points. Notice that this is a reasonable approach since, in fact, a distillation column is composed of a number of plates, each one being a first order element.

As the process is nonlinear, the response varies for different operating conditions, so different values for the parameters K, τ and τ_d were obtained changing the reflux flowrate from 3.5 mol/min to 4.5 mol/min (see table 5.1). By these tests, it was seen that variations in the process parameters as $0.107 \leq K \leq 0.112$, $15.6 \leq \tau_d \leq 16.37$ and $40.49 \leq \tau \leq 62.8$ should be considered (being τ and τ_d in minutes).

Considering a sample time of five minutes the dead time is not integer and the discrete transfer function must be that of equation (5.13). The discrete parameters can be seen in the same table. The system pole can vary between 0.8838 and 0.9234, meanwhile parameter m is going to move between 0.7381 and 0.8841.

Table 5.1: Process parameters for different operating conditions

Flow	K	τ	τ_d	a	$b_0(\times 10^{-3})$	$b_1(\times 10^{-3})$	d
4.5	0.107	62.8	15.6	0.9234	7.2402	0.9485	3
4	0.112	46.56	15.65	0.8981	9.9901	1.4142	3
3.5	0.112	40.49	16.37	0.8838	9.6041	3.4067	3

To cope with the variations of the system dynamics depending on the operating point, a gain-scheduling predictive controller can be developed. In order to do this, a set of controller parameters is obtained for each operating point, and depending on the reflux flow, the parameters in each condition will be calculated by interpolating in these sets.

For a fixed value of λ, the coefficients $k_{ij}(m)$ are used to calculate the values $l_i(a,m)$. For $\lambda = 0.8$, the following expressions can be used:

$$\begin{aligned}
k_{11} &= 0.141 * m^2 - 0.125 * m - 0.920 \\
k_{21} &= -0.061 * m^2 + 0.202 * m - 0.625 \\
k_{31} &= -0.015 * m + 1.061 \\
k_{12} &= -0.071 * m^2 + 0.054 * m + 0.180 \\
k_{22} &= -0.138 * m + 0.575 \\
k_{32} &= -0.015 * m + 1.058 \\
k_{14} &= -1.287 * m^2 + 1.747 * m + 0.117 \\
k_{24} &= -0.113 * m + 0.112 \\
k_{34} &= -0.071 * m^2 + 0.091 * m + 1.181
\end{aligned}$$

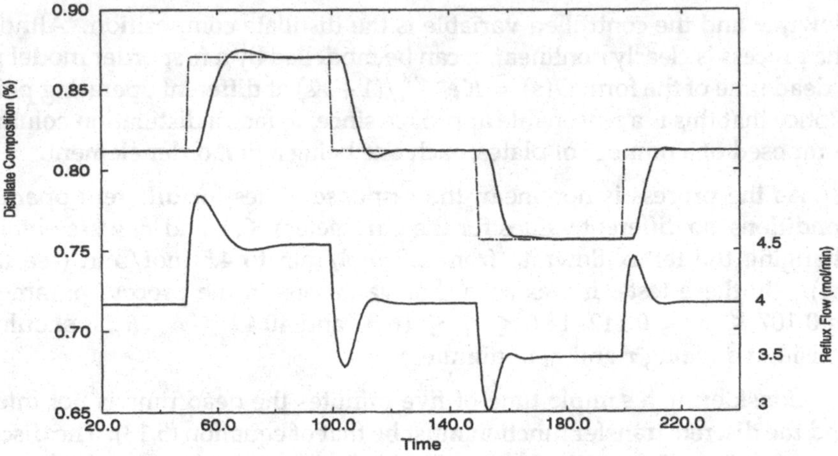

Figure 5.11: Changes in Operating Conditions

The GPC parameters for the different operating conditions are given in the table 5.2

Table 5.2: GPC parameters for different operating conditions

Flow	l_{y1}	l_{y2}	l_{r1}	l_{u1}
4.5	-4.577	3.629	0.948	-0.035
4	-3.881	2.954	0.927	-0.039
3.5	-3.637	2.745	0.892	-0.091

As the dead time is of three sampling periods, the predictor is:

$$\hat{y}(t+i \mid t) = (1+a)\hat{y}(t+i-1 \mid t) - a\hat{y}(t+i-2 \mid t) + b_0\Delta u(t+i-3) + b_1\Delta u(t+i-4)$$

and the control law,

$$u(t) = u(t-1) + (l_{y1}\hat{y}(t+3 \mid t) + l_{y2}\hat{y}(t+2 \mid t) + l_{r1}r)/G + l_{u1} \Delta u(t-1)$$

where G is the static gain and the controller and predictor parameters are obtained by interpolating the flow, once filtered by a low pass filter, in the tables given above.

To show the system behaviour, changes of the setpoint covering the complete region are produced and it can be observed in figure 5.11 that the output follows the reference, in spite of the changes in the system dynamics due to changing operating point.

5.4 Integrating Processes

In industrial practice it is easy to find some processes including an integral effect. The output of one of these processes grows infinitely when excited by a step input. This is the case of a tank where the level increases provided there is an input flow and a constant output. Also the angle of an electrical motor shaft which grows while being powered until the torque equals the load. The behaviour of these processes differs drastically from that of the ones considered up to now in this chapter.

These processes cannot be modelled by a first order plus delay transfer function, but need the addition of an $1/s$ term in order to model the integrating effect. Hence, the transfer function for this kind of processes will be:

$$G(s) = \frac{K}{s(1+\tau s)} e^{-\tau_d s} \tag{5.19}$$

In the general case of dead time being non multiple of the sampling time the equivalent discrete transfer function when a zero order hold is employed is given by:

$$G(z) = \frac{b_0 z^{-1} + b_1 z^{-2} + b_2 z^{-3}}{(1-z^{-1})(1-az^{-1})} z^{-d} \tag{5.20}$$

In the simpler case of the dead time being an integer multiple of the sampling time the term b_2 disappears.

The GPC control law for processes described by (5.19) will be calculated in this section. Notice that some formulations of MPC are unable to deal with these processes since they use the truncated impulse or step response, which is not valid for unstable processes. As GPC makes use of the transfer function, there is no problem about unstable processes.

5.4.1 Derivation of the Control Law

The procedure for obtaining the control law is analogous to the one used in previous sections, although, logically, the predictor will be different and the final expression will change slightly.

Using a CARIMA model with the noise polynomial equal to 1, the system can be written as

$$(1-z^{-1})(1-az^{-1})y(t) = (b_0 + b_1 z^{-1} + b_2 z^{-2}) z^{-d} u(t-1) + \frac{\varepsilon(t)}{\Delta}$$

which can be transformed into:

$$\begin{aligned} y(t+1) &= (2+a)y(t) - (1+2a)y(t-1) + ay(t-2) + \\ &+ b_0 \Delta u(t-d) + b_1 \Delta u(t-d-1) + b_2 \Delta u(t-d-2) + \varepsilon(t+1) \end{aligned}$$

If the values of $\hat{y}(t+d+i-1 \mid t)$, $\hat{y}(t+d+i-2 \mid t)$ and $\hat{y}(t+d+i-3 \mid t)$ are known, then the best predicted output at instant $t+d+i$ will be:

$$\hat{y}(t+d+i \mid t) = (2+a)\hat{y}(t+d+i-1 \mid t) - (1+2a)\hat{y}(t+d+i-2 \mid t) + \\ +a\hat{y}(t+d+i-3 \mid t) + b_0 \triangle u(t+i-1) + b_1 \triangle u(t+i-2) + b_2 \triangle u(t+i-3)$$

With these expressions of the predicted outputs, the cost function to be minimized will be a function of $\hat{y}(t+d \mid t)$, $\hat{y}(t+d-1 \mid t)$ and $\hat{y}(t+d-2 \mid t)$, as well as the future control signals $\Delta u(t+N-1)$, $\Delta u(t+N-2)$ ... $\Delta u(t)$, and past inputs $\Delta u(t-1)$ and $\Delta u(t-2)$ and, of course, of the reference trajectory.

Minimizing $J(N_1, N_2, N_3)$ leads to the following matrix equation for calculating $\mathbf{u}$:

$$\mathbf{M}\,\mathbf{u} = \mathbf{P}\,\mathbf{y} + \mathbf{R}\,\mathbf{w} + \mathbf{Q}_1 \triangle u(t-1) + \mathbf{Q}_2 \triangle u(t-2)$$

where $\mathbf{M}$ and $\mathbf{R}$ are matrices of dimension $N \times N$, $\mathbf{P}$ of dimension $N \times 2$ and $\mathbf{Q}_1$ and $\mathbf{Q}_2$ of $N \times 1$. As in the latter section, $\mathbf{u}$ are the future input increments and $\mathbf{y}$ the predicted outputs.

The first element of vector $\mathbf{u}$ can be obtained by:

$$\Delta u(t) = \mathbf{q}\,\mathbf{P}\,\mathbf{y} + \mathbf{q}\,\mathbf{R}\,\mathbf{w} + \mathbf{q}\,\mathbf{Q}_1 \triangle u(t-1) + \mathbf{q}\,\mathbf{Q}_2 \triangle u(t-2)$$

being $\mathbf{q}$ the first row of matrix $\mathbf{M}^{-1}$.

If the reference is considered to be constant over the prediction horizon and equal to the current setpoint:

$$\mathbf{w} = [1 \ldots 1] r(t+d)$$

the control law results as:

$$\begin{aligned} \triangle u(t) &= l_{y1}\hat{y}(t+d \mid t) + l_{y2}\hat{y}(t+d-1 \mid t) + l_{y3}\hat{y}(t+d-2 \mid t) \\ &+ l_{r1} r(t+d) + l_{u1} \triangle u(t-1) + l_{u2} \triangle u(t-2) \end{aligned} \tag{5.21}$$

Being $\mathbf{q}\,\mathbf{P} = [l_{y1}\; l_{y2}\; l_{y3}]$, $l_{r1} = \sum_{i=1}^{N}(q_i \sum_{j=1}^{N} r_{ij})$, $l_{u1} = \mathbf{q}\,\mathbf{Q}_1$ and $l_{u2} = \mathbf{q}\,\mathbf{Q}_2$.

Therefore the control law results in a linear expression depending on six coefficients which depend on the process parameters (except on the dead time) and on the control weighting factor λ. Furthermore, one of these coefficients is a linear combination of the others, since the following relation must hold so as to get a closed loop with unitary static gain:

$$l_{y1} + l_{y2} + l_{y3} + l_{r1} = 0$$

5.4.2 Controller Parameters

The control law (5.21) is very easy to implement provided the controller parameters l_{y1}, l_{y2}, l_{y3}, l_{r1}, l_{u1} and l_{u2} are known. The existence of available relationships of these parameters with process parameters is of crucial importance for a straightforward implementation of the controller. In a similar way to the previous sections, simple expressions for these relationships will be obtained.

As the process can be modelled by (5.20) four parameters (a, b_0, b_1 and b_2) are needed to describe the plant. Expressions relating the controller coefficients with these parameters can be obtained as previously, although the resulting functions are not as simple, due to the number of plant parameters involved. As the dead time can often be considered as a multiple of the sampling time, simple functions will be obtained for this case from now on. Then b_2 will be considered equal to 0.

In a similar way to the process without integrator case, the process can be considered to have $(b_0 + b_1)/(1 - a) = 1$ in order to work with normalized plants. Then the computed parameters must be divided by this value that will not be equal to 1 in general.

The controller coefficients will be obtained as a function of the pole a and a parameter:

$$n = \frac{b_0}{b_0 + b_1}$$

This parameter has a short range of variability for any process. As b_0 and b_1 are related to the continuous parameters by (see [8]):

$$b_0 = K(T + \tau(-1 + e^{-\frac{T}{\tau}})) \qquad b_1 = K(\tau - e^{-\frac{T}{\tau}}(T + \tau))$$

then

$$n = \frac{a - 1 - \log a}{(a - 1)\log a}$$

that for the usual values of the system pole is going to vary between $n = 0.5$ and $n = 0.56$. Therefore the controller parameters can be expressed as functions of the system pole, and n for a fixed value of λ.

The shape of the parameters is displayed in figure 5.12 for a fixed value of $\lambda = 1$. It can be seen that the coefficients depend mainly on the pole a, being almost independent of n except in the case of l_{u1}. Functions of the form

$$f(a, n, \lambda) = k_1(n, \lambda) + k_2(n, \lambda)\frac{a}{k_3(n, \lambda) - a}$$

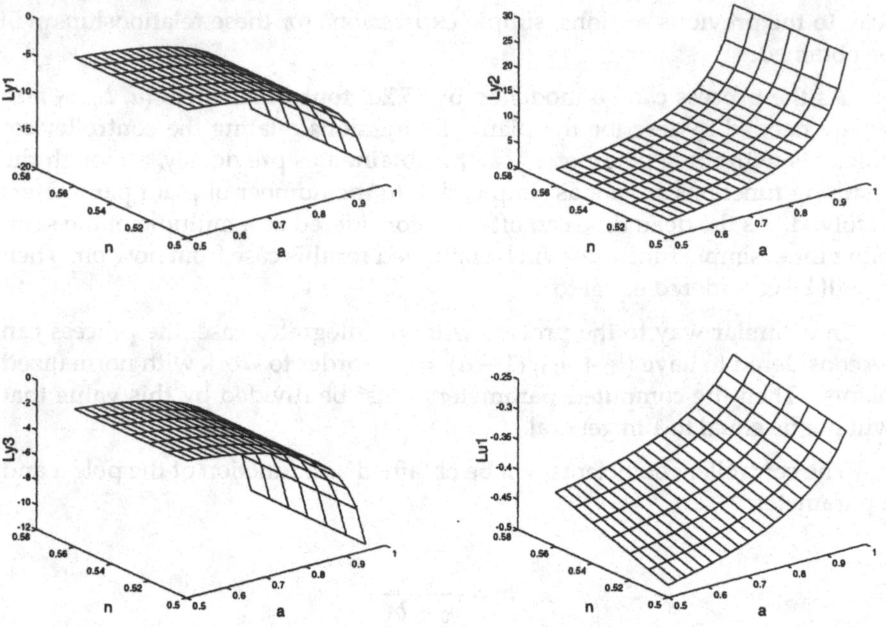

Figure 5.12: Controller coefficients l_{y1}, l_{y2}, l_{y3} and l_{u1}

where k_i can be approximated by:

$$
\begin{aligned}
k_{y1,1} &= -e^{0.955-0.559\lambda+0.135\lambda^2} \\
k_{y1,2} &= -e^{0.5703-0.513\lambda+0.138\lambda^2} \\
k_{y1,3} &= 1.0343 \\
k_{y2,1} &= e^{0.597-0.420\lambda+0.0953\lambda^2} \\
k_{y2,2} &= e^{1.016-0.4251\lambda+0.109\lambda^2} \\
k_{y2,3} &= 1.0289 \\
k_{y3,1} &= -e^{-1.761-0.422\lambda+0.071\lambda^2} \\
k_{y3,2} &= -e^{0.103-0.353\lambda+0.089\lambda^2} \\
k_{y3,3} &= 1.0258 \\
k_{u1,1} &= 1.631n - 1.468 + 0.215\lambda - 0.056\lambda^2 \\
k_{u1,2} &= -0.124n + 0.158 - 0.026\lambda + 0.006\lambda^2 \\
k_{u1,3} &= 1.173 - 0.019\lambda
\end{aligned}
\tag{5.22}
$$

provide good approximations for l_{y1}, l_{y2}, l_{y3}, l_{r1} and l_{u1} in the usual range of the plant parameter variations. Notice that an approximate function for l_{r1} is not supplied, since it is linearly dependent on the other coefficients. The functions fit the set of computed data with a maximum error of less than 1.5 percent of the nominal values. Notice that closer approximations can be obtained if developed for a concrete case where the range of variability of the process parameters is smaller.

5.4.3 Example

The control law (5.21) will be implemented in an extensively used system as a direct current motor. When the input of the process is the voltage applied to the motor (U) and the output the shaft angle (θ) it is obvious that the process has an integral effect, given that the position grows indefinitely whilst it is fed by a certain voltage. In order to obtain a model that describes the behaviour of the motor the inertia load (proportional to the angular acceleration) and the dynamic friction load (proportional to angular speed) are taken into account. Their sum is equal to the torque developed by the motor, that depends on the voltage applied to it. It is a first order system with regards to speed but a second order one if the angle is considered as the output of the process.:

$$J\frac{d^2\theta}{dt^2} + f\frac{d\theta}{dt} = M_m$$

and the transfer function will be:

$$\frac{\theta(s)}{U(s)} = \frac{K}{s(1+\tau s)}$$

where K and τ depend on electromechanical characteristics of the motor.

The controller is going to be implemented on a real motor with a feed voltage of 24 V and nominal current of 1.3 A, subjected to a constant load. The reaction curve method is used to obtain experimentally the parameters of the motor, applying a step in the feed voltage and measuring the evolution of the angular speed (which is a first order system). The parameters obtained are:

$$K = 2.5 \qquad \tau = 0.9 \text{ seconds}$$

and zero dead time. Taking a sampling time of $T = 0.06$ seconds one gets the discrete transfer function:

$$G(z) = \frac{0.004891z^{-1} + 0.004783z^{-2}}{(1-z^{-1})(1-0.935507z^{-1})}$$

If a high value of the control weighting factor is taken in order to avoid overshooting ($\lambda = 2$) the control parameters (5.21) can be calculated using expressions (5.22):

$$\begin{aligned}
l_{y1} &= -11.537\\
l_{y2} &= 19.242\\
l_{y3} &= -8.207\\
l_{u1} &= -0.118\\
l_{r1} &= 0.502
\end{aligned}$$

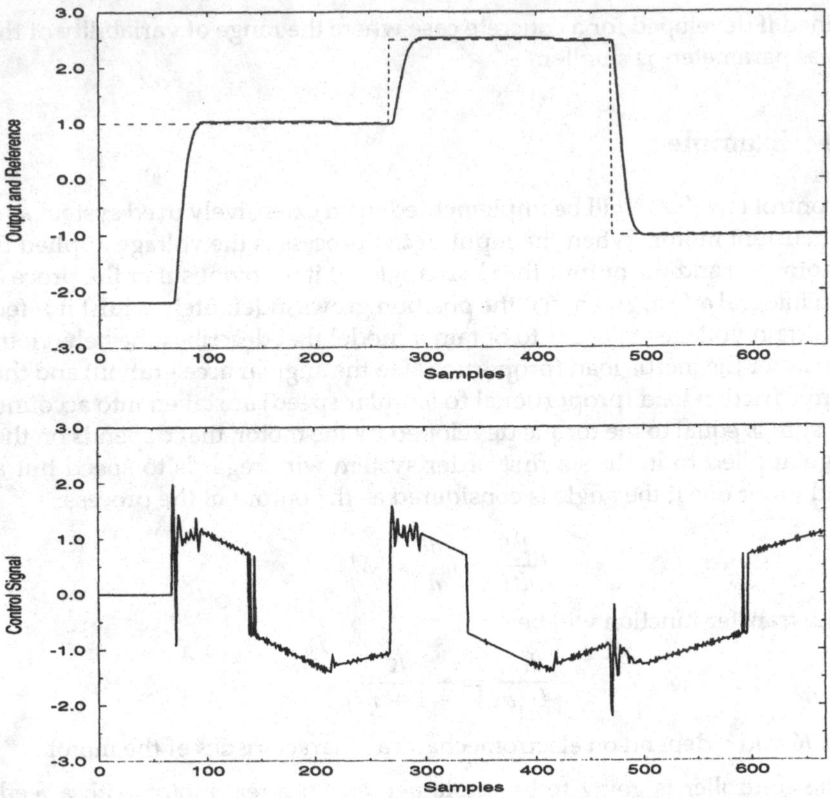

Figure 5.13: Motor response for setpoint changes

The evolution of the shaft angle when some steps are introduced in the reference can be seen in figure 5.13. It can be observed that there is no overshooting due to the high value of λ chosen. The system has a dead zone such that it is not sensitive to control signals less than 0.7 V; in order to avoid this a non-linearity is added.

It is important to remember that the sampling time is very small (0.06 seconds) which could make the implementation of the standard GPC algorithm impossible. However, due to the simple formulation used here, the implementation is reduced to the calculation of expression (5.21) and hardly takes any time in any computer.

The process is disturbed by the addition of an electromagnetic break that changes the load and the friction constant. The model parameters used for designing the GPC do not coincide with the process parameters but in spite of this, as can be seen in figure 5.14, GPC is able to control the motor reasonably well even though a slight overshoot appears.

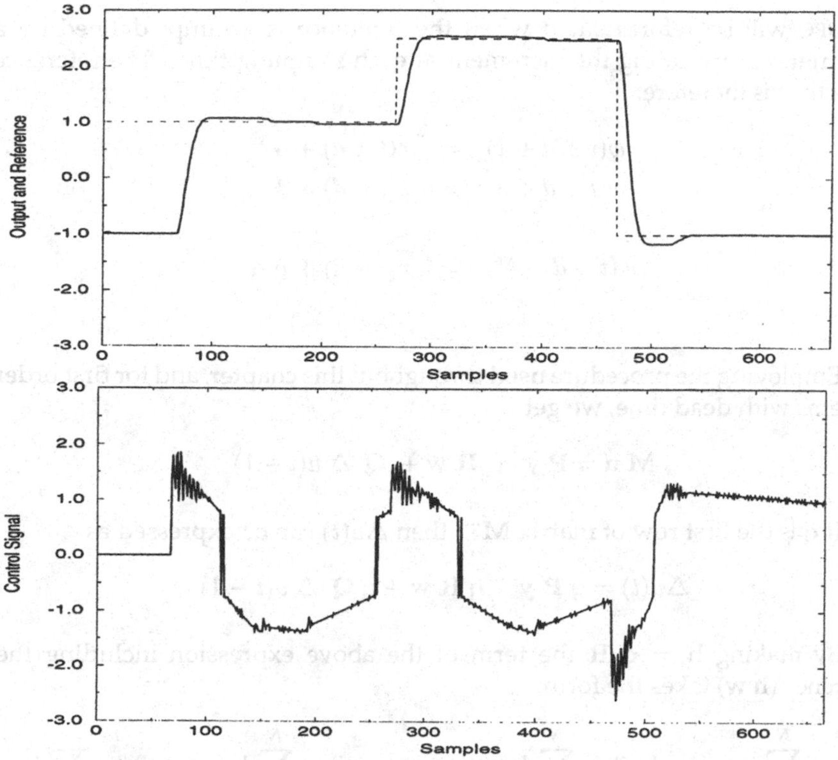

Figure 5.14: Motor response with electromagnetic break

5.5 Consideration of Ramp Setpoints

It is usual for a process reference signal to keep a certain constant value for a time and to move to other constant values by step changes during normal plant operation. This is what has been considered up to now, that is, $w(t+d+1) = w(t+d+2) \ldots = r(t)$, $r(t)$ being the setpoint at instant t which is going to maintain a fixed value.

But the reference evolution will not behave like this in all circumstances. On many occasions it can evolve as a ramp, which changes smoothly to another constant setpoint. In general it would be desirable for the process output to follow a mixed trajectory composed of steps and ramps.

This situation frequently appears in different industrial processes. In the food and pharmaceutical industries some thermal processes require the temperature to follow a profile given by ramps and steps. It is also of interest that in the control of motors and in Robotics applications the position or velocity follow evolutions of this type.

GPC will be reformulated when the reference is a ramp, defined by a parameter α indicating the increment at each sampling time. The reference trajectory is therefore:

$$\begin{aligned} w(t+d+1) &= r(t+d)+\alpha \\ w(t+d+2) &= r(t+d)+2\alpha \\ \dots & \quad \dots \\ w(t+d+N) &= r(t+d)+N\alpha \end{aligned}$$

Employing the procedure used throughout this chapter, and for first order systems with dead time, we get

$$\mathbf{M}\,\mathbf{u} = \mathbf{P}\,\mathbf{y} + \mathbf{R}\,\mathbf{w} + \mathbf{Q}\ \Delta u(t-1)$$

If $\mathbf{q}$ is the first row of matrix $\mathbf{M}^{-1}$ then $\Delta u(t)$ can be expressed as

$$\Delta u(t) = \mathbf{q}\,\mathbf{P}\,\mathbf{y} + \mathbf{q}\,\mathbf{R}\,\mathbf{w} + \mathbf{q}\,\mathbf{Q}\ \Delta u(t-1)$$

By making $\mathbf{h} = \mathbf{q}\ \mathbf{R}$ the term of the above expression including the reference ($\mathbf{h}\ \mathbf{w}$) takes the form:

$$\mathbf{h}\ \mathbf{w} = \sum_{i=1}^{N} h_i\ r(t+d+i) = \sum_{i=1}^{N} h_i\ (r(t+d)+\alpha\ i) = \sum_{i=1}^{N} h_i\ r(t+d)+\alpha\sum_{i=1}^{N} h_i\ i$$

therefore

$$\mathbf{h}\ \mathbf{w} = l_{r1}\ r(t+d) + \alpha\ l_{r2}$$

The control law can now be written as

$$\Delta u(t) = l_{y1}\hat{y}(t+d \mid t)+l_{y2}\hat{y}(t+d-1 \mid t)+l_{r1}r(t+d)+\alpha\ l_{r2}+l_{u1}\Delta u(t-1) \quad (5.23)$$

Where $\mathbf{q}\,\mathbf{P} = [l_{y1}\ \ l_{y2}]$, $l_{u1} = \mathbf{q}\,\mathbf{Q}$, $l_{r1} = \sum_{i=1}^{N}(q_i \sum_{j=1}^{N} r_{ij})$ and $l_{r2} = \alpha \sum_{i=1}^{N} h_i\ i$.

The control law is therefore linear. The new coefficient l_{r2} is due to the ramp. It can be noticed that when the ramp becomes a constant reference, the control law coincides with the one developed for the constant reference case. The only modification that needs to be made because of the ramps is the term $l_{r2}\alpha$. The predictor is the same and the resolution algorithm does not differ from the one used for the constant reference case. The new parameter l_{r2} is a function of the process parameters (a, m) and of the control weighting factor (λ). As in the previous cases an approximating function can easily be obtained. Notice that the other parameters are exactly the same as in the constant reference case, meaning that the previously obtained expressions can be used.

In what has been seen up to now (non integrating processes, integrating processes, constant reference, ramp reference), a new coefficient appeared in the control law with each new situation. All these situations can be described by the following control law:

$$\begin{aligned}\Delta u(t) &= l_{y1}\hat{y}(t+d \mid t) + l_{y2}\hat{y}(t+d-1 \mid t) + l_{y3}\hat{y}(t+d-2 \mid t) + \\ &+ l_{r1}r(t+d) + \alpha\, l_{r2} + l_{u1}\,\Delta\, u(t-1) + l_{u2}\,\Delta\, u(t-2)\end{aligned}$$

Table 5.3: Coefficients that may appear in the control law. The × indicates that the coefficient exists

Process	Reference	l_{y1}	l_{y2}	l_{y3}	l_{u1}	l_{u2}	l_{r1}	l_{r2}
$\frac{k}{1+\tau s}e^{-\tau_d s}$	Constant	×	×	0	0	0	×	0
τ_d integer	Ramp	×	×	0	0	0	×	×
$\frac{k}{1+\tau s}e^{-\tau_d s}$	Constant	×	×	0	×	0	×	0
τ_d non integer	Ramp	×	×	0	×	0	×	×
$\frac{k}{s(1+\tau s)}e^{-\tau_d s}$	Constant	×	×	×	×	0	×	0
τ_d integer	Ramp	×	×	×	×	0	×	×
$\frac{k}{s(1+\tau s)}e^{-\tau_d s}$	Constant	×	×	×	×	×	×	0
τ_d non integer	Ramp	×	×	×	×	×	×	×

Table 5.3 shows which coefficients of the above control law may be zero depending on the particular situation.

5.5.1 Example

As an application example, a GPC with ramp following capability is going to be designed for the motor described above. The reference trajectory is composed of a series of steps and ramps defined by the value of α ($\alpha = 0$ for the case of constant reference).

The same controller parameters as in the previous example are used, with the addition of the new parameter $l_{r2} = 2.674$. Considering that $(b_0+b_1)/(1-a) = 0.15$, the control law is given by:

$$\begin{aligned}\Delta u(t) &= -76.92\, y(t) + 128.29\, y(t-1) - 54.72\, y(t-2) + \\ &+ 3.35 r(t) + 17.82\, \alpha - 0.12\,\Delta\, u(t-1)\end{aligned}$$

As the dead time is zero, the predicted outputs are known at instant t.

The results obtained are shown in figure 5.15 where it can be seen that the motor is able to follow the ramp reference quite well.

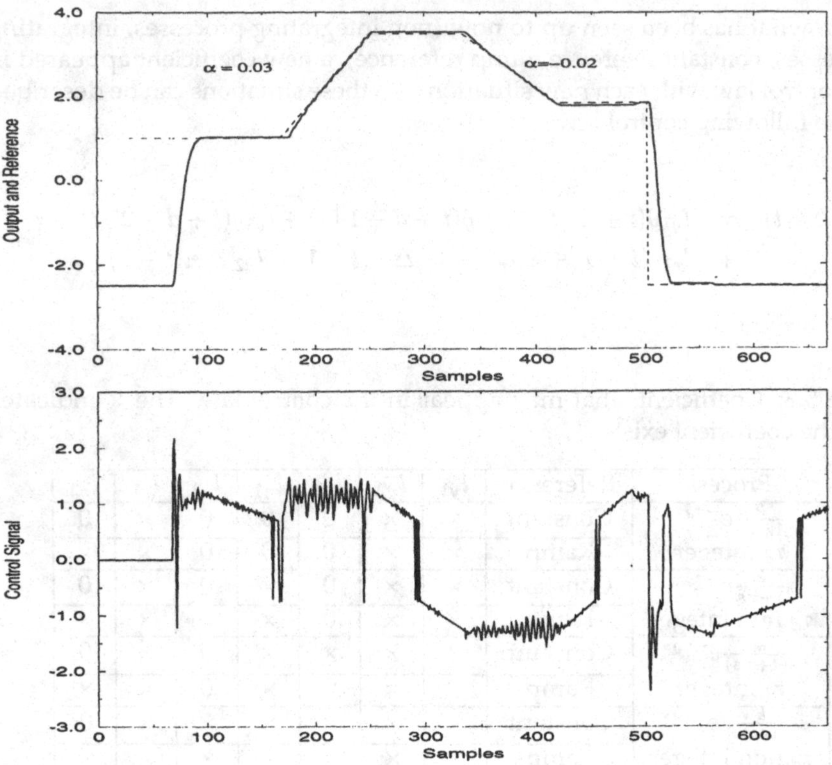

Figure 5.15: Combined steps and ramps setpoint

5.6 Comparison with Standard GPC

The approximations made in the method can affect the quality of the controlled performance. Some simulation results are presented that compare the results obtained with the proposed method with those when the standard GPC algorithm as originally proposed by Clarke *et al.* [34] is used.

Two indices are used to measure the performance: ISE (sum of the square errors during the transient) and ITAE (sum of the absolute error multiplied by discrete time). Also the number of floating point operations and the computing time needed to calculate the control law are analyzed.

First, the performance of the proposed algorithm is compared with that of the standard GPC with no modelling errors. In this situation the error is only caused by the approximative functions of the controller parameters. For the system $G(s) = \frac{1.5}{1+10s}e^{-4s}$ with a sampling time of one second, the values for the proposed algorithm when the process is perturbed by a white noise uniformly distributed in the interval ± 0.015 are : ISE= 7.132, ITAE= 101.106 and for the standard controller: ISE= 7.122, ITAE=100.536. The plot comparing

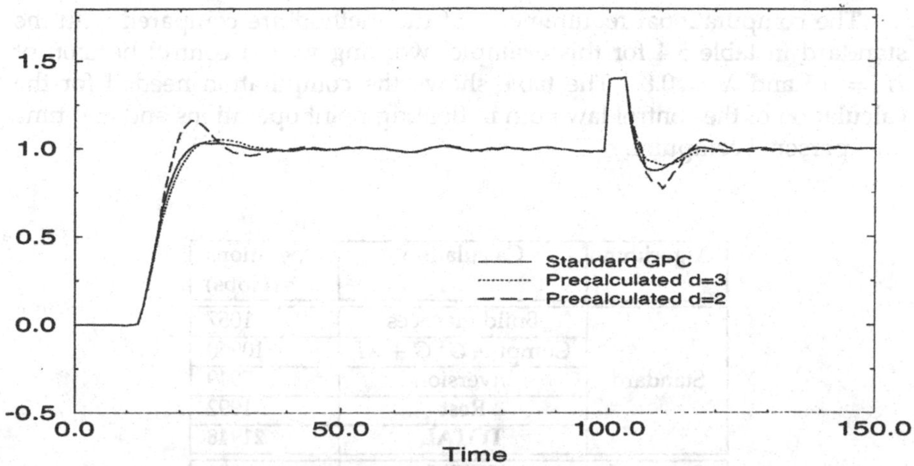

Figure 5.16: System performance. Standard GPC: ISE=6.809, ITAE= 395.608 , proposed algorithm $d = 3$ ISE=6.987, ITAE= 402.544 proposed algorithm $d = 2$ ISE=7.064, ITAE=503.687

both responses is not shown because practically there is no difference.

The plant model is supposed to be first order plus dead-time. If the process behaviour can be reasonably described by this model, there will not be a substantial loss of performance. Consider for instance the process modelled by:

$$G_p(z) = \frac{0.1125z^{-1} - 0.225z^{-2}}{1 - z^{-1} + 0.09z^{-2}} z^{-3}$$

For control purposes it is approximated by the following first order model, obtained from data generated by the process G_p:

$$G_m(z) = \frac{0.1125z^{-1}}{1 - 0.8875z^{-1}} z^{-3}$$

That is, the precomputed GPC is working in the presence of unmodelled dynamics. From the previous studies of robust stability, it can be deduced that the closed loop system is going to be stable. The performance in this situation is shown in figure 5.16 where the system response for both controllers is shown; notice that for t=100 a disturbance is added to the output. The figure also shows the behaviour of the precomputed GPC when an additional dead-time mismatch is included, that is, the controller uses a model with $d = 2$ instead of the true value $d = 3$.

Logically, there is a slight loss of performance due to the uncertainties, that must be considered in conjunction with the benefits in the calculation. Besides, consider that in a real case the uncertainties (such as dead-time mismatch) can also affect the standard GPC since high frequency effects are usually very difficult to model.

The computational requirements of the method are compared with the standard in table 5.4 for this example, working with a control horizon of $N = 15$ and $\lambda = 0.8$. The table shows the computation needed for the calculation of the control law both in floating point operations and CPU time on a personal computer.

Algorithm	Calculation	Operations (Flops)
Standard	Build matrices	1057
	Compute $G^T G + \lambda I$	10950
	Inversion	7949
	Rest	1992
	TOTAL	**21948**
Proposed	**TOTAL**	**79**

Table 5.4: Computational requirements for the standard and the precalculated GPC

As can be seen, these examples show that although a little performance is lost, there is a great improve in real-time implementability, reaching a computing effort around 275 times smaller. This advantage can represent a crucial factor for the implementation of this strategy in small size controllers with low computational facilities, considering that the impact on the performance is negligible. The simulations also show the robustness of the controller in the presence of structured uncertainties.

5.7 Stability Robustness Analysis

The elaboration of mathematical models of processes in real life requires simplifications to be adopted. In practice no mathematical model capable of exactly describing a physical process exists. It is always necessary to bear in mind that modelling errors may adversely affect the behaviour of the control system. The aim is that the controller should be insensitive to these uncertainties in the model, that is, that it should be robust. The aim here is to deal with the robustness of the controller presented. In any case, developments of predictive controller design using robust criteria can be found for instance in [22], [126].

The modeling errors, or uncertainties, can be represented in different forms, reflecting in certain ways the knowledge of the physical mechanisms which cause the discrepancy between the model and the process as well as the capacity to formalize these mechanisms so that they can be handled. Uncertainties can, in many cases, be expressed in a structured way, as expressions in function of determined parameters which can be considered in the transfer function [41]. However, there are usually residual errors particularly dominant at high frequencies which cannot be modelled in this way, which constitute unstructured uncertainties [39]. In this section a study of the precalculated GPC stability in the presence of both types of uncertainties is made; that is, the stability robustness of the method will be studied.

This section aims to study the influence of uncertainties on the behaviour of the process working with a controller which has been developed for the nominal model. That is, both the predictor and the controller parameters are calculated for a model which does not exactly coincide with the real process to be controlled. The following question is asked: what discrepancies are permissible between the process and the model in order for the controlled system to be stable?

The controller parameters l_{y1}, l_{y2}, l_{r1} and l_{u1} that appear in the control law:

$$\Delta u(t) = l_{y1}\hat{y}(t+d \mid t) + l_{y2}\hat{y}(t+d-1 \mid t) + l_{r1}r(t) + l_{u1} \Delta u(t-1)$$

have been precalculated for the model (not for the process as this is logically unknown). Likewise the predictor works with the parameters of the model, although it keeps up to date with the values taken from the output produced by the real process.

5.7.1 Structured Uncertainties

A first order model with pure delay, in spite of its simplicity, describes the dynamics of most plants in the process industry. However it is fundamental to consider the case where the model is unable to completely describe all the dynamics of the real process. Two types of structured uncertainties are considered: parametric uncertainties and unmodelled dynamic uncertainties. In the first case, the order of the control model is supposed to be identical to the order of the plant but the parameters are considered to be within an uncertainty region around the nominal parameters (these parameters will be the pole, the gain, and the coefficient $m = b_0/(b_0 + b_1)$ that measures the fractional delay between d and $d+1$). The other type of uncertainty will take into account the existence of process dynamics not included in the control model as an additional unmodelled pole and delay estimation error. This will be reflected in differences between the plant and model orders.

The uncertainty limits have been obtained numerically for the range of variation of the process parameters ($0.5 < a < 0.98, 0 \leq m \leq 1$) with a delay $0 \leq d \leq 10$ obtaining the following results (for more details, see [22]):

- **Uncertainty at the pole:** for a wide working zone ($a < 0.75$) and for normal values of the delay an uncertainty of more than $\pm 20\%$ is allowed. For higher poles the upper limit decreases due almost exclusively to the fact that the open loop would now be unstable. The stable area only becomes narrower for very slow systems with large delays. Notice that this uncertainty refers to the time constant (τ) uncertainty of the continuous process ($a = \exp(-T/\tau)$) and thus the time constant can vary around 500% of the nominal one in many cases.

- **Gain uncertainty:** When the gain of the model is G_e and that of the process is $\gamma \times G_e$, γ will be allowed to move between 0.5 and 1.5, that is, uncertainties in the value of the gain of about 50% are permitted. For small delays $(1, 2)$ the upper limit is always above the value $\gamma = 2$ and only comes close to the value 1.5 for delays of about 10. It can thus be concluded that the controller is very robust when faced with this type of error.
- **Uncertainty in** m**:** The effect of this parameter can be ignored since a variation of 300% is allowed without reaching instability.
- **Unmodelled pole:** The real process has another less dominant pole $(k \times a)$ apart from the one appearing in the model (a), and the results show that the system is stable even for values of k close to 1; stability is only lost for systems with very large delays.
- **Delay estimation error:** From the results obtained in a numerical study, it is deduced that for small delays stability is guaranteed for errors of up to two units through all the range of the pole, but when bigger poles are dealt with this only happens for small values of a, and even for delay 10 only a delay mismatch of one unit is permitted. It can be concluded, therefore, that a good delay estimation is fundamental to GPC, because for errors of more than one unit the system can become unstable if the process delay is high.

5.7.2 Unstructured Uncertainties

In order to consider unstructured uncertainties, it will be assumed from now on that the dynamic behaviour of a determined process is described not by an invariant time linear model but by a family of linear models. Thus the real possible processes (G) will be in a vicinity of the nominal process ($\widetilde{G}$), which will be modelled by a first order plus delay system.

A family $\mathcal{F}$ of processes in the frequency domain will therefore be defined which in the Nyquist plane will be represented by a region about the nominal plant for each ω frequency. If this family is defined as

$$\mathcal{F} = \{G :\mid G(i\omega) - \widetilde{G}(i\omega) \mid \leq \bar{l}_a(\omega)\}$$

the region consists of a disc with its centre at $G(i\omega)$ and radius $\bar{l}_a(\omega)$. Therefore any member of the family fulfils the condition

$$G(i\omega) = \widetilde{G}(i\omega) + l_a(i\omega) \qquad \mid l_a(i\omega) \mid \leq \bar{l}_a(\omega)$$

This region will change with ω because l_a does and, therefore, in order to describe family $\mathcal{F}$ we will have a zone formed by the discs at different frequencies. If one wishes to work with multiplicative uncertainties the family of processes can be described by:

$$\mathcal{F} = \{G : \left| \frac{G(i\omega) - \widetilde{G}(i\omega)}{\widetilde{G}(i\omega)} \right| \leq \bar{l}_m(\omega)\} \tag{5.24}$$

simply considering

$$l_m(i\omega) = l_a(i\omega)/\widetilde{G}(i\omega) \qquad \bar{l}_m(i\omega) = \bar{l}_a(\omega)/\mid \widetilde{G}(i\omega) \mid$$

Therefore any member of family $\mathcal{F}$ satisfies that:

$$G(i\omega) = \widetilde{G}(i\omega)(1 + l_m(i\omega)) \qquad \mid l_m(i\omega) \mid \leq \bar{l}_m(i\omega)$$

This representation of uncertainties in the Nyquist plane as a disc around the nominal process can encircle any set of structured uncertainties, although some times it can result in a rather conservative attitude [77].

The measurement of the robustness of the method can be tackled using the robust stability theorem [77], that for discrete systems states:

Suppose that all processes G of family $\mathcal{F}$ have the same number of unstable poles, which do not become unobservable for the sampling, and that a controller $C(z)$ stabilizes the nominal process $\widetilde{G}(s)$. Then the system has robust stability with controller C if, and only if, the complementary sensitivity function for the nominal process satisfies the following relation:

$$\mid \widetilde{T}(e^{i\omega T}) \mid \bar{l}_m(\omega) < 1 \qquad 0 \leq \omega \leq \pi/T \tag{5.25}$$

Using this condition the robustness limits will be obtained for systems that can be described by (5.13), and for all the values of the parameters that describe the system (a, m and d). For each value of the frequency ω the limits can be calculated as:

$$\bar{l}_m = \left| \frac{1 + \widetilde{G}C}{\widetilde{G}C} \right| \qquad \bar{l}_a = \bar{l}_m \mid \widetilde{G} \mid$$

In figure 5.17 the form taken by the limits in function of ωT can be seen for some values of a, and fixed values of m and d. Both limits are practically constant and equal to unity at low frequencies and change (the additive limit $\bar{l}_a$ decreases and the multiplicative $\bar{l}_m$ increases) at a certain point. Notice that these curves show the great degree of robustness that the GPC possesses since $\bar{l}_m$ is relatively big at high frequencies, where multiplicative uncertainties are normally smaller than unity, and increases with frequency as uncertainties

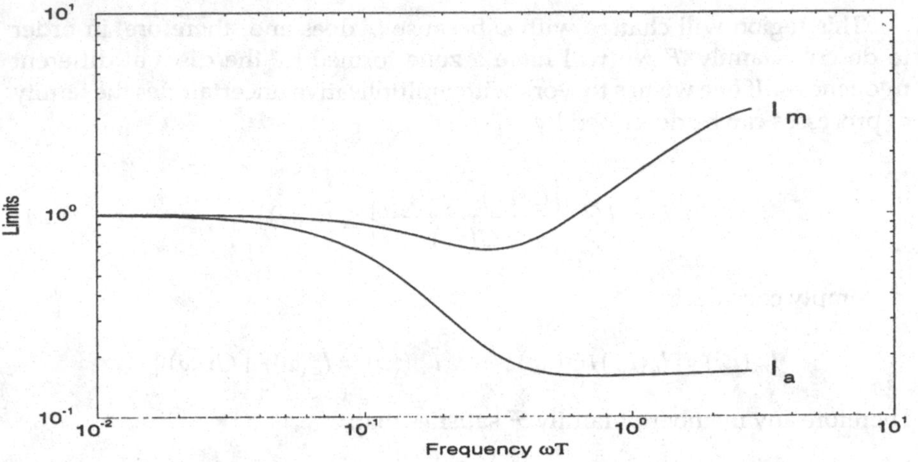

Figure 5.17: General shape of $\bar{l}_a$ and $\bar{l}_m$

do. The small value of $\bar{l}_a$ at high frequencies is due to the fact that the process itself has a small gain at those frequencies; remember that both limits are dependent and related by $\bar{l}_a = \bar{l}_m \mid \tilde{G} \mid$.

Figure 5.18 shows the frequency response of the nominal process alone and with the controller, as well as the discs of radius $\bar{l}_a$ and $\bar{l}_m \mid \widetilde{GC} \mid$ for a certain frequency. All the G processes belonging to the $\mathcal{F}$ family maintaining the stability of the closed loop can be found inside the disc of radius $\bar{l}_a$. The shape of the frequency response leads to limits $\bar{l}_a$ and $\bar{l}_m$. Thus, $\widetilde{GC}(\omega)$ has a big modulus (due to the integral term) at low frequencies, leading to a value of $\bar{l}_m$ close to unity. When ω increases, $\widetilde{GC}(\omega)$ separates from -1 (without decreasing in modulus) and therefore the limit can safely grow.

It can be seen that the most influential parameters are pole a and delay d. The evolution of limit $\bar{l}_a$ with frequency (ωT) for parameter a changing between 0.5 and 0.98 is presented in figure 5.19 for a concrete value of delay, $d = 1$, and for an average value of m, $m = 0.5$. As was to be expected, the limit decreases for greater poles because with open loop poles near to the limit of the unit circle the uncertainty allowed is less, as it would be easier to enter the open loop unstable zone.

5.7.3 General Comments

The results obtained for both types of uncertainties are qualitatively the same. It can be concluded that the factor that mainly affects robustness is delay uncertainty, because of its effect at high frequencies. The robustness zone decreases when the open loop pole increases whilst the parameter m hardly

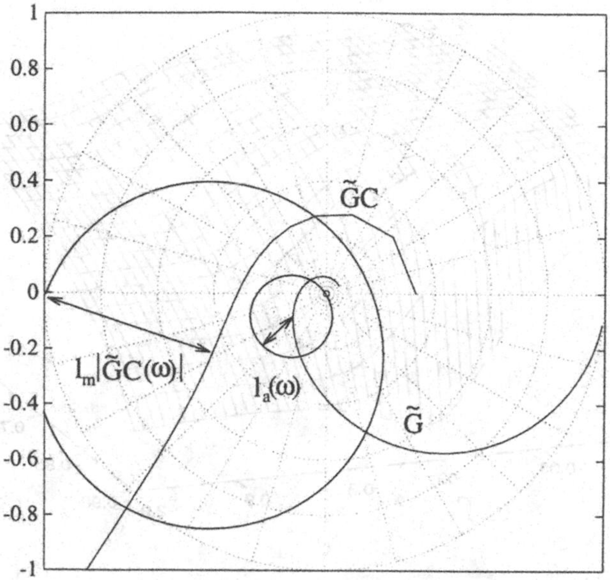

Figure 5.18: Polar diagram of the process $\tilde{G}$ and $\tilde{G}C$ showing the limits for a given frequency

has any influence. As the analysis has been performed based on a particular choice of parameters in the GPC formulation the conclusions depend on these values. The influence of the choice of these parameters on the closed loop stability is studied in [85].

In any case, the GPC algorithm presented has shown itself to be very robust against the types of uncertainties considered. For small delays the closed loop is stable for static gain mismatch of more than 100% and time constant mismatch of more than 200%.

The stability robustness of GPC can be improved with the use of an observer polynomial, the so-called $T(z^{-1})$ polynomial. In [33] a reformulation of the standard GPC algorithm including this polynomial can be found. In order to do this, the CARIMA model is expressed in the form:

$$A(z^{-1})y(t) = B(z^{-1})u(t-1) + \frac{T(z^{-1})}{\Delta}\xi(t)$$

Up to now the $T(z^{-1})$ has been considered equal to 1, describing the most common disturbances or as the colouring polynomial $C(z^{-1})$. But it can also be considered as a design parameter. In consequence the predictions will

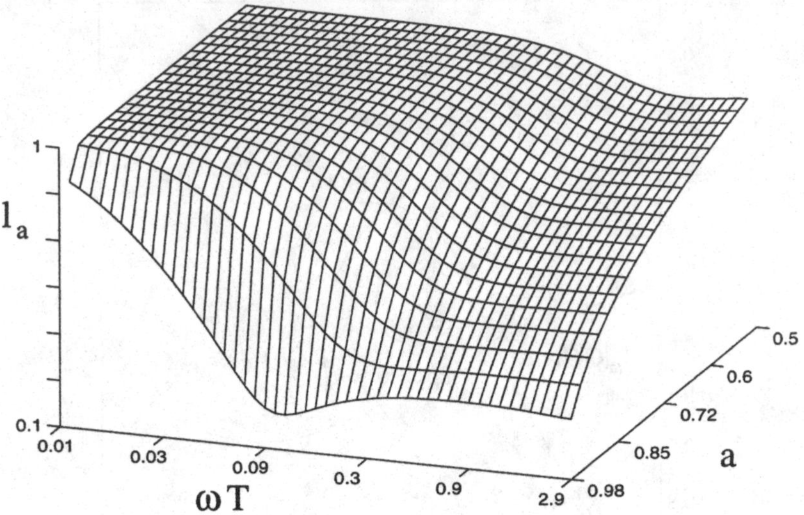

Figure 5.19: Limit $\bar{l}_a$ for dead time 1

not be optimal but on the other hand robustness in the face of uncertainties can be achieved, in a similar interpretation as that used by Ljung [68]. Then this polynomial can be considered as a prefilter as well as an observer. The effective use of observers is known to play an essential role in the robust realization of predictive controllers (see [33] for the effect of prefiltering on robustness and [127] for guidelines for the selection of T).

This polynomial can be easily added to the proposed formulation, computing the prediction with the values of inputs and outputs filtered by $T(z^{-1})$. Then, the predictor works with $y^f(t) = y(t)/T(z^{-1})$ and $u^f(t) = u(t)/T(z^{-1})$. The actual prediction for the control law is computed as $\hat{y}(t+d) = T(z^{-1})\hat{y}^f(t+d)$.

5.8 Composition Control in an Evaporator

This chapter ends presenting an application of the method to a typical process. An evaporator, a very common process in industry, has been chosen as a testing bed for the GPC. This process involves a fair number of interrelated variables and although it may appear to be rather simple compared to other processes of greater dimension, it allows the performance of any control technique to be checked. The results presented in this section have been obtained by simulation on a non-linear model of the process.

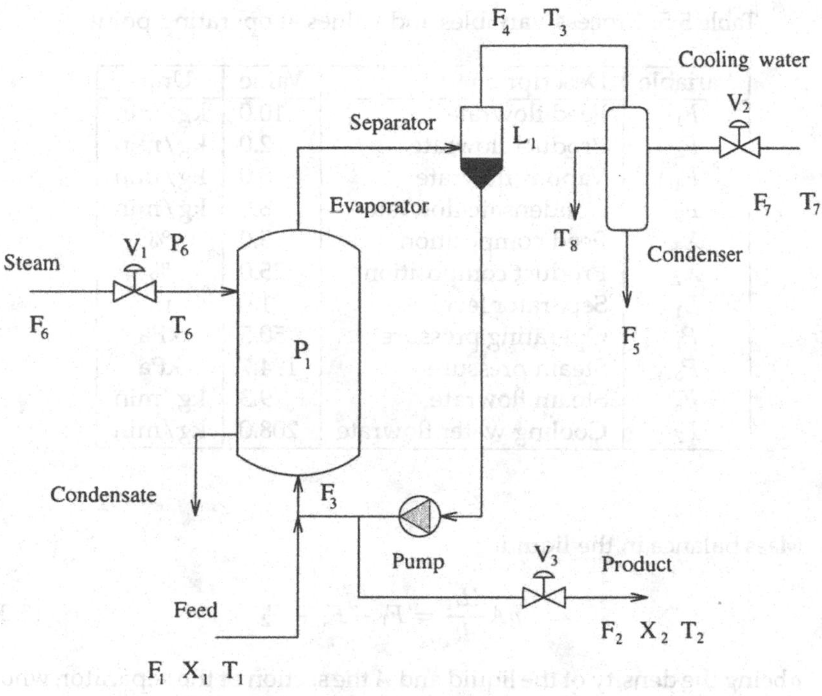

Figure 5.20: Diagram of the evaporator

5.8.1 Description of the Process

The process in question is a forced circulation evaporator in which the raw material is mixed with an extraction of the product and is pumped through a vertical heat exchanger though which water steam is circulating, which condenses in the tubes. The mix evaporates and passes through a separating vessel where the liquid and the vapour are separated. The former is made to recycle and a part is extracted as the final product whilst the vapour is condensed with cooling water. Figure 5.20 shows the diagram of this process, used in many production sectors, such as, for example, the sugar industry.

The process behaviour can be modelled by a series of equations obtained from the mass and energy balance equations, as well as by making some realistic assumptions. The equations describing the process behaviour can be found in [83]. The main variables, together with their values at the point of operation are grouped in table 5.5.

The system dynamics is mainly dictated by the differential equations modelling the mass balances:

Table 5.5: Process Variables and values at operating point

Variable	Description	Value	Units
F_1	Feed flowrate	10.0	kg/min
F_2	Product flowrate	2.0	kg/min
F_4	Vapour flowrate	8.0	kg/min
F_5	Condensate flowrate	8.0	kg/min
X_1	Feed composition	5.0	%
X_2	Product composition	25.0	%
L_1	Separator level	1.0	m
P_1	Operating pressure	50.5	kPa
P_6	Steam pressure	194.7	kPa
F_6	Steam flowrate	9.3	kg/min
F_7	Cooling water flowrate	208.0	kg/min

- Mass balance in the liquid:

$$\rho A \frac{dL_1}{dt} = F_1 - F_4 - F_2 \tag{5.26}$$

  ρ being the density of the liquid and A the section of the separator, whose product can be considered constant.

- Mass balance in the solute:

$$M \frac{dX_2}{dt} = F_1 X_1 - F_2 X_2 \tag{5.27}$$

  M is the total quantity of liquid in the evaporator.

- Mass balance in the process vapour: the total amount of water vapour can be expressed in function of the pressure existing in the system according to:

$$C \frac{dP_1}{dt} = F_4 - F_5 \tag{5.28}$$

  C being a constant that converts the steam mass into an equivalent pressure.

The dynamics of the interchanger and the condenser can be considered to be very fast compared to the previous ones.

Degrees of freedom

Twelve equations can be found for twenty variables so there are eight degrees of freedom. Eight more equations must therefore be considered in order to close the problem; these will be the ones which provide the values of the manipulated variables and of the disturbances:

- Three manipulated variables: the steam pressure P_6, which depends on the opening of valve V_1, the cooling water flowrate F_7, controlled by valve V_2 and the product flowrate F_2 with V_3.
- Five disturbances: feed flowrate F_1, circulating flowrate F_3, composition and temperature of feed X_1 and T_1 and cooling water temperature T_7.

A single solution can be obtained with these considerations which allows the value of the remaining variables to be calculated.

5.8.2 Obtaining the Linear Model

As can be deduced from equations (5.26)-(5.28) the process is a non-linear system with a strong interrelationship amongst the variables. Even so, a linear model with various independent loops will be used to design the controller. It is clear that these hypothesis of work are incorrect as will be made obvious on putting the control to work. The control of the evaporator includes maintaining certain stable working conditions as well as obtaining a product of a determined quality. In order to achieve the first objective it is necessary to control the mass and the energy of the system, which can be achieved by keeping the level in the separator L_1 and the process pressure P_1 constant . In order to do this two PI type local controllers will be used, so that the level L_1 is controlled by acting on the product flowrate F_2 and the process pressure P_1 is controlled by the cooling water F_7. The justification of the choice of the couplings and the tuning of these loops can be found in [83].

The other objective is to obtain a determined product composition. This is achieved by acting on the remaining manipulated variable, the steam pressure which supplies the energy to the evaporator, P_6. The interaction amongst the variables is very strong, as can be seen by using Bristol method [15] and it would even be possible to have coupled X_2 with F_7 and P_1 with P_6. To obtain the linear model of the composition loop a step is applied at the input (P_6) and the effect on the output (X_2) is studied. As was to be expected and can be seen from figure 5.21 which shows the evolution of the more significant variables for a 10% step, the interaction amongst the variables is considerable.

The evolution of the composition does not therefore follow the pattern of a first order system, due mainly to the fact that the experiment was not done in open loop because the level and pressure regulators are functioning which, as has been indicated, need to be activated for stable functioning of the evaporator and indirectly affect the composition. The approximation of loop $X_2 - P_6$ to that of a first order model with delay of the form

$$G(s) = \frac{K}{1+\tau s} e^{-\tau_d s}$$

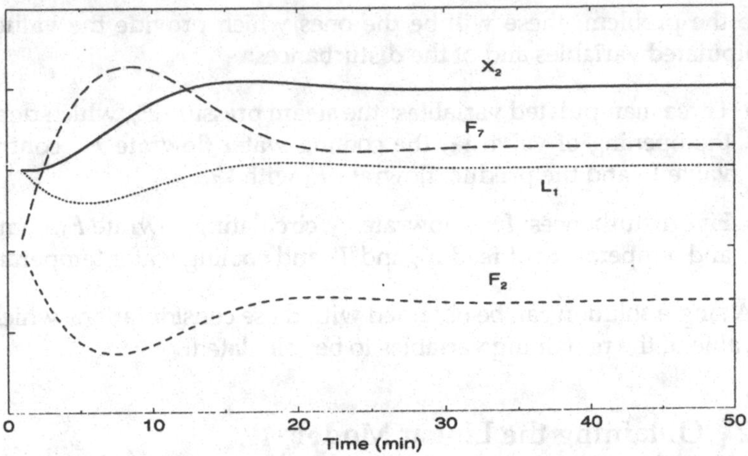

Figure 5.21: Evaporator response to step input

that, as is known, in spite of its simplicity is much used in practice, will be attempted. The reaction curve method will be used to obtain the model parameters; this provides the values of K, τ and τ_d starting from the graph of the system response at a step input. Due to the non-linearity the system behaviour will be different for inputs of different value and different sign. By making various experiments for steps of different sign and magnitude the following parameters can be considered to be appropriate:

$$K = 0.234\,\%/\mathrm{KPa} \qquad \tau = 4.5\ \mathrm{min} \qquad \tau_d = 3.5\ \mathrm{min}$$

By taking a sampling time of one minute the delay is not integer so that the discrete transfer function about the working point will be (see conversion expressions in chapter 3):

$$G(z^{-1}) = \frac{0.02461z^{-1} + 0.02202z^{-2}}{1 - 0.8007374z^{-1}} z^{-3}$$

This transfer function will be used for the design of the controller in spite of its limitations because of the existence of the previously mentioned phenomena.

5.8.3 Controller Design

Once a linear model of the process is obtained, the design of the controller is direct if the precalculated GPC is used. It is only necessary to calculate the parameters which appear in the control law.

$$\triangle u(t) = (l_{y1}\hat{y}(t+d \mid t) + l_{y2}\hat{y}(t+d-1 \mid t) + l_{r1}r(t))/K + l_{u1} \triangle u(t-1) \quad (5.29)$$

If one wants to design a fixed (non adaptive) regulator it is only necessary to calculate these parameters once. In the case of the evaporator with $a = 0.8007374, m = b_0/(b_0 + b_1) = 0.528$ and for a value of λ of 1.2 one has:

$$\begin{aligned} l_{y1} &= -2.2748 \\ l_{y2} &= 1.5868 \\ l_{r1} &= 0.6879 \\ l_{u1} &= -0.1862 \end{aligned}$$

the control signal at each instant being therefore

$$u(t) = 0.814u(t-1)+0.186u(t-2)-9.721\hat{y}(t+d \mid t)+6.781\hat{y}(t+d-1 \mid t)+2.939r(t)$$

In order to complete the computations of the control law (5.29) the predicted values of the output at instants $t + d$ and $t + d - 1$ are necessary. This computation is easy to do given the simplicity of the model. It is enough to project the equation of the model towards the future:

$$\begin{aligned} \hat{y}(t+i) &= (1+a)\hat{y}(t+i-1) - a\hat{y}(t+i-2) \\ &+b_0(u(t-d+i-1) - u(t-d+i-2)) \\ &+b_1(u(t-d+i-2) - u(t-d+i-3)) \quad i = 1 \ldots d \end{aligned}$$

the elements $\hat{y}(t) = y(t)$ and $\hat{y}(t-1) = y(t-1)$ being known values at instant t. Note that if one wanted to make the controller adaptive it would be enough to just calculate the new value of l_i when the parameters of the system change. The simplicity of the control law obtained is obvious, being comparable to that of a digital PID, and is therefore easy to implant in any control system however simple it is.

5.8.4 Results

In the following some results of applying the previous control law to the evaporator are presented (simulated to a non-linear model). Even though the simplifications which were employed in the design phase (monovariable system, first order linear model) were not very realistic, a reasonably good behaviour of the closed loop system is obtained.

In figure 5.22 the behaviour of the process in the presence of changes in the reference of the composition is shown. It can be observed that the output clearly follows the reference although with certain initial overshoot. It should be taken into account that the loops considered to be independent are greatly interrelated amongst themselves and in particular that the composition is very disturbed by the variations in the cooling water flow rate F_7, which is constantly changing in order to maintain the process pressure constant.

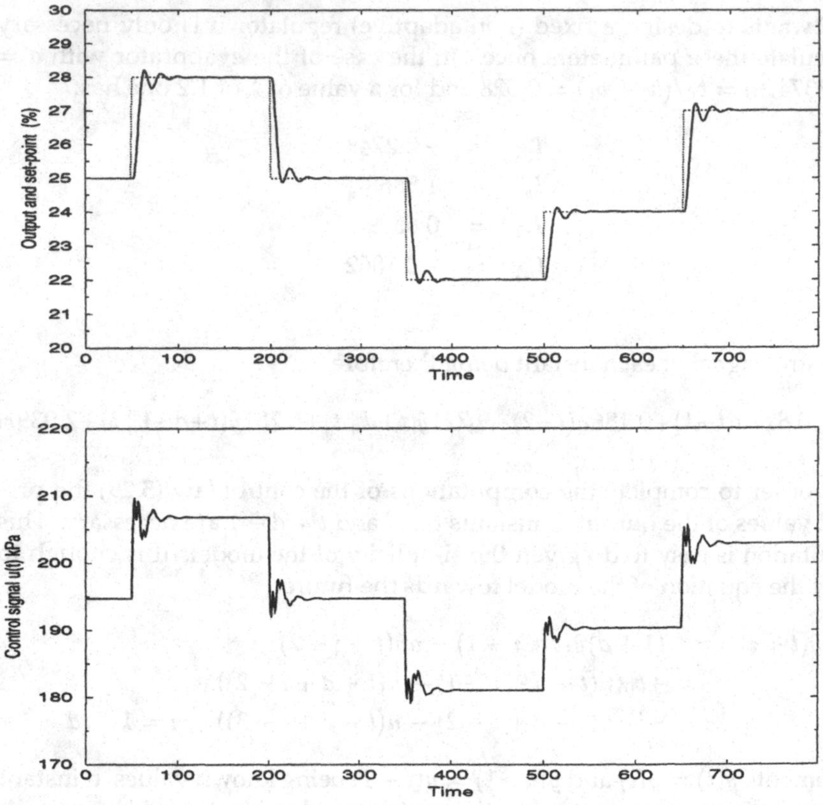

Figure 5.22: GPC behaviour in the evaporator

In spite of the overshoot the behaviour can be considered to be good. It can be compared to that obtained with a classical controller such as PI. Some good values for adjusting this controller are those calculated in [83]

$$K = 1.64\ \mathrm{kPa}/\% \qquad T_I = 3.125\ \mathrm{min}$$

In figure 5.23 both regulators are compared for a change in the reference from 28 to 25 percent. The GPC is seen to be faster and overshoots less than the PI as well as not introducing great complexity in the design, as has been seen in the previous section. The responses of both controllers in the presence of changes in the feed flowrate are reflected in figure 5.24, where at instant $t = 50$ the feed flowrate changes from 10 to 11 kg/min and at $t = 200$ the composition at the input brusquely changes from 5 to 6 percent. It can be seen that these changes considerably affect the composition of the product and although both controllers return the output to the reference value, the GPC does it sooner and with less overshoot, reducing the peaks by about 30%.

Tests can also be made with regard to the study of robustness. It is already

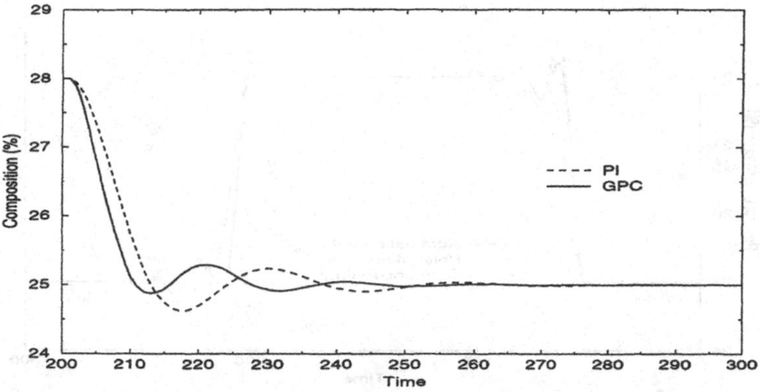

Figure 5.23: Comparison of GPC and PI for a setpoint change

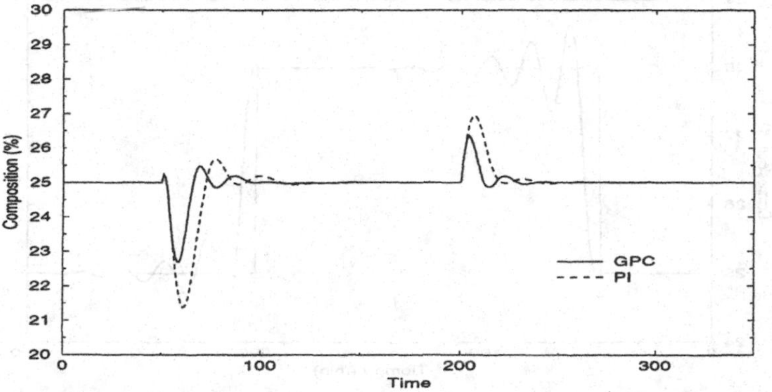

Figure 5.24: Comparison of GPC and PI for feed changes

known that the model used does not correspond to the real one (which is neither first order nor linear) and therefore the controller already possesses certain robustness. However to corroborate the robustness results previously presented, instead of using as a control model the linear model that best fits the nonlinear process, a model with estimation errors is going to be used. For example, when working in a model with an error on the pole estimation of the form that $\hat{a} = \alpha \times a$ with $\alpha = 0.9$, that is an error of 10%, it is necessary to calculate the new l_i parameters using $\hat{a} = 0.9 \times a = 0.9 \times 0.8007374 = 0.7206$ (supposing that $a = 0.8007374$ is the "real"value). Thus, by recalculating the control law (including the predictor) for this new value and leaving the gain unaltered, the response shown in figure 5.25 is obtained. As can be seen, the composition is hardly altered by this modelling error and similar response to the initial model is obtained. The same can be done by changing the system gain. For the model values considered to be "good", uncertainties of up to 100% in the gain can be seen to be permissible without any problem. Thus

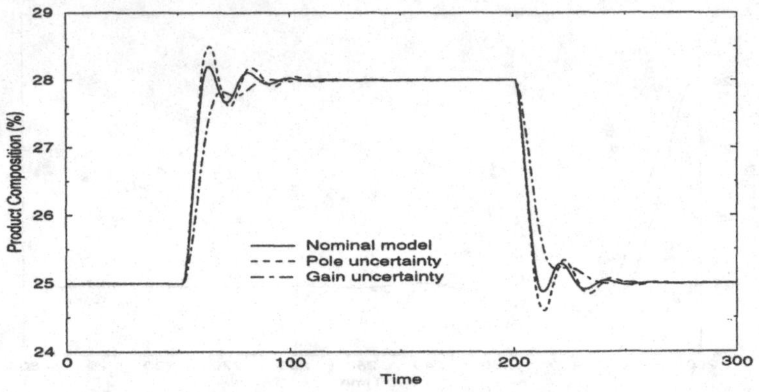

Figure 5.25: Influence of errors on the estimation of gain and delay

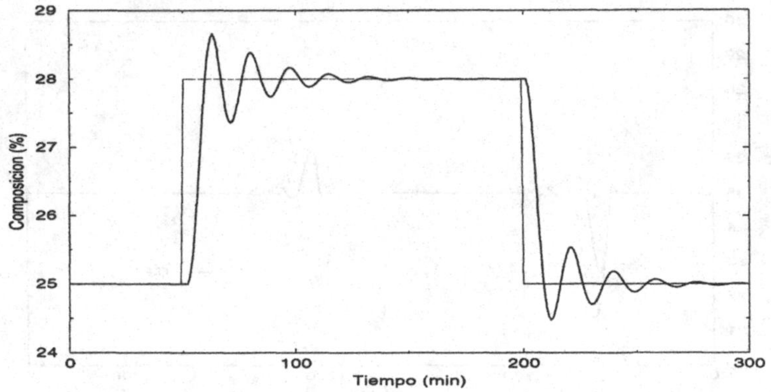

Figure 5.26: Influence of error on the estimation of delay

by doubling the gain of the model used and calculating the new control law, slower but not less satisfactory behaviour is obtained, as is shown in the same figure.

Knowledge of delay is a fundamental factor of model based predictive methods, to such an extent that large errors in estimation can give rise to instability. Whilst for a difference of one unit the response hardly varies, the same is not true if the discrepancy is of two or more units. Figure 5.26 shows the effect of using a model with delay 1 the "real" being equal to 3. The response is defective but does not reach instability. It should be noted that once a mistake in the value of the delay is detected it is very easy to correct as it is enough to calculate higher or lower values of $\hat{y}(t+d)$ using the model equation whose other parameters are unchanged. Furthermore it is not necessary to change the values of the l_i coefficients as they are independent of the delay.

Chapter 6

Multivariable MPC

Most industrial plants have many variables that have to be controlled (outputs) and many manipulated variables or variables used to control the plant (inputs). In certain cases a change in one of the manipulated variables mainly affects the corresponding controlled variable and each of the input-output pairs can be considered as a single-input single-output (SISO) plant and controlled by independent loops. In many cases, when one of the manipulated variables is changed, it not only affects the corresponding controlled variable but also upsets the other controlled variables. These interactions between process variables may result in poor performance of the control process or even instability. When the interactions are not negligible, the plant must be considered to be a process with multiple inputs and outputs (MIMO) instead of a set of SISO processes. The control of MIMO processes has been extensively treated in literature; perhaps the most popular way of controlling MIMO processes is by designing decoupling compensators to suppress or diminish the interactions and then designing multiple SISO controllers. This first requires determining how to pair the input and output variables, that is, which manipulated variable will be used to control each of the output variables, and also that the plant have the same number of manipulated and controlled variables. Total decoupling is very difficult to achieve for processes with complex dynamics or exhibiting dead times.

One of the advantages of Model Predictive Control is that multivariable processes can be handled in a straightforward manner [112], [122]. This chapter is dedicated to showing how GPC can be implemented on MIMO processes.

6.1 Derivation of Multivariable GPC

A CARIMA model for an n-output, m-input multivariable process can be expressed as:

$$\mathbf{A}(z^{-1})y(t) = \mathbf{B}(z^{-1})u(t-1) + \frac{1}{\Delta}\mathbf{C}(z^{-1})e(t) \tag{6.1}$$

where $\mathbf{A}(z^{-1})$ and $\mathbf{C}(z^{-1})$ are $n \times n$ monic polynomial matrices and $\mathbf{B}(z^{-1})$ is an $n \times m$ polynomial matrix defined as:

$$\begin{aligned}
\mathbf{A}(z^{-1}) &= I_{n\times n} + A_1 z^{-1} + A_2 z^{-2} + \cdots + A_{n_a} z^{-n_a} \\
\mathbf{B}(z^{-1}) &= B_0 + B_1 z^{-1} + B_2 z^{-2} + \cdots + B_{n_b} z^{-n_b} \\
\mathbf{C}(z^{-1}) &= I_{n\times n} + C_1 z^{-1} + C_2 z^{-2} + \cdots + C_{n_c} z^{-n_c}
\end{aligned}$$

The operator Δ is defined as $\Delta = 1 - z^{-1}$. The variables $y(t)$, $u(t)$ and $e(t)$ are the $n \times 1$ output vector, the $m \times 1$ input vector and the $n \times 1$ noise vector at time t. The noise vector is supposed to be a white noise with zero mean.

Let us consider the following finite horizon quadratic criterion:

$$J(N_1, N_2, N_3) = \sum_{j=N_1}^{N_2} \|\hat{y}(t+j \mid t) - w(t+j)\|_R^2 + \sum_{j=1}^{N_3} \| \Delta u(t+j-1)\|_Q^2 \tag{6.2}$$

where $\hat{y}(t+j \mid t)$ is an optimum j-step ahead prediction of the system output on data up to time t; that is, the expected value of the output vector at time t if the past input and output vectors and the future control sequence are known. N_1 and N_2 are the minimum and maximum prediction horizons and $w(t+j)$ is a future setpoint or reference sequence for the output vector. R and Q are positive definite weighting matrices.

6.1.1 White Noise Case

We shall first consider the most usual case when matrix $\mathbf{C}(z^{-1}) = I_{n\times n}$. The reason for this is that the colouring polynomials are very difficult to estimate with sufficient accuracy in practice, especially in the multivariable case. In fact, many predictive controllers use colouring polynomials as design parameters. The optimal prediction for the output vector can be generated as in the monovariable case as follows:

Consider the following Diophantine equation:

$$I_{n\times n} = \mathbf{E}_j(z^{-1})\tilde{\mathbf{A}}(z^{-1}) + z^{-j}\mathbf{F}_j(z^{-1}) \tag{6.3}$$

where $\tilde{\mathbf{A}}(z^{-1}) = \mathbf{A}(z^{-1})\Delta$, $\mathbf{E}_j(z^{-1})$ and $\mathbf{F}_j(z^{-1})$ are unique polynomial matrices of order $j-1$ and n_a respectively. If (6.1) is multiplied by $\Delta \mathbf{E}_j(z^{-1})z^j$:

$$\mathbf{E}_j(z^{-1})\tilde{\mathbf{A}}(z^{-1})y(t+j) = \mathbf{E}_j(z^{-1})\mathbf{B}(z^{-1}) \Delta u(t+j-1) + \mathbf{E}_j(z^{-1})e(t+j)$$

By using (6.3) and after some manipulation we get:

$$y(t+j) = \mathbf{F}_j(z^{-1})y(t) + \mathbf{E}_j(z^{-1})\mathbf{B}(z^{-1}) \triangle u(t+j-1) + \mathbf{E}_j(z^{-1})e(t+j) \quad (6.4)$$

Notice that because the degree of $\mathbf{E}_j(z^{-1})$ is $j-1$, the noise terms of equation (6.4) are all in the future. By taking the expectation operator and considering that $E[e(t)] = 0$, the expected value for $y(t+j)$ is given by:

$$\hat{y}(t+j|t) = E[y(t+j)] = \mathbf{F}_j(z^{-1})y(t) + \mathbf{E}_j(z^{-1})\mathbf{B}(z^{-1}) \triangle u(t+j-1) \quad (6.5)$$

Notice that the prediction can easily be extended to the non zero mean noise case by adding vector $\mathbf{E}_j(z^{-1})E[e(t)]$ to prediction $\hat{y}(t+j|t)$.

Recursion of the Diophantine Equation

Let us consider that a solution $(\mathbf{E}_j(z^{-1}), \mathbf{F}_j(z^{-1}))$ for the Diophantine equation has been obtained. That is:

$$I_{n\times n} = \mathbf{E}_j(z^{-1})\tilde{\mathbf{A}}(z^{-1}) + z^{-j}\mathbf{F}_j(z^{-1}) \quad (6.6)$$

with

$$\tilde{\mathbf{A}}(z^{-1}) = \mathbf{A}(z^{-1})\triangle = I_{n\times n} + \widetilde{A}_1 z^{-1} + \widetilde{A}_2 z^{-2} + \cdots + \widetilde{A}_{n_a} z^{-n_a} + \widetilde{A}_{n_a+1} z^{-(n_a+1)}$$
$$= I_{n\times n} + (A_1 - I_{n\times n})z^{-1} + (A_2 - A_1)z^{-2} + \cdots + (A_{n_a} - A_{n_a-1})z^{-n_a} - A_{n_a} z^{-(n_a+1)}$$

$$\mathbf{E}_j(z^{-1}) = E_{j,0} + E_{j,1}z^{-1} + E_{j,2}z^{-2} + \cdots + E_{j,j-1}z^{j-1}$$
$$\mathbf{F}_j(z^{-1}) = F_{j,0} + F_{j,1}z^{-1} + F_{j,2}z^{-2} + \cdots + F_{j,n_a}z^{-n_a}$$

Now consider the Diophantine equation corresponding to the prediction for $\hat{y}(t+j+1|t)$

$$I_{n\times n} = \mathbf{E}_{j+1}(z^{-1})\tilde{\mathbf{A}}(z^{-1}) + z^{-(j+1)}\mathbf{F}_{j+1}(z^{-1}) \quad (6.7)$$

Let us subtract equation (6.6) from equation (6.7)

$$\mathbf{0}_{n\times n} = (\mathbf{E}_{j+1}(z^{-1}) - \mathbf{E}_j(z^{-1}))\tilde{\mathbf{A}}(z^{-1}) + z^{-j}(z^{-1}\mathbf{F}_{j+1}(z^{-1}) - \mathbf{F}_j(z^{-1})) \quad (6.8)$$

Matrix $(\mathbf{E}_{j+1}(z^{-1}) - \mathbf{E}_j(z^{-1}))$ is of degree j. Let us make

$$(\mathbf{E}_{j+1}(z^{-1}) - \mathbf{E}_j(z^{-1})) = \tilde{\mathbf{R}}(z^{-1}) + R_j z^{-j}$$

where $\mathbf{R}(z^{-1})$ is an $n\times n$ polynomial matrix of degree smaller or equal to $j-1$ and R_j is an $n \times n$ real matrix. By substituting in equation (6.8):

$$\mathbf{0}_{n\times n} = \tilde{\mathbf{R}}(z^{-1})\tilde{\mathbf{A}}(z^{-1}) + z^{-j}(R_j\tilde{\mathbf{A}}(z^{-1}) + z^{-1}\mathbf{F}_{j+1}(z^{-1}) - \mathbf{F}_j(z^{-1})) \quad (6.9)$$

As $\tilde{\mathbf{A}}(z^{-1})$ is monic, it is easy to see that $\tilde{\mathbf{R}}(z^{-1}) = \mathbf{0}_{n\times n}$. That is, matrix $\mathbf{E}_{j+1}(z^{-1})$ can be computed recursively by:

$$\mathbf{E}_{j+1}(z^{-1}) = \mathbf{E}_j(z^{-1}) + R_j z^{-j}$$

The following expressions can easily be obtained from (6.9):

$$\begin{aligned} R_j &= F_{j,0} \\ F_{j+1,i} &= F_{j,i+1} - R_j \widetilde{A}_{i+1} \text{ for } i = 0 \cdots \delta(\mathbf{F}_{j+1}) \end{aligned}$$

It can easily be seen that the initial conditions for the recursion equation are given by:

$$\begin{aligned} \mathbf{E}_1 &= I \\ \mathbf{F}_1 &= z(I - \tilde{\mathbf{A}}) \end{aligned}$$

By making the polynomial matrix $\mathbf{E}_j(z^{-1})\mathbf{B}(z^{-1}) = \mathbf{G}_j(z^{-1}) + z^{-j}\mathbf{G}_{jp}(z^{-1})$, with $\delta(\mathbf{G}_j(z^{-1})) < j$, the prediction equation can now be written as:

$$\hat{y}(t+j|t) = \mathbf{G}_j(z^{-1})\,\Delta u(t+j-1) + \mathbf{G}_{jp}(z^{-1})\,\Delta u(t-1) + \mathbf{F}_j(z^{-1})y(t) \quad (6.10)$$

Notice that the last two terms of the right hand side of equation (6.10) depend on past values of the process output and input variables and correspond to the free response of the process considered if the control signals are kept constant, while the first term only depends on future values of the control signal and can be interpreted as the forced response. That is, the response obtained when the initial conditions are zero $y(t-j) = 0$, $\Delta u(t-j) = 0$ for $j = 0, 1 \cdots$. Equation (6.10) can be rewritten as:

$$\hat{y}(t+j|t) = \mathbf{G}_j(z^{-1})\,\Delta\, u(t+j-1) + \mathbf{f}_j$$

with $\mathbf{f}_j = \mathbf{G}_{jp}(z^{-1})\,\Delta\, u(t-1) + \mathbf{F}_j(z^{-1})y(t)$. Let us now consider a set of N j-ahead predictions:

$$\begin{aligned} \hat{y}(t+1|t) &= \mathbf{G}_1(z^{-1})\,\Delta\, u(t) + \mathbf{f}_1 \\ \hat{y}(t+2|t) &= \mathbf{G}_2(z^{-1})\,\Delta\, u(t+1) + \mathbf{f}_2 \\ &\vdots \\ \hat{y}(t+N|t) &= \mathbf{G}_N(z^{-1})\,\Delta\, u(t+N-1) + \mathbf{f}_N \end{aligned} \quad (6.11)$$

Because of the recursive properties of the $\mathbf{E}_j$ polynomial matrix described above, expressions (6.11) can be rewritten as:

$$\begin{bmatrix} \hat{y}(t+1|t) \\ \hat{y}(t+2|t) \\ \vdots \\ \hat{y}(t+j|t) \\ \vdots \\ \hat{y}(t+N|t) \end{bmatrix} = \begin{bmatrix} G_0 & 0 & \cdots & 0 & \cdots & 0 \\ G_1 & G_0 & \cdots & 0 & \cdots & 0 \\ \vdots & \vdots & \ddots & \vdots & \vdots & \vdots \\ G_{j-1} & G_{j-2} & \cdots & G_0 & \vdots & 0 \\ \vdots & \vdots & \vdots & \vdots & \ddots & \vdots \\ G_{N-1} & G_{N-2} & \cdots & \cdots & \cdots & G_0 \end{bmatrix} \begin{bmatrix} \Delta u(t) \\ \Delta u(t+1) \\ \vdots \\ \Delta u(t+j-1) \\ \vdots \\ \Delta u(t+N-1) \end{bmatrix} + \begin{bmatrix} \mathbf{f}_1 \\ \mathbf{f}_2 \\ \vdots \\ \mathbf{f}_j \\ \vdots \\ \mathbf{f}_N \end{bmatrix}$$

where $\mathbf{G}_j(z^{-1}) = \sum_{i=0}^{j-1} G_i z^{-i}$. The predictions can be expressed in condensed form as:

$$\mathbf{y} = \mathbf{G}\mathbf{u} + \mathbf{f}$$

Notice that if all initial conditions are zero, the free response $\mathbf{f}$ is also zero. If a unit step is applied to the first input at time t; that is

$$\Delta u(t) = [1, 0, \cdots, 0]^T, \Delta u(t+1) = 0, \cdots, \Delta u(t+N-1) = 0$$

the expected output sequence $[\hat{y}(t+1)^T, \hat{y}(t+2)^T, \cdots, \hat{y}(t+N)^T]^T$ is equal to the first column of matrix $\mathbf{G}$ or the first columns of matrices $G_0, G_1, \cdots, G_{N-1}$. That is, the first column of matrix $\mathbf{G}$ can be calculated as the step response of the plant when a unit step is applied to the first control signal. Column i can be obtained in a similar manner by applying a unit step to the i-input. In general, matrix G_k can be obtained as follows:

$$(G_k)_{i,j} = y_{i,j}(t+k+1)$$

where $(G_k)_{i,j}$ is the (i,j) element of matrix G_k and $y_{i,j}(t+k+1)$ is the i-output of the system when a unit step has been applied to control input j at time t.

The free response term can be calculated recursively by:

$$\mathbf{f}_{j+1} = z(I - \tilde{\mathbf{A}}(z^{-1}))\mathbf{f}_j + \mathbf{B}(z^{-1}) \Delta u(t+j)$$

with $\mathbf{f}_0 = y(t)$ and $\Delta u(t+j) = 0$ for $j \geq 0$.

Notice that if matrix $\mathbf{A}(z^{-1})$ is diagonal, matrices $\mathbf{E}_j(z^{-1})$ and $\mathbf{F}_j(z^{-1})$ are also diagonal matrices and the problem is reduced to the recursion of n scalar Diophantine equations which are much simpler to program and require less computation. The computation of $\mathbf{G}_j(z^{-1})$ and $\mathbf{f}_j$ is also considerably simplified.

If the control signal is kept constant after the first N_3 control moves, the set of predictions affecting the cost function (6.2) $\mathbf{y}_{N_{12}} = [\hat{y}(t+N_1|t)^T \cdots \hat{y}(t+N_2|t)^T]$ can be expressed as:

$$\mathbf{y}_{N_{12}} = \mathbf{G}_{N_{123}}\mathbf{u}_{N_3} + \mathbf{f}_{N_{12}}$$

where $\mathbf{u}_{N_3} = [\Delta u(t)^T \cdots \Delta u(t+N_3-1)^T]^T$, $\mathbf{f}_{N_{12}} = [\mathbf{f}_{N_1}^T \cdots \mathbf{f}_{N_2}^T]^T$ and $\mathbf{G}_{N_{123}}$ is the following submatrix of $\mathbf{G}$

$$\mathbf{G}_{N_{123}} = \begin{bmatrix} G_{N_1-1} & G_{N_1-2} & \cdots & G_{N_1-N_3} \\ G_{N_1} & G_{N_1-1} & \cdots & G_{N_1+1-N_3} \\ \vdots & \ddots & \ddots & \vdots \\ G_{N_2-1} & G_{N_2-2} & \cdots & G_{N_2-N_3} \end{bmatrix}$$

with $G_i = 0$ for $i < 0$. Equation (6.2) can be rewritten as:

$$J = (\mathbf{G}_{N_{123}}\mathbf{u}_{N_3} + \mathbf{f}_{N_{12}} - \mathbf{w})^T \overline{R} (\mathbf{G}_{N_{123}}\mathbf{u}_{N_3} + \mathbf{f}_{N_{12}} - \mathbf{w}) + \mathbf{u}_{N_3}^T \overline{Q} \mathbf{u}_{N_3}$$

where $\overline{R} = diag(R, \cdots, R)$ and $\overline{Q} = diag(Q, \cdots, Q)$

If there are no constraints, the optimum can be expressed as:

$$\mathbf{u} = (\mathbf{G}_{N_{123}}^T \overline{R} \mathbf{G}_{N_{123}} + \overline{Q})^{-1} \mathbf{G}_{N_{123}}^T \overline{R} (\mathbf{w} - \mathbf{f}_{N_{12}})$$

Because of the receding control strategy, only $\Delta u(t)$ is needed at instant t. Thus only the first m rows of $(\mathbf{G}_{N_{123}}^T \overline{R} \mathbf{G}_{N_{123}} + \overline{Q})^{-1} \mathbf{G}_{N_{123}}^T \overline{R}$, say K, have to be computed. This can be done beforehand for the non-adaptive case. The control law can then be expressed as $\Delta u(t) = K(\mathbf{w} - \mathbf{f})$. That is a linear gain matrix that multiplies the predicted errors between the predicted references and the predicted free response of the plant.

In the case of adaptive control, matrix $\mathbf{G}_{N_{123}}$ has to be computed every time the estimated parameters change and the way of computing the control action increment would be by solving the linear set of equations: $(\mathbf{G}_{N_{123}}^T \overline{R} \mathbf{G}_{N_{123}} + \overline{Q})\mathbf{u} = \mathbf{G}_{N_{123}}^T \overline{R} (\mathbf{w} - \mathbf{f}_{N_{12}})$. Again only the first m components of $\mathbf{u}$ have to be found and as matrix $(\mathbf{G}_{N_{123}}^T \overline{R} \mathbf{G}_{N_{123}} + \overline{Q})$ is positive definite, Cholesky's algorithm [117] can be used to find the solution.

6.1.2 Coloured Noise Case

When the noise is coloured, $\mathbf{C}(z^{-1}) \neq I$, and provided that the colouring matrix $\mathbf{C}(z^{-1})$ is stable, the optimal predictions needed can be generated as follows [44]:

First solve the Diophantine equation:

$$\mathbf{C}(z^{-1}) = \mathbf{E}_j(z^{-1})\tilde{\mathbf{A}}(z^{-1}) + z^{-j}\mathbf{F}_j(z^{-1}) \tag{6.12}$$

where $\mathbf{E}_j(z^{-1})$ and $\mathbf{F}_j(z^{-1})$ are unique polynomial matrices of order $j-1$ and n_a respectively. Note that the Diophantine equation (6.12) can be solved recursively. Consider the Diophantine equations for j and $j+1$:

$$\begin{aligned} \mathbf{C}(z^{-1}) &= \mathbf{E}_j(z^{-1})\tilde{\mathbf{A}}(z^{-1}) + z^{-j}\mathbf{F}_j(z^{-1}) \\ \mathbf{C}(z^{-1}) &= \mathbf{E}_{j+1}(z^{-1})\tilde{\mathbf{A}}(z^{-1}) + z^{-(j+1)}\mathbf{F}_{j+1}(z^{-1}) \end{aligned}$$

By differentiating them we get equation (6.8), hence $\mathbf{E}_{j+1}(z^{-1})$ and $\mathbf{F}_{j+1}(z^{-1})$ can be computed recursively by using the same expressions obtained in the case of $\mathbf{C}(z^{-1}) = I$ with initial conditions: $\mathbf{E}_1(z^{-1}) = I$ and $\mathbf{F}_1(z^{-1}) = z(\mathbf{C}(z^{-1}) - \tilde{\mathbf{A}}(z^{-1}))$.

Define the polynomial matrices $\overline{\mathbf{E}}_j(z^{-1})$ and $\overline{\mathbf{C}}_j(z^{-1})$ such that

$$\overline{\mathbf{E}}_j(z^{-1})\mathbf{C}(z^{-1}) = \overline{\mathbf{C}}_j(z^{-1})\mathbf{E}_j(z^{-1}) \tag{6.13}$$

with $\overline{E}_0 = I$ and $\det(\overline{\mathbf{C}}_j(z^{-1}))=\det(\mathbf{C}_j(z^{-1}))$. Note that

$$\overline{\mathbf{E}}_j(z^{-1})^{-1}\overline{\mathbf{C}}_j(z^{-1}) = \mathbf{C}(z^{-1})\mathbf{E}_j(z^{-1})^{-1}$$

and thus, matrices $\overline{\mathbf{E}}_j(z^{-1})$ and $\overline{\mathbf{C}}_j(z^{-1})$ can be interpreted (and computed) as a left fraction matrix description of the process having a right matrix fraction description given by matrices $\mathbf{E}(z^{-1})$ and $\mathbf{C}(z^{-1})$. Define

$$\overline{\mathbf{F}}_j(z^{-1}) \quad = \quad z^j(\overline{\mathbf{C}}_j(z^{-1}) - \overline{\mathbf{E}}_j(z^{-1})\tilde{\mathbf{A}}(z^{-1})) \tag{6.14}$$

Premultiplying equation (6.1) by $\overline{\mathbf{E}}_j(z^{-1})\Delta$

$$\overline{\mathbf{E}}_j(z^{-1})\tilde{\mathbf{A}}(z^{-1})y(t+j) = \overline{\mathbf{E}}_j(z^{-1})\mathbf{B}(z^{-1}) \triangle u(t+j-1) + \overline{\mathbf{E}}_j(z^{-1})e(t+j)$$

Using equations (6.13) and (6.14) we get

$$\overline{\mathbf{C}}_j(z^{-1})(y(t+j)-\overline{\mathbf{E}}_j(z^{-1})e(t+j)) = \overline{\mathbf{E}}_j(z^{-1})\mathbf{B}(z^{-1})\triangle u(t+j-1)+\overline{\mathbf{F}}_j(z^{-1})y(t)$$

By taking the expected value $E[\overline{\mathbf{C}}_j(z^{-1})y(t+j)-\overline{\mathbf{E}}_j(z^{-1})e(t+j)] = \hat{y}(t+j|t)$. The optimal predictions $\hat{y}(t+j|t)$ can be generated by the equation:

$$\overline{\mathbf{C}}_j(z^{-1})\hat{y}(t+j|t) = \overline{\mathbf{E}}_j(z^{-1})\mathbf{B}(z^{-1}) \triangle u(t+j-1) + \overline{\mathbf{F}}_j(z^{-1})y(t)$$

Now solving the Diophantine equation:

$$I = \mathbf{J}_j(z^{-1})\overline{\mathbf{C}}_j(z^{-1}) + z^{-j}\mathbf{K}_j(z^{-1}) \tag{6.15}$$

with $\delta(\mathbf{J}(z^{-1})) < j$. Multiplying by $\mathbf{J}_j(z^{-1})^{-1}$ and using equation (6.15)

$$(I - z^{-j}\mathbf{K}_j(z^{-1}))y(t+j|t) = \mathbf{J}_j(z^{-1})\overline{\mathbf{E}}_j(z^{-1})\mathbf{B}(z^{-1}) \triangle u(t+j-1)+$$
$$+\mathbf{J}_j(z^{-1})\overline{\mathbf{F}}_j(z^{-1})y(t)$$

or,

$$y(t+j|t) = \mathbf{J}_j(z^{-1})\overline{\mathbf{E}}_j(z^{-1})\mathbf{B}(z^{-1})\triangle u(t+j-1)+(\mathbf{K}_j(z^{-1})+\mathbf{J}_j(z^{-1})\overline{\mathbf{F}}_j(z^{-1}))y(t)$$

If $\mathbf{J}_j(z^{-1})\overline{\mathbf{E}}_j(z^{-1})\mathbf{B}(z^{-1}) = \mathbf{G}_j(z^{-1}) + z^{-j}\mathbf{G}p_j(z^{-1})$, with $\delta(\mathbf{G}_j(z^{-1})) < j$, the optimal j-step ahead prediction can be expressed as:

$$y(t+j|t) = \mathbf{G}_j(z^{-1}) \triangle u(t+j-1) + \mathbf{G}p_j(z^{-1}) \triangle u(t-1) + (\mathbf{K}_j(z^{-1})+$$
$$+\mathbf{J}_j(z^{-1})\overline{\mathbf{F}}_j(z^{-1}))y(t)$$

The first term of the prediction corresponds to the forced response due to future control increments, while the last two terms correspond to the free response $\mathbf{f}c_j$ and are generated by past input increments and past output.

The set of pertinent j-ahead predictions can be written as: $\mathbf{y}_{N_{12}} = \mathbf{G}_{N_{123}}\mathbf{u}_{N_3}+$ $\mathbf{f}c_{N_{12}}$ generated as before.

The objective function can be expressed as:

$$J = (\mathbf{G}_{N_{123}}\mathbf{u}_{N_3} + \mathbf{f}c_{N_{12}} - \mathbf{w})^T\overline{R}(\mathbf{G}_{N_{123}}\mathbf{u}_{N_3} + \mathbf{f}c_{N_{12}} - \mathbf{w}) + \mathbf{u}_{N_3}^T\overline{Q}\mathbf{u}_{N_3}$$

And the optimal solution can be found solving a set of linear equations as in the white noise case but notice that the computation required is now more complex.

It is in reality very difficult to obtain the colouring polynomial matrix $\mathbf{C}(z^{-1})$ and in most cases this matrix is chosen arbitrarily by the user in order to gain robustness. If matrices $\mathbf{C}(z^{-1})$ and $\mathbf{A}(z^{-1})$ are chosen to be diagonal, the problem is transformed into generating a set of optimal predictions for a series of multi input single output processes, which is an easier problem to solve and the computation required can be substantially simplified.

Consider a CARIMA multivariable process with $\mathbf{A}(z^{-1}) = \text{diag}(A_{ii}(z^{-1}))$ and $\mathbf{C}(z^{-1}) = \text{diag}(C_{ii}(z^{-1}))$. The model equation corresponding to the ith-output variable can be expressed as:

$$A_{ii}(z^{-1})y_i(t) = \sum_{j=1}^{m} B_{ij}(z^{-1})u_j(t-1) + C_{ii}(z^{-1})\frac{e_i(t)}{\Delta} \tag{6.16}$$

Solve the scalar Diophantine equation:

$$C_{ii}(z^{-1}) = E_{i_k}(z^{-1})\tilde{A}_{ii}(z^{-1}) + z^{-k}F_{i_k}(z^{-1}) \tag{6.17}$$

with $\delta(E_{i_k}(z^{-1})) = k-1$ and $\delta(F_{i_k}(z^{-1})) = \delta(\tilde{A}_{ii}(z^{-1})) - 1$. Multiplying equation (6.16) by $\Delta E_{i_k}(z^{-1})$ and using (6.17)

$$C_{ii}(z^{-1})(y_i(t+j) - E_{i_k}(z^{-1})e_i(t+j)) =$$

$$= E_{i_k}(z^{-1})\sum_{j=1}^{m} B_{ij}(z^{-1}) \Delta u_j(t+j-1) + F_{i_k}(z^{-1})y_i(t)$$

As the noise terms are all in the future, the expected value of the left hand side of the above equation is: $E[C_{ii}(z^{-1})(y_i(t+k) - E_{i_k}(z^{-1})e_i(t+k))] = C_{ii}(z^{-1})\hat{y}_i(t+k|t)$

The expected value of the output can be generated by the equation:

$$C_{ii}(z^{-1})\hat{y}_i(t+k|t) = E_{i_k}(z^{-1})\sum_{j=1}^{m} B_{ij}(z^{-1}) \Delta u_j(t+k-1) + F_{i_k}(z^{-1})y_i(t) \tag{6.18}$$

Notice that this prediction equation could be used to generate the predictions in a recursive way. An explicit expression for the optimal k-step ahead prediction, can be obtained by solving the Diophantine equation:

$$1 = C_{ii}(z^{-1})M_{i_k}(z^{-1}) + z^{-k}N_{i_k}(z^{-1}) \tag{6.19}$$

with $\delta(M_{i_k}(z^{-1})) = k-1$ and $\delta(N_{i_k}(z^{-1})) = \delta(C_{ii}(z^{-1})) - 1$.

Multiplying equation (6.18) by $M_{i_k}(z^{-1})$ and using (6.19),

$$\hat{y}_i(t+k|t) = M_{i_k} E_{i_k}(z^{-1}) \sum_{j=1}^{m} B_{ij}(z^{-1}) \triangle u_j(t+k-1) +$$

$$+ M_{i_k}(z^{-1}) F_{i_k}(z^{-1}) y_i(t) + N_{i_k}(z^{-1}) y_i(t)$$

which can be expressed as:

$$\begin{aligned} \hat{y}_i(t+k|t) &= \sum_{j=1}^{m} G_{ij}(z^{-1}) \triangle u_j(t+k-1) + \sum_{j=1}^{m} Gp_{ij}(z^{-1}) \triangle u_j(t+k-1) + \\ &+ (M_{i_k}(z^{-1}) F_{i_k}(z^{-1}) + N_{i_k}(z^{-1})) y_i(t) \end{aligned}$$

with $\delta(G_{ij}(z^{-1})) < k$. These predictions can be substituted in the cost function which can be minimized as previously. Note that the amount of computation required has been considerably reduced in respect to the case of a non diagonal colouring matrix.

6.1.3 Measurable Disturbances

The measurable disturbances can be handled for the MIMO case in the same way as for SISO processes. It will be seen that only the *free* response has to be changed to take into account the measurable disturbances. Consider a multivariable process described by the following CARIMA model:

$$\mathbf{A}(z^{-1}) y(t) = \mathbf{B}(z^{-1}) u(t) + \mathbf{D}(z^{-1}) v(t) + \frac{1}{\triangle} \mathbf{C}(z^{-1}) e(t) \qquad (6.20)$$

where the variables $v(t)$ is a $n \times 1$ vector of measured disturbances at time t and $\mathbf{D}(z^{-1})$ is a $n \times n$ polynomial matrix defined as:

$$\mathbf{D}(z^{-1}) = D_0 + D_1 z^{-1} + D_2 z^{-2} + \cdots + D_{n_d} z^{-n_d}$$

Multiplying equation (6.20) by $\Delta \mathbf{E}_j(z^{-1}) z^j$:

$$\begin{aligned} \mathbf{E}_j(z^{-1}) \tilde{\mathbf{A}}(z^{-1}) y(t+j) &= \mathbf{E}_j(z^{-1}) \mathbf{B}(z^{-1}) \triangle u(t+j-1) + \\ &+ \mathbf{E}_j(z^{-1}) \mathbf{D}(z^{-1}) \triangle v(t+j) + \mathbf{E}_j(z^{-1}) e(t+j) \end{aligned}$$

By using (6.3) and after some manipulation we get:

$$\begin{aligned} y(t+j) &= \mathbf{F}_j(z^{-1}) y(t) + \mathbf{E}_j(z^{-1}) \mathbf{B}(z^{-1}) \triangle u(t+j-1) + \\ &+ \mathbf{E}_j(z^{-1}) \mathbf{D}(z^{-1}) \triangle v(t+j) + \mathbf{E}_j(z^{-1}) e(t+j) \end{aligned} \qquad (6.21)$$

Notice that because the degree of $\mathbf{E}_j(z^{-1})$ is $j-1$, the noise terms of equation (6.4) are all in the future. By taking the expectation operator and considering that $E[e(t)] = 0$, the expected value for $y(t+j)$ is given by:

$$\hat{y}(t+j|t) = E[y(t+j)] = \mathbf{F}_j(z^{-1})y(t) + \mathbf{E}_j(z^{-1})\mathbf{B}(z^{-1}) \triangle u(t+j-1) + \\ + \mathbf{E}_j(z^{-1})\mathbf{D}(z^{-1}) \triangle v(t+j) \tag{6.22}$$

By making the polynomial matrix

$$\mathbf{E}_j(z^{-1})\mathbf{D}(z^{-1}) = \mathbf{H}_j(z^{-1}) + z^{-j}\mathbf{H}_{jp}(z^{-1}),$$

with $\delta(\mathbf{H}_j(z^{-1})) = j-1$, the prediction equation can now be written as:

$$\hat{y}(t+j|t) = \mathbf{G}_j(z^{-1}) \triangle u(t+j-1) + \mathbf{H}_j(z^{-1}) \triangle v(t+j) + \mathbf{G}_{jp}(z^{-1}) \triangle u(t-1) + \\ + \mathbf{H}_{jp}(z^{-1}) \triangle v(t) + \mathbf{F}_j(z^{-1})y(t) \tag{6.23}$$

Notice that the last three terms of the right hand side of equation (6.23) depend on past values of the process output measured disturbances and input variables and correspond to the free response of the process considered if the control signals and measured disturbances are kept constant, while the first term only depends on future values of the control signal and can be interpreted as the forced response. That is, the response obtained when the initial conditions are zero $y(t-j) = 0, \Delta u(t-j-1) = 0, \Delta v(t-j)$ for $j > 0$.

The second term of equation (6.23) depends on the future deterministic disturbances. In some cases, when they are related to the process load, future disturbances are known. In other cases, they can be predicted using trends or by other means. If this is the case, the term corresponding to future deterministic disturbances can be computed. If the future load disturbances are supposed to be constant and equal to the last measured value (i.e. $v(t+j) = v(t)$), then $\Delta v(t+j) = 0$ and the second term of this equation vanishes.

Equation (6.23) can be rewritten as:

$$\hat{y}(t+j|t) = \mathbf{G}_j(z^{-1}) \triangle u(t+j-1) + \mathbf{H}_j(z^{-1}) \triangle v(t+j) + \mathbf{f}_j$$

with $\mathbf{f}_j = \mathbf{G}_{jp}(z^{-1}) \triangle u(t-1) + \mathbf{H}_{jp}(z^{-1}) \triangle v(t) + \mathbf{F}_j(z^{-1})y(t)$.

Let us now consider a set of N j-ahead predictions:

$$\begin{aligned} \hat{y}(t+1|t) &= \mathbf{G}_1(z^{-1}) \triangle u(t) + \mathbf{H}_j(z^{-1}) \triangle v(t+1) + \mathbf{f}_1 \\ \hat{y}(t+2|t) &= \mathbf{G}_2(z^{-1}) \triangle u(t+1) + \mathbf{H}_j(z^{-1}) \triangle v(t+2) + \mathbf{f}_2 \\ &\vdots \\ \hat{y}(t+N|t) &= \mathbf{G}_N(z^{-1}) \triangle u(t+N-1) + \mathbf{H}_j(z^{-1}) \triangle v(t+N) + \mathbf{f}_N \end{aligned} \tag{6.24}$$

Because of the recursive properties of the $\mathbf{E}_j$ polynomial matrix described above, expressions (6.24) can be rewritten as:

$$\begin{bmatrix} \hat{y}(t+1|t) \\ \hat{y}(t+2|t) \\ \vdots \\ \hat{y}(t+j|t) \\ \vdots \\ \hat{y}(t+N|t) \end{bmatrix} = \begin{bmatrix} G_0 & 0 & \cdots & 0 & \cdots & 0 \\ G_1 & G_0 & \cdots & 0 & \cdots & 0 \\ \vdots & \vdots & \ddots & \vdots & \vdots & \vdots \\ G_{j-1} & G_{j-2} & \cdots & G_0 & \vdots & 0 \\ \vdots & \vdots & \vdots & \vdots & \ddots & \vdots \\ G_{N-1} & G_{N-2} & \cdots & \cdots & \cdots & G_0 \end{bmatrix} \begin{bmatrix} \Delta u(t) \\ \Delta u(t+1) \\ \vdots \\ \Delta u(t+j-1) \\ \vdots \\ \Delta u(t+N-1) \end{bmatrix} +$$

$$+ \begin{bmatrix} H_0 & 0 & \cdots & 0 & \cdots & 0 \\ H_1 & H_0 & \cdots & 0 & \cdots & 0 \\ \vdots & \vdots & \ddots & \vdots & \vdots & \vdots \\ H_{j-1} & \cdots & H_1 & H_0 & \vdots & 0 \\ \vdots & \vdots & \vdots & \ddots & \vdots & \vdots \\ H_{N-1} & \cdots & \cdots & \cdots & H_1 & H_0 \end{bmatrix} \begin{bmatrix} \Delta v(t+1) \\ \Delta v(t+2) \\ \vdots \\ \Delta v(t+j-1) \\ \vdots \\ \Delta v(t+N) \end{bmatrix} + \begin{bmatrix} \mathbf{f}_1 \\ \mathbf{f}_2 \\ \vdots \\ \mathbf{f}_j \\ \vdots \\ \mathbf{f}_N \end{bmatrix}$$

where $\mathbf{H}_j(z^{-1}) = \sum_{i=1}^{j} H_i z^{-i}$. The predictions can be expressed in condensed form as:

$$\mathbf{y} = \mathbf{G}\mathbf{u} + \mathbf{H}\mathbf{v} + \mathbf{f}$$

Notice that if all initial conditions and future control moves are zero, the free response f and force response are also zero. If a unit step is applied to the first disturbance at time $t+1$; that is

$$\Delta v(t+1) = [1, 0, \cdots, 0]^T, \Delta v(t+2) = 0, \cdots, \Delta v(t+N) = 0$$

the expected output sequence $[\hat{y}(t+2)^T, \hat{y}(t+3)^T, \cdots, \hat{y}(t+N)^T]^T$ is equal to the first column of matrix $\mathbf{H}$ or the first columns of matrices $H_1, H_2, \cdots, H_{N-1}$. That is, the first columns of matrix $\mathbf{H}$ can be interpreted as the step response of the plant when a unit step is applied to the first disturbance signal. Column i can be obtained in a similar manner by applying a unit step to the i-disturbance. In general, matrix H_k could be obtained as follows:

$$(H_k)_{i,j} = y_{i,j}(t+k+1)$$

where $(H_k)_{i,j}$ is the (i,j) element of matrix H_k and $y_{i,j}(t+k+1)$ is the i-output of the system when a unit step has been applied to the disturbance input j at time $t+1$. Notice that to do this test in practice external deterministic variables need to be manipulated, and this is not the usual case. However, they can be computed from the nominal model of the plant by simulation.

Notice that if matrix $\mathbf{A}(z^{-1})$ is diagonal, matrices $\mathbf{E}_j(z^{-1})$ and $\mathbf{F}_j(z^{-1})$ are also diagonal matrices and the problem is reduced to the recursion of n scalar Diophantine equations which are much simpler to program and require less computation. The computation of $\mathbf{G}_j(z^{-1})$, $\mathbf{H}_j(z^{-1})$ and $\mathbf{f}_j$ is also considerably simplified.

By making $\mathbf{f}' = \mathbf{H}\mathbf{v} + \mathbf{f}$, the prediction equation is now:

$$\mathbf{y} = \mathbf{G}\mathbf{u} + \mathbf{f}'$$

Which has the same shape as the prediction equation used for the case of zero external measured disturbances. The future control signal can now be found in the same way, but using as free response the response of the process due to initial conditions (including external disturbances) and future "known" disturbances.

6.2 Obtaining a Matrix Fraction Description

6.2.1 Transfer Matrix Representation

The transfer matrix is the most popular representation of multivariable processes. The reason for this is that transfer matrices can very easily be obtained by a frequency analysis or by applying pulses or steps to the plant, as in the case of the Reaction Curve method. For most plants in the process industry, any column of the plant transfer matrix can be obtained by applying a step to the corresponding input and measuring the static gain, time constant and equivalent delay time for each of the outputs. If the process is repeated for all the inputs, the full transfer matrix is obtained.

The input-output transfer matrix of the CARIMA multivariable model described by equation (6.1) is given by the following $n \times m$ rational matrix:

$$\mathbf{T}(z^{-1}) = \mathbf{A}(z^{-1})^{-1}\mathbf{B}(z^{-1})z^{-1} \tag{6.25}$$

Given a rational matrix $\mathbf{T}(z^{-1})$, the problem consists of finding two polynomial matrices $\mathbf{A}(z^{-1})$ and $\mathbf{B}(z^{-1})$ so that equation (6.25) holds. The most simple way of accomplishing this task is by making $\mathbf{A}(z^{-1})$ a diagonal matrix with its diagonal elements equal to the least common multipliers of the denominators of the corresponding row of $\mathbf{T}(z^{-1})$. Matrix $\mathbf{B}(z^{-1})$ is then equal to $\mathbf{B}(z^{-1}) = \mathbf{A}(z^{-1})\mathbf{T}(z^{-1})z$.

Matrices $\mathbf{A}(z^{-1})$ and $\mathbf{B}(z^{-1})$ obtained this way do not have to be left coprime in general. A left coprime representation can be obtained [44] as follows:

Find a right matrix fraction description $\mathbf{T}(z^{-1}) = \mathbf{N}_R(z^{-1})\mathbf{D}_R(z^{-1})^{-1}$ by making $\mathbf{D}_R(z^{-1})$ a diagonal matrix with its diagonal elements equal to the least common denominator of the corresponding column and form $\mathbf{N}_R(z^{-1})$ accordingly. Note that these polynomial matrices do not have to be right coprime in general.

Find a unimodular matrix $\mathbf{U}(z^{-1})$ such that

$$\begin{bmatrix} \mathbf{U}_{11} & \mathbf{U}_{12} \\ \mathbf{U}_{21} & \mathbf{U}_{22} \end{bmatrix} \begin{bmatrix} \mathbf{D}_R(z^{-1}) \\ \mathbf{N}_R(z^{-1}) \end{bmatrix} = \begin{bmatrix} \mathbf{R}(z^{-1}) \\ \mathbf{0} \end{bmatrix} \tag{6.26}$$

where $\mathbf{R}(z^{-1})$ is the greatest right common divisor of $\mathbf{D}_R(z^{-1})$ and $\mathbf{N}_R(z^{-1})$. That is, $\mathbf{R}(z^{-1})$ is a right divisor of $\mathbf{D}_R(z^{-1})$ and $\mathbf{N}_R(z^{-1})$ ($\mathbf{D}_R(z^{-1}) = \mathbf{D}'_R(z^{-1})\mathbf{R}(z^{-1})$, $\mathbf{N}_R(z^{-1}) = \mathbf{N}'_R(z^{-1})\mathbf{R}(z^{-1})$) and if there is another right divisor $\mathbf{R}'(z^{-1})$ then $\mathbf{R}(z^{-1}) = \mathbf{W}(z^{-1})\mathbf{R}'(z^{-1})$ where $\mathbf{W}(z^{-1})$ is a polynomial matrix.

The greatest right common divisor can be obtained by using the following algorithm (Goodwin and Sin [44]):

1. Form matrix

$$\mathbf{P}(z^{-1}) = \begin{bmatrix} \mathbf{D}_R(z^{-1}) \\ \mathbf{N}_R(z^{-1}) \end{bmatrix}$$

2. Make zero by elementary row transformation all the elements of the first column of $\mathbf{P}(z^{-1})$ below the main diagonal as follows:

   Choose the entry of the first column with smallest degree and interchange the corresponding rows to leave this element in position $(1,1)$ of the matrix (now $\tilde{\mathbf{P}}(z^{-1})$). Obtain $g_{i1}(z^{-1})$ and $r_{i1}(z^{-1})$ for all the elements of the first column such that $\widetilde{P}_{i1}(z^{-1}) = \widetilde{P}_{11}(z^{-1})g_{i1}(z^{-1}) + r_{i1}(z^{-1})$, with $\delta(r_{i1}(z^{-1})) < \delta(\widetilde{P}_{i1}(z^{-1}))$. For all rows below the main diagonal subtract the first row multiplied by $g_{i1}(z^{-1})$, leaving $r_{i1}(z^{-1})$. Repeat the procedure until all the elements below the main diagonal are zero.

3. For the remaining columns use the same procedure described in step 2 to make zero all the elements below the main diagonal but now using the element (i,i) and at the same time reducing the order of elements on the right of the main diagonal as much as possible.

4. Apply the same elementary transformations to an identity matrix, the resulting unimodular matrix will be matrix $\mathbf{U}(z^{-1})$.

The submatrices $\mathbf{U}_{21}(z^{-1})$ and $\mathbf{U}_{22}(z^{-1})$ are left coprime, and $\mathbf{U}_{22}(z^{-1})$ is nonsingular and from (6.26) $\mathbf{N}_R(z^{-1})\mathbf{D}_R(z^{-1})^{-1} = -\mathbf{U}_{22}(z^{-1})^{-1}\mathbf{U}_{21}(z^{-1})$

That is, $\mathbf{A}(z^{-1}) = \mathbf{U}_{22}(z^{-1})$ and $\mathbf{B}(z^{-1}) = -\mathbf{U}_{21}(z^{-1})$. Although $\mathbf{A}(z^{-1})$ and $\mathbf{B}(z^{-1})$ do not have to be left coprime for implementing a GPC, they will in general have higher degrees and may in some cases result in a less efficient algorithm.

Example

In order to illustrate how to obtain a matrix fraction description and how to apply GPC to a MIMO process given by its transfer matrix, consider the small

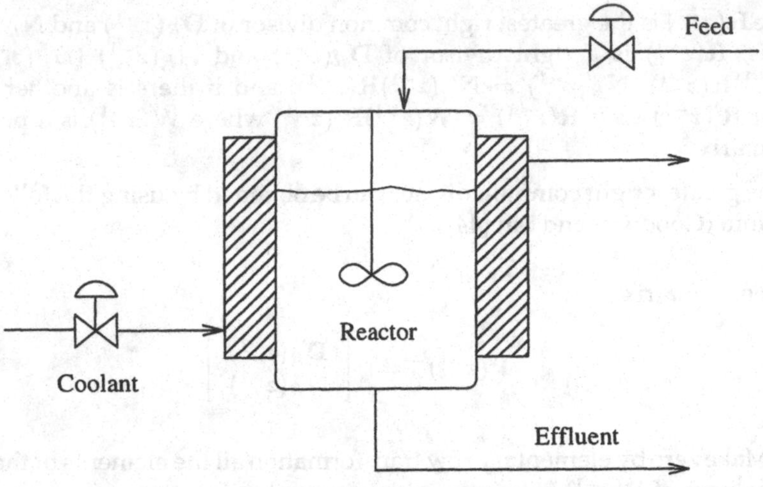

Figure 6.1: Stirred tank reactor

signal model of a stirred tank reactor, figure 6.1 described by the following transfer matrix (the time constants are expressed in minutes)

$$\begin{bmatrix} Y_1(s) \\ Y_2(s) \end{bmatrix} = \begin{bmatrix} \dfrac{1}{1+0.7s} & \dfrac{5}{1+0.3s} \\ \dfrac{1}{1+0.5s} & \dfrac{2}{1+0.4s} \end{bmatrix} \begin{bmatrix} U_1(s) \\ U_2(s) \end{bmatrix}$$

where the manipulated variables $U_1(s)$ and $U_2(s)$ are the feed flowrate and the flow of coolant in the jacket respectively. The controlled variables $Y_1(s)$ and $Y_2(s)$ are the effluent concentration and the reactor temperature respectively.

The discretized model for a sampling time of 0.03 minutes is

$$\begin{bmatrix} y_1(t) \\ y_2(t) \end{bmatrix} = \begin{bmatrix} \dfrac{0.0420z^{-1}}{1-0.9580z^{-1}} & \dfrac{0.4758z^{-1}}{1-0.9048z^{-1}} \\ \dfrac{0.0582z^{-1}}{1-0.9418z^{-1}} & \dfrac{0.1445z^{-1}}{1-0.9277z^{-1}} \end{bmatrix} \begin{bmatrix} u_1(t) \\ u_2(t) \end{bmatrix}$$

A left matrix fraction description can be obtained by making matrix $\mathbf{A}(z^{-1})$ equal to a diagonal matrix with diagonal elements equal to the least common multiple of the denominators of the corresponding row of the transfer function, resulting in:

$$\mathbf{A}(z^{-1}) = \begin{bmatrix} 1-1.8629z^{-1}+0.8669z^{-2} & 0 \\ 0 & 1-1.8695z^{-1}+0.8737z^{-2} \end{bmatrix}$$

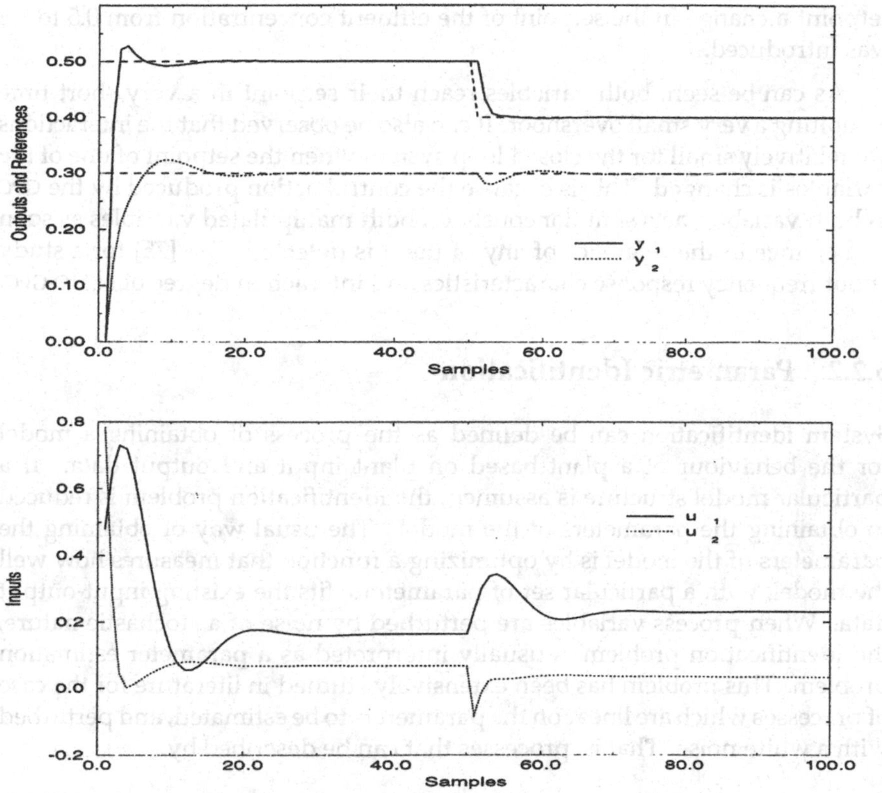

Figure 6.2: Evolution of reactor temperature and effluent concentration

$$\mathbf{B}(z^{-1}) = \begin{bmatrix} 0.0420 - 0.0380z^{-1} & 0.4758 - 0.4559z^{-1} \\ 0.0582 - 0.0540z^{-1} & 0.1445 - 0.1361z^{-1} \end{bmatrix}$$

For a prediction horizon $N_2 = 3$, a control horizon $N_3 = 2$ and a control weight $\lambda = 0.05$, matrix $\mathbf{G}_{N_{123}}$ results as:

$$\mathbf{G}_{N_{123}} = \begin{bmatrix} 0.0420 & 0.4758 & 0 & 0 \\ 0.0582 & 0.1445 & 0 & 0 \\ 0.0821 & 0.9063 & 0.0420 & 0.4758 \\ 0.1131 & 0.2786 & 0.0582 & 0.1445 \\ 0.1206 & 1.2959 & 0.0821 & 0.9063 \\ 0.1647 & 0.4030 & 0.1131 & 0.2786 \end{bmatrix}$$

The evolution of the reactor temperature and of the effluent concentration obtained when the GPC is applied without a prior knowledge of the references and can be seen in figure 6.2. The setpoints were increased by 0.5 and 0.3 at the beginning of the simulation. Once the variables reached the initial setpoint a change in the setpoint of the effluent concentration from 0.5 to 0.4 was introduced.

As can be seen, both variables reach their setpoint in a very short time exhibiting a very small overshoot. It can also be observed that the interactions are relatively small for the closed loop system when the setpoint of one of the variables is changed. This is because the control action produced by the GPC in both variables acts simultaneously on both manipulated variables as soon as a change in the reference of any of them is detected. See [75] for a study about frequency response characteristics and interaction degree of MIMO GPC.

6.2.2 Parametric Identification

System identification can be defined as the process of obtaining a model for the behaviour of a plant based on plant input and output data. If a particular model structure is assumed, the identification problem is reduced to obtaining the parameters of the model. The usual way of obtaining the parameters of the model is by optimizing a function that measures how well the model, with a particular set of parameters, fits the existing input-output data. When process variables are perturbed by noise of a stochastic nature, the identification problem is usually interpreted as a parameter estimation problem. This problem has been extensively studied in literature for the case of processes which are linear on the parameters to be estimated, and perturbed with a white noise. That is, processes that can be described by:

$$\mathbf{z}_k = \Theta \, \Phi_k + \mathbf{e}_k \tag{6.27}$$

where Θ is the vector of parameters to be estimated, Φ_k is a vector of past input and output measures, $\mathbf{z}_k$ is a vector of the latest output measures and $\mathbf{e}_k$ is a white noise.

A multivariable CARIMA model described by equation (6.1) can easily be expressed as (6.27). Multiply equation (6.1) by $\triangle$:

$$\tilde{\mathbf{A}}(z^{-1})y(t) = \mathbf{B}(z^{-1}) \triangle u(t-1) + e(t)$$

which can be rewritten as

$$y(t) = \tilde{\mathbf{A}}'(z^{-1})y(t-1) + \mathbf{B}(z^{-1}) \triangle u(t-1) + e(t)$$

with

$$\tilde{\mathbf{A}}'(z^{-1}) = (I_{n\times n} - \tilde{\mathbf{A}}(z^{-1}))z =$$
$$= -\tilde{A}_1 - \tilde{A}_2 z^{-2} - \cdots - \tilde{A}_{n_a} z^{-(n_a-1)} - \tilde{A}_{n_a+1} z^{-n_a}$$

Which can be expressed as (6.27) by making

$$\Theta \;=\; \left[\tilde{A}_1 \, \tilde{A}_2 \cdots \tilde{A}_{n_a} \, \tilde{A}_{n_a+1} \, B_0 \, B_1 \cdots B_{n_b}\right]$$

$$\Phi_k = \begin{bmatrix} -y(t-1) \\ -y(t-2) \\ \vdots \\ -y(t-na) \\ \Delta u(t-1) \\ \Delta u(t-2) \\ \vdots \\ \Delta u(t-n_b) \end{bmatrix}$$

The parameter can be identified by using a least squares identification algorithm [68],[131].

Notice that estimated parameters correspond to the coefficient matrices of polynomial matrices $\tilde{\mathbf{A}}(z^{-1})$ and $\mathbf{B}(z^{-1})$ which are used for the recursion of the Diophantine equation and for the prediction of forced and free responses.

Notice that if some knowledge about the structure of matrices $A(z^{-1})$ and $B(z^{-1})$ is available, the number of parameters to be identified can be reduced substantially, resulting in greater efficiency of the identification algorithms. For example, if matrix $A(z^{-1})$ is considered to be diagonal, only the parameters of the diagonal elements need to be identified and thus appear in Θ. The form of vectors Θ and Φ_k has to be changed accordingly.

6.3 State Space Formulation

Let us consider a multivariable process with n outputs and m inputs described by the following state space model:

$$\begin{aligned} x(t+1) &= Mx(t) + N \Delta u(t) + Pv(t) \\ y(t) &= Qx(t) + w(t) \end{aligned} \tag{6.28}$$

where $x(t)$ is the state vector, $v(t)$ and $w(t)$ are the noise affecting the process and the output respectively and are assumed to be white stationary random processes with $E[v(t)] = 0$, $E[w(t)] = 0$, $E[v(t)v(t)^T] = \Gamma_v$, $E[w(t)w(t)^T] = \Gamma_w$, and $E[v(t)w(t)^T] = \Gamma_{vw}$.

The output of the model for instant $t+j$, assuming that the state at instant t and future control increments are known, can be computed by recursively applying equation (6.28), resulting in:

$$y(t+j) = QM^j x(t) + \sum_{i=0}^{j-1} QM^{j-i-1} N \Delta u(t+i) +$$

$$+ \sum_{i=0}^{j-1} QM^{j-i-1} Pv(t+i) + w(k+j)$$

Taking the expected value:

$$\begin{aligned}\hat{y}(t+j|t) &= E[y(t+j)] = QM^j E[x(t)] + \sum_{i=0}^{j-1} QM^{j-i-1} N \triangle u(t+i) + \\ &+ \sum_{i=0}^{j-1} QM^{j-i-1} P E[v(t+i)] + E[w(k+j)]\end{aligned}$$

As $E[v(t+i)] = 0$ and $E[w(t+j)] = 0$, the optimal j-ahead prediction is given by:

$$\hat{y}(t+j|t) = QM^j E[x(t)] + \sum_{i=0}^{j-1} QM^{j-i-1} N \triangle u(t+i)$$

Let us now consider a set of N_2 j-ahead predictions:

$$\mathbf{y} = \begin{bmatrix} \hat{y}(t+1|t) \\ \hat{y}(t+2|t) \\ \vdots \\ \hat{y}(t+N_2|t) \end{bmatrix} = \begin{bmatrix} QME[x(t)] + QN \triangle u(t) \\ QM^2 E[x(t)] + \sum_{i=0}^{1} QM^{1-i} N \triangle u(t+i) \\ \vdots \\ QM^{N_2} E[x(t)] + \sum_{i=0}^{N_2-1} QM^{N_2-1-i} N \triangle u(t+i) \end{bmatrix}$$

which can be expressed as:

$$\mathbf{y} = \mathbf{F}\hat{x}(t) + \mathbf{H}\mathbf{u}$$

where $\hat{x}(t) = E[x(t)]$, $\mathbf{H}$ is a block lower triangular matrix with its non-null elements defined by $(\mathbf{H})_{ij} = QM^{i-j}N$ and matrix $\mathbf{F}$ is defined as:

$$\mathbf{F} = \begin{bmatrix} QM \\ QM^2 \\ \vdots \\ QM^{N_2} \end{bmatrix}$$

The prediction equation (6.3) requires an unbiased estimation of the state vector $x(t)$. If the state vector is not accessible, a Kalman filter [8] is required.

Let us now consider a set of j-ahead predictions affecting the cost function: $\mathbf{y}_{N_{12}} = [\hat{y}(t+N_1|t)^T \cdots \hat{y}(t+N_2|t)^T]^T$ and the vector of N_3 future control moves $\mathbf{u}_{N_3} = [\triangle u(t)^T \cdots \triangle u(t+N_3-1)^T]^T$. Then,

$$\mathbf{y}_{N_{12}} = \mathbf{F}_{N_{12}}\hat{x}(t) + \mathbf{H}_{N_{123}}\mathbf{u}_{N_3}$$

where matrices $\mathbf{F}_{N_{12}}$ and $\mathbf{H}_{N_{123}}$ are formed by the corresponding submatrices in $\mathbf{F}$ and $\mathbf{H}$ respectively. Equation (6.2) can be rewritten as:

$$J = (\mathbf{H}_{N_{123}}\mathbf{u}_{N_3} + \mathbf{F}_{N_{12}}\hat{x}(t) - \mathbf{w})^T \overline{R} (\mathbf{H}_{N_{123}}\mathbf{u}_{N_3} + \mathbf{F}_{N_{12}}\hat{x}(t) - \mathbf{w}) + \mathbf{u}_{N_3}^T \overline{Q} \mathbf{u}_{N_3}$$

If there are no constraints, the optimum can be expressed as:

$$\mathbf{u} = ((\mathbf{H}_{N_{123}}^T \overline{R} \mathbf{H}_{N_{123}}) + \overline{Q})^{-1} \mathbf{H}_{N_{123}}^T \overline{R} (\mathbf{w} - \mathbf{F}_{N_{12}}\hat{x}(t))$$

6.3.1 Matrix Fraction and State Space Equivalences

The output signal of processes described by equations (6.28) and (6.1), with zero initial conditions, can be expressed as:

$$\begin{aligned} y(t) &= Q(zI - M)^{-1}N \triangle u(t) + Q(zI - M)^{-1}Pv(t) + w(t) \\ y(t) &= \tilde{\mathbf{A}}(z^{-1})^{-1}\mathbf{B}(z^{-1})z^{-1} \triangle u(t) + \tilde{\mathbf{A}}(z^{-1})^{-1}\mathbf{C}(z^{-1})e(t) \end{aligned}$$

By comparing these equations it is clear that both representation are equivalent if

$$\begin{aligned} Q(zI - M)^{-1}N &= \tilde{\mathbf{A}}(z^{-1})^{-1}\mathbf{B}(z^{-1})z^{-1} \\ Q(zI - M)^{-1}Pv(t) + w(t) &= \tilde{\mathbf{A}}(z^{-1})^{-1}\mathbf{C}(z^{-1})e(t) \end{aligned}$$

This can be achieved by making $w(t) = 0, v(t) = e(t)$ and finding a left matrix fraction description of $Q(zI - M)^{-1}N$ and $Q(zI - M)^{-1}P$ with the same left matrix $\tilde{\mathbf{A}}(z^{-1})^{-1}$.

The state space description can be obtained from the matrix fraction description of equation (6.1), used in the previous section, as follows:

Consider the state vector $x(t) = [y(t)^T \cdots y(t-n_a)^T \triangle u(t-1)^T \cdots \triangle u(t-n_b)^T e(t)^T \cdots e(t-n_c)^T]$ and the noise vector $v(t) = e(t+1)$. Equation (6.1) can now be expressed as equation (6.28) with:

$$M = \left[\begin{array}{cccc|cccc|cccc} \tilde{A}'_1 & \cdots & \tilde{A}'_{n_a} & \tilde{A}'_{n_a+1} & B_1 & \cdots & B_{n_b-1} & B_{n_b} & C_1 & \cdots & C_{n_c-1} & C_{n_c} \\ I & \cdots & 0 & 0 & 0 & \cdots & 0 & 0 & 0 & \cdots & 0 & 0 \\ 0 & \ddots & \vdots & \vdots & \vdots & & \vdots & \vdots & \vdots & \vdots & \vdots & \vdots \\ 0 & \cdots & I & 0 & 0 & \cdots & 0 & 0 & 0 & \cdots & 0 & 0 \\ \hline 0 & \cdots & 0 & 0 & 0 & \cdots & 0 & 0 & 0 & \cdots & 0 & 0 \\ 0 & \cdots & 0 & 0 & I & \cdots & 0 & 0 & 0 & \cdots & 0 & 0 \\ \vdots & \vdots & \vdots & \vdots & 0 & \ddots & \vdots & \vdots & \vdots & \vdots & \vdots & \vdots \\ 0 & \cdots & 0 & 0 & 0 & \cdots & I & 0 & 0 & \cdots & 0 & 0 \\ \hline 0 & \cdots & 0 & 0 & 0 & \cdots & 0 & 0 & 0 & \cdots & 0 & 0 \\ 0 & \cdots & 0 & 0 & 0 & \cdots & 0 & 0 & I & \cdots & 0 & 0 \\ \vdots & \vdots & \vdots & \vdots & \vdots & \vdots & \vdots & \vdots & 0 & \ddots & \vdots & \vdots \\ 0 & \cdots & 0 & 0 & 0 & \cdots & 0 & 0 & 0 & \cdots & I & 0 \end{array}\right]$$

$$\begin{aligned} N &= \left[B_0^T \, 0 \cdots 0 \middle| I \, 0 \cdots 0 \middle| 0 \cdots 0 \, 0 \right]^T \\ P &= [\, I \, 0 \cdots 0 | 0 \, 0 \cdots 0 | I \, 0 \cdots 0]^T \end{aligned}$$

The measurement error vector $w(t)$ has to be made zero for both descriptions to coincide. If the colouring polynomial matrix is the identity matrix, the state vector is only composed of past inputs and outputs $x(t) = [y(t)^T \cdots y(t-n_a)^T \triangle u(t-1)^T \cdots \triangle u(t-n_b)^T]$ and only the first two column blocks of matrices M, N and P and the first two row blocks of matrix M have to be considered.

Notice that no Kalman filter is needed to implement the GPC because the state vector is composed of past inputs and outputs. However, the description does not have to be minimal in terms of the state vector dimension. If there is a big difference in the degrees of the polynomials $(\tilde{\mathbf{A}}(z^{-1}))_{ij}$ and $(\mathbf{B}(z^{-1}))_{ij}$ it is better to consider only the past inputs and outputs that are really needed to compute future output signals. In order to do this, consider the i-component of the output vector:

$$\begin{aligned} y_i(t+1) = &-\tilde{\mathbf{A}}_{i1}(z^{-1})y_1(t) - \tilde{\mathbf{A}}_{i2}(z^{-1})y_2(t) - \cdots - \tilde{\mathbf{A}}_{in}(z^{-1})y_n(t) + \\ &+\mathbf{B}_{i1}(z^{-1}) \triangle u_1(t) + \mathbf{B}_{i2}(z^{-1}) \triangle u_2(t) + \cdots + \mathbf{B}_{im}(z^{-1}) \triangle u_m(t) + \\ &+\mathbf{C}_{i1}(z^{-1})e_1(t+1) + \mathbf{C}_{i2}(z^{-1})e_2(t+1) + \cdots + \mathbf{C}_{in}(z^{-1})e_n(t+1) \end{aligned}$$

where $\tilde{\mathbf{A}}_{ij}(z^{-1})$, $\mathbf{B}_{ij}(z^{-1})$ and $\mathbf{C}_{ij}(z^{-1})$ are the ij entries of polynomial matrices $\tilde{\mathbf{A}}(z^{-1})$, $\mathbf{B}(z^{-1})$ and $\mathbf{C}(z^{-1})$ respectively.

The state vector can be defined as:

$$\begin{aligned} x(t) \;=\; & [y_1(t) \cdots y_1(t-n_{y_1}), y_2(t) \cdots y_2(t-n_{y_2}), \cdots, y_n(t) \cdots y_n(t-n_{y_n}), \\ & \Delta u_1(t-1) \cdots \Delta u_1(t-n_{u_1}), \Delta u_2(t-1) \cdots \Delta u_2(t-n_{u_2}), \cdots, \\ & \Delta u_m(t-1) \cdots \Delta u_m(t-n_{u_m}), \\ & e_1(t) \cdots e_1(t-n_{e_1}), e_2(t) \cdots e_2(t-n_{e_2}), \cdots, e_n(t) \cdots e_n(t-n_{e_n})]^T \end{aligned}$$

where $n_{y_i} = \max_j \delta(\tilde{\mathbf{A}}_{ij}(z^{-1}))$, $n_{u_j} = \max_i \delta(\mathbf{B}_{ij}(z^{-1}))$ and $n_{e_j} = \max_j \delta(\mathbf{C}_{ij}(z^{-1}))$. Matrices M,N and P can be expressed as:

$$M = \left[\begin{array}{ccc|ccc|ccc} Myy_{11} & \cdots & Myy_{1n} & Myu_{11} & \cdots & Myu_{1m} & Mye_{11} & \cdots & Mye_{1n} \\ \vdots & \vdots & \vdots & \vdots & \vdots & \vdots & \vdots & \vdots & \vdots \\ Myy_{n1} & \cdots & Myy_{nn} & Myu_{n1} & \cdots & Myu_{nm} & Mye_{n1} & \cdots & Mye_{nn} \\ \hline 0 & \cdots & 0 & Muu_{11} & \cdots & Muu_{1m} & 0 & \cdots & 0 \\ \vdots & \vdots & \vdots & \vdots & \vdots & \vdots & \vdots & \vdots & \vdots \\ 0 & \cdots & 0 & Muu_{m1} & \cdots & Muu_{mm} & 0 & \cdots & 0 \\ \hline 0 & \cdots & 0 & 0 & \cdots & 0 & Mee_{11} & \cdots & Mee_{1n} \\ \vdots & \vdots & \vdots & \vdots & \vdots & \vdots & \vdots & \vdots & \vdots \\ 0 & \cdots & 0 & 0 & \cdots & 0 & Mee_{n1} & \cdots & Mee_{nn} \end{array}\right]$$

$$\begin{aligned} N &= [Ny_1^T \cdots Ny_n^T | Nu_1^T \cdots Nu_m^T | 0 \cdots 0]^T \\ P &= [Py_1^T \cdots Py_n^T | 0 \cdots 0 | 0 \cdots 0]^T \end{aligned}$$

where the submatrices Myy_{ij}, Myu_{ij}, Mye_{ij}, Muu_{ij} and Ny_i have the following form:

$$Myy_{ij} = \begin{bmatrix} -\tilde{a}_{ij_1} & -\tilde{a}_{ij_2} & \cdots & \cdots & -\tilde{a}_{ij_{n_{y_i}}} \\ 1 & 0 & \cdots & \cdots & 0 \\ 0 & 1 & 0 & \cdots & 0 \\ \vdots & \ddots & \ddots & \ddots & \vdots \\ 0 & \cdots & \cdots & 1 & 0 \end{bmatrix} \quad Myu_{ij} = \begin{bmatrix} b_{ij_1} & \cdots & b_{ij_{n_{u_j}}} \\ 0 & \cdots & 0 \\ \vdots & \vdots & \vdots \\ 0 & \cdots & 0 \end{bmatrix}$$

$$Mye_{ij} = \begin{bmatrix} c_{ij_1} & \cdots & c_{ij_{n_{uj}}} \\ 0 & \cdots & 0 \\ \vdots & \vdots & \vdots \\ 0 & \cdots & 0 \end{bmatrix} \quad Muu_{ij} = \begin{bmatrix} 0 & 0 & 0 & \cdots & 0 \\ 1 & 0 & 0 & \cdots & 0 \\ 0 & 1 & 0 & \cdots & 0 \\ \vdots & \ddots & \ddots & \ddots & \vdots \\ 0 & 0 & \cdots & 1 & 0 \end{bmatrix}$$

$$Ny_i = \begin{bmatrix} b_{il_0} & \cdots & b_{im_0} \\ 0 & \cdots & 0 \\ \vdots & \vdots & \vdots \\ 0 & \cdots & 0 \end{bmatrix}$$

Matrix Mee_{ij} has the same form as matrix Muu_{ij}. Matrices Nu_j and Py_j have all elements zero except the j element of the first row which is 1. The noise vectors are $v(t) = e(t+1)$ and $w(t) = 0$.

This state space description corresponds to the one used in [91] for the SISO case. Other state space descriptions have been proposed in literature in the MPC context. Albertos and Ortega [1] proposed a state space description involving an artificial sampling interval equal to the prediction horizon multiplied by the sampling time. The vectors of predicted inputs and outputs over the control horizon are used as input and output signals. The GPC costing function, for the noise free case, is then transformed into a one-step performance index. A state space description based on the step response of the plant has been proposed in [62]. Models based on the step response of the plant are widely used in industry because they are very intuitive and require less *a priori* information to identify the plant. The main disadvantages are that more parameters are needed and only stable processes can be modelled, although the description proposed in [62] allows for the modelling of processes containing integrators.

6.4 Dead Time Problems

Most plants in industry, especially in the process industry, exhibit input-output delays or dead times. That is, the effect of a change in the manipulated variable is not felt on the process output until the dead time has elapsed. Dead times are mainly caused by transport delays or sometimes as the result of processes with dynamics composed of multiple chained lags. The difficulties of controlling processes with significant dead time are well known and are due to the fact that a dead time produces a phase lag that deteriorates the phase margin. As a result, low gain controllers producing sluggish responses (which have to be added to the dead time of the process) have to be used in order to avoid high oscillations. There are different techniques to cope with delays. The most popular is, perhaps, the Smith predictor [114] which basically consists of getting the delay out of the closed loop by generating a prediction of the process output and designing a controller for the process minus the dead time. The error between process output and predictions is fed back to the controller to cope with plant and model mismatch.

Because of the predictive nature of model predictive controllers, time delays are inherently considered by them. Process input-output dead times are reflected in the polynomial matrix $\mathbf{B}(z^{-1})$. The dead time from the j-input to the i-output, expressed in sampling time units, is the maximum integer d_{ij} such that the entry $(\mathbf{B}(z^{-1}))_{ij}$ of polynomial matrix $\mathbf{B}(z^{-1})$ can be expressed as $(\mathbf{B}(z^{-1}))_{ij} = z^{-d_{ij}}(\mathbf{B}'(z^{-1}))_{ij}$. Let us define $d_{min} = \min_{i,j} d_{ij}$, and $d_{max} = \max_{i,j} d_{ij}$. Although process dead time is implicitly considered in the previous section by the first coefficient matrices of polynomial matrix $\mathbf{B}(z^{-1})$ being zero, the computation will not be efficient if precautions are not taken.

The natural extension of the dead time to multivariable processes is the interactor matrix [44] which represents the time delay structure of a multivariable process. The interactor matrix always exists if the transfer matrix $\mathbf{T}(z)$ is strictly proper with $\det(\mathbf{T}(z)) \neq 0$ for almost all z. It is defined as a polynomial matrix $\xi(z)$ such that:

$$\begin{aligned}\det(\xi(z)) &= z^k \\ \lim_{z\to\infty} \xi(z)\mathbf{T}(z) &= K\end{aligned}$$

where k is an integer and K is a nonsingular matrix. The interactor matrix can be made to have the following structure: $\xi(z) = M(z)D(z)$ where $D(z) = \text{diag}(z^{d_1} \cdots z^{d_n})$ and $M(z)$ is a lower triangular matrix with the elements on the main diagonal equal to unity and the elements below the main diagonal either zero or divisible by z. The interactor matrix can be used to design precompensators as indicated in [119] by making the control signal $u(t) = \xi_r(z^{-1})z^{-d}v(t)$, with $\xi_r(z^{-1})$ equal to the right interactor matrix. The output vector is then equal to

$$\begin{aligned}y(t) &= \mathbf{T}(z^{-1})u(t) = \mathbf{T}(z^{-1})\xi_r(z^{-1})z^{-d}v(t) = \\ &= [\mathbf{T}'(z^{-1})\xi_r(z^{-1})^{-1}]\xi_r(z^{-1})z^{-d}v(t) = \mathbf{T}'(z^{-1})z^{-d}v(t)\end{aligned}$$

The process can now be interpreted as a process with a common delay d for all the variables. Notice that the precompensator consists of adding delays to the process. Model predictive control, as pointed out in [112], does not require the use of this type of pre- or post-compensation and the unwanted effects caused by adding extra delays at the input or output are avoided.

In most cases the interactor matrix will take a diagonal form, one corresponding to a single delay d_{min} for every output and the other with a delay d_i for each output. These two cases will be discussed in the following.

First consider the case where there is not much difference between d_{max} and d_{min} and a single delay d_{min} is associated to all output variables. The output of the process will not be affected by $\Delta u(t)$ until the time instant $t + d_{min} + 1$, the previous outputs will be a part of the free response and there is no point in considering them as part of the objective function. The lower limit of the prediction horizon N_1 can therefore be made equal to $d_{min} + 1$. Note that there is no point in making it smaller, and furthermore if it is made bigger, the first predictions, the ones predicted with greater certainty, will not be considered in the objective function. If the difference between d_{max} and d_{min} is not significant, and there is not much difference in the dynamics of the process variables, a common lower ($N_1 = d_{min} + 1$) and upper ($N_2 = N_1 + N - 1$) limit can be chosen for the objective function. Computation can be simplified by considering $\mathbf{B}(z^{-1}) = z^{-d_{min}}\mathbf{B}'(z^{-1})$ and computing the predictions as:

$$\hat{y}(t + N_1 + j|t) = \mathbf{E}_{N_1+j}(z^{-1})\mathbf{B}(z^{-1}) \triangle u(t + N_1 + j - 1) + \mathbf{F}_{N_1+j}(z^{-1})y(t)$$

$$= \mathbf{E}_{N_1+j}(z^{-1})\mathbf{B}'(z^{-1}) \triangle u(t + N_1 + j - 1 - d_{min}) + \mathbf{F}_{N_1+j}(z^{-1})y(t)$$

$$= \mathbf{E}_{N_1+j}(z^{-1})\mathbf{B}'(z^{-1}) \triangle u(t + j) + \mathbf{F}_{N_1+j}(z^{-1})y(t)$$

By making the polynomial matrix

$$\mathbf{E}_{N_1+j}(z^{-1})\mathbf{B}'(z^{-1}) = \mathbf{G}_{N1_j}(z^{-1}) + z^{-(j+1)}\mathbf{G}p_{N1_j}(z^{-1})$$

the prediction equation can now be written as:

$$\hat{y}(t + N_1 + j|t) = \mathbf{G}_{N1_j}(z^{-1}) \triangle u(t + j) +$$

$$+ \mathbf{G}p_{N1_j}(z^{-1}) \triangle u(t - 1) + \mathbf{F}_{N_1+j}(z^{-1})y(t) \tag{6.29}$$

Notice that the last two terms of the right hand side of equation (6.29) depend on past values of the process output and process input and correspond to the free response of the process when the control signals are kept constant, while the first term only depends on the future values of the control signal and can be interpreted as the forced response. That is, the response obtained when the initial conditions are zero. Equation (6.29) can be rewritten as:

$$\hat{y}(t + N_1 + j|t) = \mathbf{G}_{N1_j}(z^{-1}) \triangle u(t + j) + \mathbf{f}_{N_1+j}$$

If there is a significant difference between d_{max} and d_{min}, there will be a lot of zero entries in the coefficient matrices of polynomial matrix $\mathbf{B}(z^{-1})$ resulting in low computational efficiency. Costing horizons should be defined independently in order to obtain higher efficiency.

Let us consider the polynomial matrix $\mathbf{A}(z^{-1})$ to be diagonal (this can easily be done from the process transfer matrix and as shown previously has many advantages). The minimum delay from the input variables to the i-output variable d_i is given by: $d_i = \min_j d_{ij}$. The minimum meaningful value for the lower limit of the prediction horizon for output variable y_i is $N_{1_i} = d_i + 1$. The upper limit $N_{2_i} = N_{1_i} + N_i - 1$ will mainly be dictated by polynomial $\mathbf{A}_{ii}(z^{-1})$. Let us define the pertinent set of optimal j-ahead output predictions $\mathbf{y} = [\mathbf{y}_1^T\ \mathbf{y}_2^T \cdots \mathbf{y}_n^T]^T$ with

$$\mathbf{y}_i = [\hat{y}_i(t+N_{1_i}|t)\ \hat{y}_i(t+N_{1_i}+1|t)\cdots\hat{y}_i(t+N_{2_i}|t)]^T$$

Notice that the set of optimal j-ahead predictions for the i-output variable can be computed by solving a one dimension Diophantine equation:

$$1 = E_{ik}(z^{-1})\tilde{A}_{ii}(z^{-1}) + z^{-k}F_{ik}(z^{-1})$$

with $\tilde{A}_{ii}(z^{-1}) = A_{ii}(z^{-1})\Delta$. The optimum prediction for the i component of the output variable vector is then given by,

$$y_i(t+N_{1_i}+k|t) = \sum_{j=1}^{m} E_{ik}(z^{-1})B_{ij}(z^{-1})\,\Delta\, u_j(t+N_{1_i}+k-1) + F_{ik}(z^{-1})y_i(t)$$

If we make $B_{ij}(z^{-1}) = z^{-d_i}B'_{ij}(z^{-1})$

$$y_i(t+N_{1_i}+k|t) = \sum_{j=1}^{m} E_{ik}(z^{-1})B'_{ij}(z^{-1})\,\Delta\, u_j(t+k) + F_{ik}(z^{-1})y_i(t)$$

which can be expressed as:

$$y_i(t+N_{1_i}+k|t) = \sum_{j=1}^{m} G_{ij_k}(z^{-1})\,\Delta\, u_j(t+k) + \sum_{j=1}^{m} Gp_{ij_k}(z^{-1})\,\Delta\, u_j(t-1) + F_{ik}(z^{-1})y_i(t)$$

where

$$E_{ik}(z^{-1})B'_{ij}(z^{-1}) = G_{ij_k}(z^{-1}) + z^{-(k+1)}Gp_{ij_k}(z^{-1})$$

Let us define $\mathbf{f}_i$ as the free response of $y_i(t)$.

$$\mathbf{f}_i = [f_i(t+N_{1_i})\cdots f_i(t+N_{2_i})]^T$$

with

$$f_i(t+N_{1_i}+k) = \sum_{j=1}^{m} Gp_{ij_k}(z^{-1})\,\Delta\, u_j(t-1) + F_{ik}(z^{-1})y_i(t)$$

The output prediction affecting the objective function can be expressed as:

$$\begin{bmatrix} \mathbf{y}_1 \\ \mathbf{y}_2 \\ \vdots \\ \mathbf{y}_n \end{bmatrix} = \begin{bmatrix} G_{11} & G_{12} & \cdots & G_{1m} \\ G_{21} & G_{22} & \cdots & G_{2m} \\ \vdots & \vdots & \ddots & \vdots \\ G_{n1} & G_{n2} & \cdots & G_{nm} \end{bmatrix} \begin{bmatrix} \mathbf{u}_1 \\ \mathbf{u}_2 \\ \vdots \\ \mathbf{u}_m \end{bmatrix} + \begin{bmatrix} \mathbf{f}_1 \\ \mathbf{f}_2 \\ \vdots \\ \mathbf{f}_n \end{bmatrix}$$

with $\mathbf{u}_j = [\Delta u_j(t)\ \Delta u_j(t+1) \cdots \Delta u_j(t+Nu_j)]^T$ and $Nu_j = \max_i(N_i - d_{ij} - 1)$. The $N_i \times Nu_j$ block matrix G_{ij} has the following form:

$$G_{ij} = \left[\begin{array}{cccc|ccc} 0 & 0 & \cdots & 0 & 0 & \cdots & 0 \\ \vdots & \vdots & \vdots & \vdots & \vdots & \vdots & \vdots \\ 0 & 0 & \cdots & 0 & 0 & \cdots & 0 \\ \hline g_{ij_0} & 0 & \cdots & 0 & 0 & \cdots & 0 \\ g_{ij_1} & g_{ij_0} & \ddots & 0 & 0 & \cdots & 0 \\ \cdots & \cdots & \ddots & \vdots & \vdots & \vdots & \vdots \\ g_{ij_l} & g_{ij_{l-1}} & \cdots & g_{ij_0} & 0 & \cdots & 0 \end{array}\right]$$

where the number of leading zero rows of matrix G_{ij} is $d_{ij} - N_{1_i}$ and the number of trailing zero columns is $Nu_j - N_i + d_{ij}$. Note that the dimension of matrix **G** is $(\sum_{i=1}^{i=n} N_i) \times (\sum_{j=1}^{m} \max_i(N_i - d_{ij} - 1))$, while for the single delay case it is $(N \times n) \times ((N - d_{min} - 1) \times m)$ with $N \geq N_i$ and $d_{min} \leq d_{ij}$ in general. The reduction of the matrix dimension, and hence the computation required depends on how the delay terms are structured.

6.5 Example: Distillation Column

In order to illustrate the problem of controlling multivariable processes with different dead times for the output variables we are going to consider the control of a distillation column.

The model chosen corresponds to a heavy oil fractionator and is referred to in literature as the Shell Oil's heavy oil fractionator [27], [7]. The model was first described by Prett and Morari [95] and has been widely used to try different control strategies for distillation columns.

The process, shown in figure 6.3, has three variables that have to be controlled: The top and side product compositions, which are measured by analyzers, and the bottom temperature. The manipulated variables are the top draw rate, the side draw rate and the bottom reflux duty. The feed provides all heat requirements for the column. Top and side product specifications are fixed by economic and operational goals and must be kept within a 0.5 percent of their setpoint at steady state. The bottom temperature must be controlled within limits fixed by operational constraints. The top end point must be maintained within the maximum and minimum value of -0.5 and 0.5. The manipulated variables are also constrained as follows: all draws must be within hard minimum and maximum bounds of -0.5 and 0.5. The bottom reflux duty is also constrained by -0.5 and 0.5. The maximum allowed slew rates for all manipulated variables is of 0.05 per minute. The dynamics of the process can be described by the following

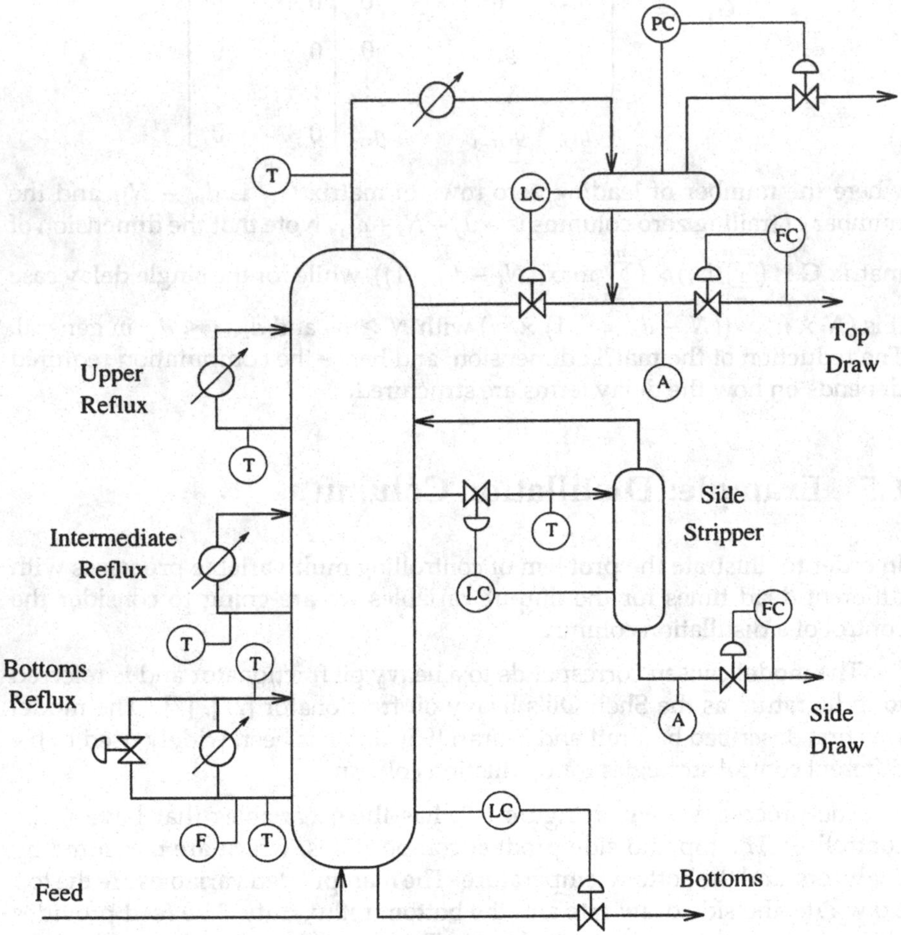

Figure 6.3: Heavy oil Fractionator

$$\begin{bmatrix} Y_1(s) \\ Y_2(s) \\ Y_3(s) \end{bmatrix} = \begin{bmatrix} \frac{4.05e^{-27s}}{1+50s} & \frac{1.77e^{-28s}}{1+60s} & \frac{5.88e^{-27s}}{1+50s} \\ \frac{5.39e^{-18s}}{1+50s} & \frac{5.72e^{-14s}}{1+60s} & \frac{6.9e^{-15s}}{1+40s} \\ \frac{4.38e^{-20s}}{1+33s} & \frac{4.42e^{-22s}}{1+44s} & \frac{7.2}{1+19s} \end{bmatrix} \begin{bmatrix} U_1(s) \\ U_2(s) \\ U_3(s) \end{bmatrix}$$

where $U_1(s), U_2(s)$ and $U_3(s)$ correspond to the top draw, side draw and bottom reflux duties and $Y_1(s), Y_2(s)$ and $Y_3(s)$ correspond to the top end point composition, side end point compositions and bottom reflux temperature respectively.

Notice that the minimum dead time for the three output variables are 27, 14 and 0 minutes respectively.

The discrete transfer matrix for a sampling time of 4 minutes is:

$$\begin{bmatrix} \frac{0.08(z^{-1}+2.88z^{-2})}{1-0.923z^{-1}}z^{-6} & \frac{0.114z^{-1}}{1-0.936z^{-1}}z^{-7} & \frac{0.116(z^{-1}+2.883z^{-2})}{1-0.923z^{-1}}z^{-6} \\ \frac{0.211(z^{-1}+0.96z^{-2})}{1-0.923z^{-1}}z^{-4} & \frac{0.187(z^{-1}+0.967z^{-2})}{1-0.936z^{-1}}z^{-3} & \frac{0.17(z^{-1}+2.854z^{-2})}{1-0.905z^{-1}}z^{-3} \\ \frac{0.5z^{-1}}{1-0.886z^{-1}}z^{-5} & \frac{0.196z^{-1}+0.955z^{-2}}{1-0.913z^{-1}}z^{-5} & \frac{1.367z^{-1}}{1-0.81z^{-1}} \end{bmatrix}$$

A left matrix fraction description can be obtained by making matrix $A(z^{-1})$ equal to a diagonal matrix with diagonal elements equal to the least common multiple of the denominators of the corresponding row of the transfer function, resulting in:

$$\begin{aligned}
A_{11}(z^{-1}) &= 1-1.859z^{-1}+0.8639z^{-2} \\
A_{22}(z^{-1}) &= 1-2.764z^{-1}+2.5463z^{-2}-0.7819z^{-3} \\
A_{33}(z^{-1}) &= 1-2.609z^{-1}+2.2661z^{-2}-0.6552z^{-3} \\
B_{11}(z^{-1}) &= (0.08+0.155z^{-1}-0.216z^{-2})z^{-6} \\
B_{12}(z^{-1}) &= (0.114-0.105z^{-1})z^{-7} \\
B_{13}(z^{-1}) &= (0.116+0.226z^{-1}-0.313z^{-2})z^{-6} \\
B_{21}(z^{-1}) &= (0.211-0.186z^{-1}-0.194z^{-2}+0.172z^{-3})z^{-4} \\
B_{22}(z^{-1}) &= (0.187-0.161z^{-1}-0.174z^{-2}+0.151z^{-3})z^{-3} \\
B_{23}(z^{-1}) &= (0.17+0.169z^{-1}-0.755z^{-2}+0.419z^{-3})z^{-4} \\
B_{31}(z^{-1}) &= (0.5-0.8615z^{-1}+0.369z^{-2})z^{-5} \\
B_{32}(z^{-1}) &= (0.196+0.145z^{-1}-1.77z^{-2}+0.134z^{-3})z^{-5} \\
B_{33}(z^{-1}) &= 1.367-2.459z^{-1}+1.105z^{-2}
\end{aligned}$$

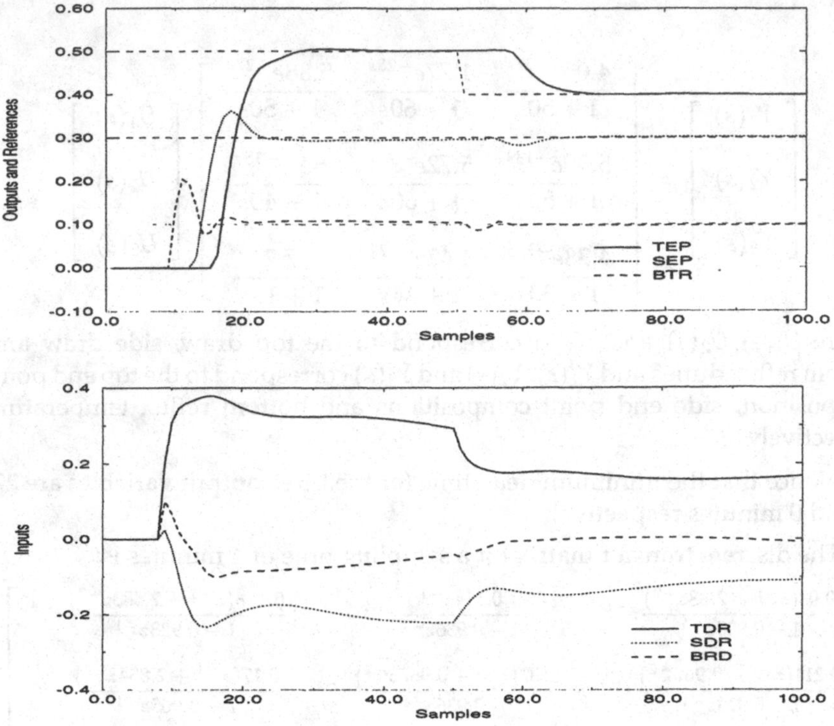

Figure 6.4: Multivariable control of the distillation column

The minimum pure delay time for each of the output variables expressed in sampling time units are 6, 3 and 0 respectively. The results obtained when applying the multivariable GPC can be seen in figure 6.4 for the case of a common prediction horizon of 30 and a control horizon of 5 for all the variables. The weighting matrices were chosen to be $Q = I$ and $R = 2\,I$. The reference trajectories are supposed to be equal to the actual setpoints: 0.5, 0.3 and 0.1 respectively. A change was produced in the setpoint of the top end point composition from 0.5 to 0.4 in the middle of the simulation.

The control increments needed could be, however, too big to be applied in reality and all the manipulated variables were saturated to the hard bounds described above. As can be seen, all the variables reach the setpoint quite rapidly and only the bottom reflux temperature exhibits a significant overshoot. The perturbations produced in the side end point composition and on the bottom reflux temperature due to the setpoint change of the top end point composition are quite small, indicating a low closed loop interaction degree between the setpoint and controlled variables, in spite of the highly coupled open loop dynamics.

The upper and intermediate reflux duties are considered to act as unmea-

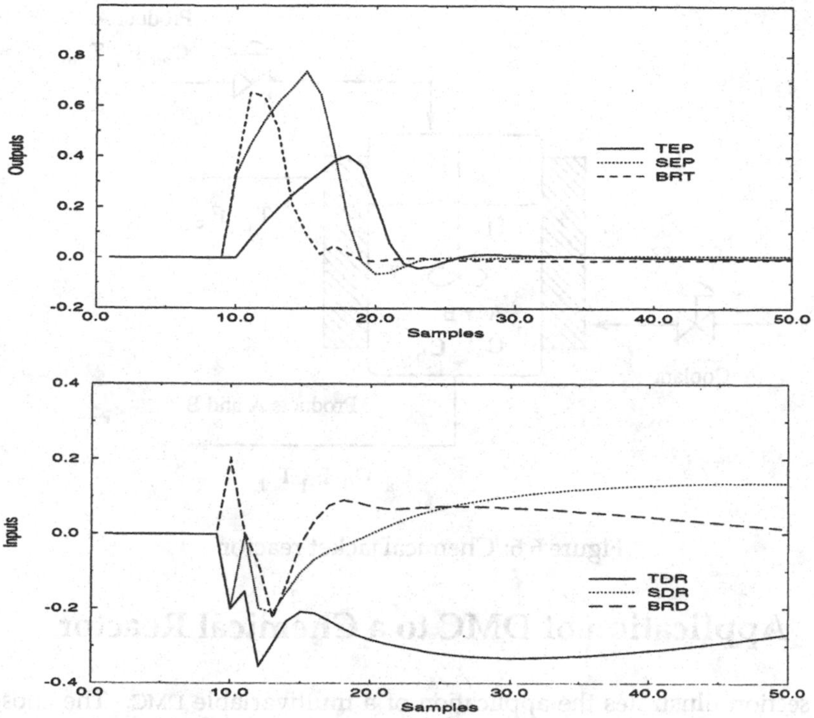

Figure 6.5: Effect of a perturbation in the upper reflux duty

surable disturbances. The small signal dynamic load model for the upper reflux duty is given by the following transfer functions.

$$\begin{bmatrix} Yp_1(s) \\ Yp_2(s) \\ Yp_3(s) \end{bmatrix} = \begin{bmatrix} \dfrac{1.44e^{-27s}}{1+40s} \\ \dfrac{1.83e^{-15s}}{1+20s} \\ \dfrac{1.26}{1+32s} \end{bmatrix} \begin{bmatrix} U_1(s) \\ U_2(s) \\ U_3(s) \end{bmatrix}$$

A step perturbation of 0.5 is introduced in the upper reflux duty, keeping all the setpoints at zero. The results obtained when applying the GPC with the previous weighting matrices are shown in figure 6.5.

As can be seen the perturbations are very rapidly cancelled in spite of the high load perturbations (the steady state value for the load perturbation in the side end point composition is 0.91).

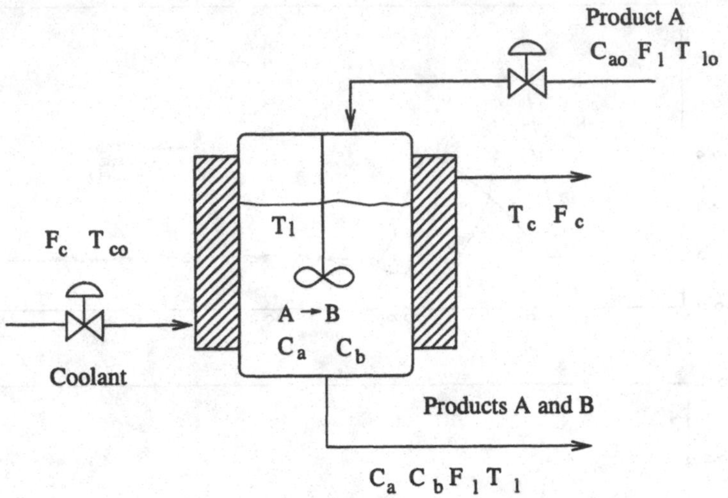

Figure 6.6: Chemical jacket reactor

6.6 Application of DMC to a Chemical Reactor

This section illustrates the application of a multivariable DMC. The chosen system is a chemical jacket reactor. The results have been obtained on a system simulation using the non-linear differential equations which model its behaviour. The model used has been taken from [5] and can be considered to be a very precise representation of this type of process.

6.6.1 Plant Description

The decomposition of a product A into a product B is produced in the reactor (see figure 6.6). This reaction is exothermic and therefore the interior temperature must be controlled by means of cold water circulating through the jacket round the tank walls.

The variables which come into play are:

- A: feed product arriving at the reactor
- B: product arising from the transformation of product A in the tank interior
- C_{a0}: concentration of product A arriving at the reactor
- T_{l0}: temperature of liquid containing product A
- F_l : flow of liquid passing through the reactor. At the inlet it only contains product A and at the outlet it contains A and B

- T_l: temperature of the liquid at the outlet of the reactor
- C_b: concentration of product B at the outlet of the reactor and in the interior
- C_a: concentration of A. The inequality $C_a < C_{a0}$ is always fulfilled and at stationary state $C_a + C_b = C_{a0}$
- T_{c0}: temperature of coolant on entering the jacket
- T_c: temperature of coolant in the interior and at the outlet of the jacket
- F_c: coolant flow

The concentrations are given by kmol/m^3, the flows by m^3/h and the temperatures in °C.

By applying the conservation laws of mass and energy the differential equations defining the dynamics of the system can be obtained. To do this it is presumed that there is no liquid accumulated in the reactor, that the concentrations and temperature are homogeneous and that the energy losses to the exterior are insignificant.

The mass balance equations are as follows:

$$\frac{d(V_l C_a)}{dt} = F_l C_{a0} - V_l k C_a - F_l C_a$$
$$\frac{d(V_l C_b)}{dt} = V_l k C_a - F_l C_b$$

and the energy balance ones are:

$$\frac{d(V_l \rho_l C_{pl} T_l)}{dt} = F_l \rho_l C_p l T_{l0} - F_l \rho_l C_p l T_l - Q + V_l k C_a H$$
$$\frac{d(V_c \rho_c C_{pc} T_c)}{dt} = F_c \rho_c C_p c (T_{c0} - T_c) + Q$$

Table 6.1 gives the meaning and nominal value of the parameters appearing in the equations.

The aim is to regulate the temperature in the tank interior (T_l) and the concentration at the reactor outlet of product B (C_b), the control variables being the flows of the liquid (F_l) and the cooling fluid (F_c). It is, therefore, a system with two inlets and two outlets.

6.6.2 Obtaining the Plant Model

The design of the controller calls for a knowledge of the system dynamics to be controlled. To achieve this, step inputs are produced in the manipulated variables and the behaviour of the process variables is studied.

Table 6.1: Process Variables and values at operating point

Variable	Description	Value	Unit
k	Speed of reaction $k = \alpha e^{-E_a/R(272+T_l)}$		h^{-1}
α	Coefficient of speed of reaction	59.063	h^{-1}
R	Constant of ideal gas	8.314	kJ/kg kmol
E_a	Activation energy	2100	kJ/kmol
H	Enthalpy of reaction	2100	kJ/kmol
Q	Heat absorbed by coolant		kJ
U	Global heat transmission coefficient	4300	kJ/(h m^2 K)
ρ_l	Liquid density	800	kg/m^3
ρ_c	Coolant density	1000	kg/m^3
C_pl	Specific heat of liquid	3	kJ/(kg K)
C_pc	Specific heat of coolant	4.1868	kJ/(kg K)
S	Effective heat interchange surface	24	m^2
V_l	Tank volume	24	m^3
V_c	Jacket volume	8	m^3

On the left hand side of figure 6.7 the response to a change in the feed flow from 25 to 26 m^3/h is shown. It can be seen that the concentrations present a fairly fast response and of opposite sign. The temperatures, however, vary more slowly. The right hand column of figure 6.7 shows the effect of a step change of 1 m^3/h in the cooling flow. Due to thermic inertia the variation in temperatures is slow and dampened, presenting opposite sign those corresponding to feeding and cooling fluid. A great deal of interaction is observed, therefore, between the circuits for feeding and cooling and furthermore the dynamics of the controlled variables are very different depending on the control variable operating, to which is added slight effects of the non minimum phase. All of this justifies the use of a multivariable controller instead of two monovariables ones.

Although the system is non linear it is possible to work with a model linearized about the operating point. The model is obtained from the response to steps shown in figure 6.7.

One has that:

$$y_1(t) = \sum_{i=1}^{N_{11}} g_i^{11} \triangle u_1(t) + \sum_{i=1}^{N_{12}} g_i^{12} \triangle u_2(t)$$

$$y_2(t) = \sum_{i=1}^{N_{21}} g_i^{21} \triangle u_1(t) + \sum_{i=1}^{N_{22}} g_i^{22} \triangle u_2(t)$$

where y_1 and y_2 correspond to the concentration of product B and the temperature in the interior of the reactor and u_1 and u_2 correspond to the flow

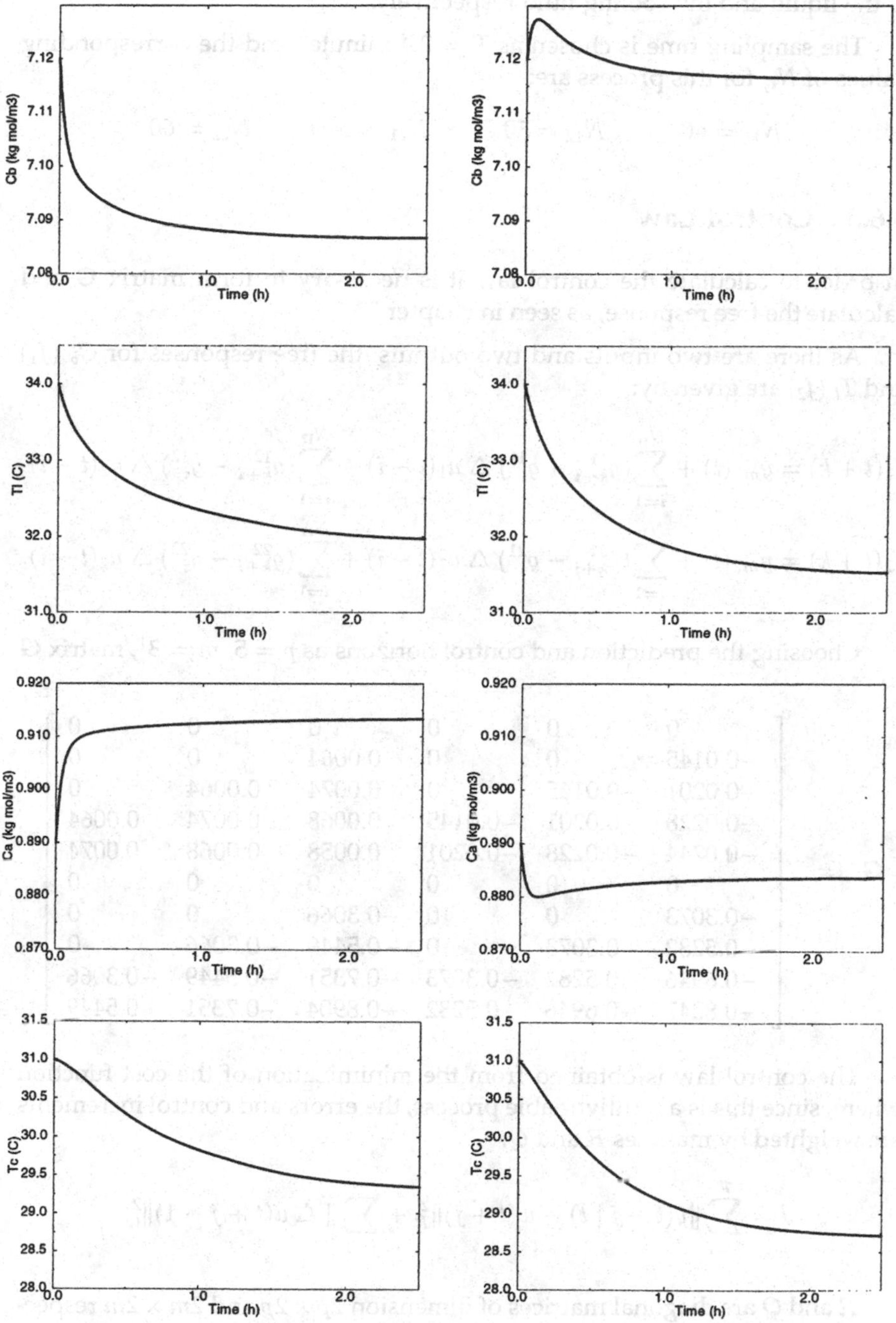

Figure 6.7: System response to changes in the feed flow (left) and the coolant flow (right)

of the liquid and the cooling fluid respectively.

The sampling time is chosen as $T = 2.4$ minutes and the corresponding values of N_{ij} for this process are:

$$N_{11} = 40 \qquad N_{12} = 50 \qquad N_{21} = 55 \qquad N_{22} = 60$$

6.6.3 Control Law

In order to calculate the control law it is necessary to form matrix G and calculate the free response, as seen in chapter 3.

As there are two inputs and two outputs, the free responses for C_b (f_1) and T_l (f_2) are given by:

$$f_1(t+k) = y_{m1}(t) + \sum_{i=1}^{N_{11}} (g_{k+i}^{11} - g_i^{11}) \triangle u_1(t-i) + \sum_{i=1}^{N_{12}} (g_{k+i}^{12} - g_i^{12}) \triangle u_2(t-i)$$

$$f_2(t+k) = y_{m2}(t) + \sum_{i=1}^{N_{21}} (g_{k+i}^{21} - g_i^{21}) \triangle u_1(t-i) + \sum_{i=1}^{N_{22}} (g_{k+i}^{22} - g_i^{22}) \triangle u_2(t-i)$$

Choosing the prediction and control horizons as $p = 5$, $m = 3$[1], matrix G is:

$$\mathbf{G} = \begin{bmatrix} 0 & 0 & 0 & 0 & 0 & 0 \\ -0.0145 & 0 & 0 & 0.0064 & 0 & 0 \\ -0.0201 & -0.0145 & 0 & 0.0074 & 0.0064 & 0 \\ -0.0228 & -0.0201 & -0.0145 & 0.0068 & 0.0074 & 0.0064 \\ -0.0244 & -0.0228 & -0.0201 & 0.0058 & 0.0068 & 0.0074 \\ 0 & 0 & 0 & 0 & 0 & 0 \\ -0.3073 & 0 & 0 & -0.3066 & 0 & 0 \\ -0.5282 & -0.3073 & 0 & -0.5449 & -0.3066 & 0 \\ -0.6946 & -0.5282 & -0.3073 & -0.7351 & -0.5449 & -0.3066 \\ -0.8247 & -0.6946 & -0.5282 & -0.8904 & -0.7351 & -0.5449 \end{bmatrix}$$

The control law is obtained from the minimization of the cost function where, since this is a multivariable process, the errors and control increments are weighted by matrices R and Q:

$$J = \sum_{j=1}^{p} \|\hat{y}(t+j \mid t) - w(t+j)\|_R^2 + \sum_{j=1}^{m} \| \triangle u(t+j-1)\|_Q^2$$

R and Q are diagonal matrices of dimension $2p \times 2p$ and $2m \times 2m$ respectively. In this application the first m elements of R are taken equal to 1 and

[1]Notice that best results can be obtained for bigger values of the horizon, although these small values have been used in this example for the sake of simplicity.

the second part equal to 10 in order to compensate for the different range of values in temperature and concentration. The control weightings are taken as 0.1 for both manipulated variables.

The solution is given by:

$$\mathbf{u} = (\mathbf{G}^T\mathbf{R}\mathbf{G} + \mathbf{Q})^{-1}\mathbf{G}^T\mathbf{R}(\mathbf{w} - \mathbf{f})$$

and the control increment at instant t is calculated multiplying the first row of $(\mathbf{G}^T\mathbf{R}\mathbf{G} + \mathbf{Q})^{-1}\mathbf{G}^T\mathbf{R}$ by the difference between the reference trajectory and the free response:

$$\Delta u(t) = \mathbf{l}(\mathbf{w} - \mathbf{f})$$

with

$$\mathbf{l} = [0\ -0.1045\ -0.1347\ -0.1450\ -0.1485\, 0\ -1.3695\ -0.1112\ -0.1579\, 0.1381]$$

6.6.4 Simulation Results

In this section some results of applying the controller to a non linear model of the reactor are presented. Although the controller has been designed using a linear model and the plant is non linear, the results obtained are satisfactory.

The charts on the left of figure 6.8 shows the behaviour of the process in the presence of changes in the composition reference (C_b). As can be observed, the output follows the reference by means of the contribution of the two manipulated variables F_l and F_c. It can also be seen that any change affects all the variables, such as the concentration of A (C_a) and the coolant temperature (T_c), and also the other output T_l, which is slightly moved from its reference during the transient stage.

The response to a change in the temperature reference is drawn in the charts on the right of figure 6.8. As can be seen, the temperature reference is followed satisfactorily but the concentration is affected and separated from its setpoint.

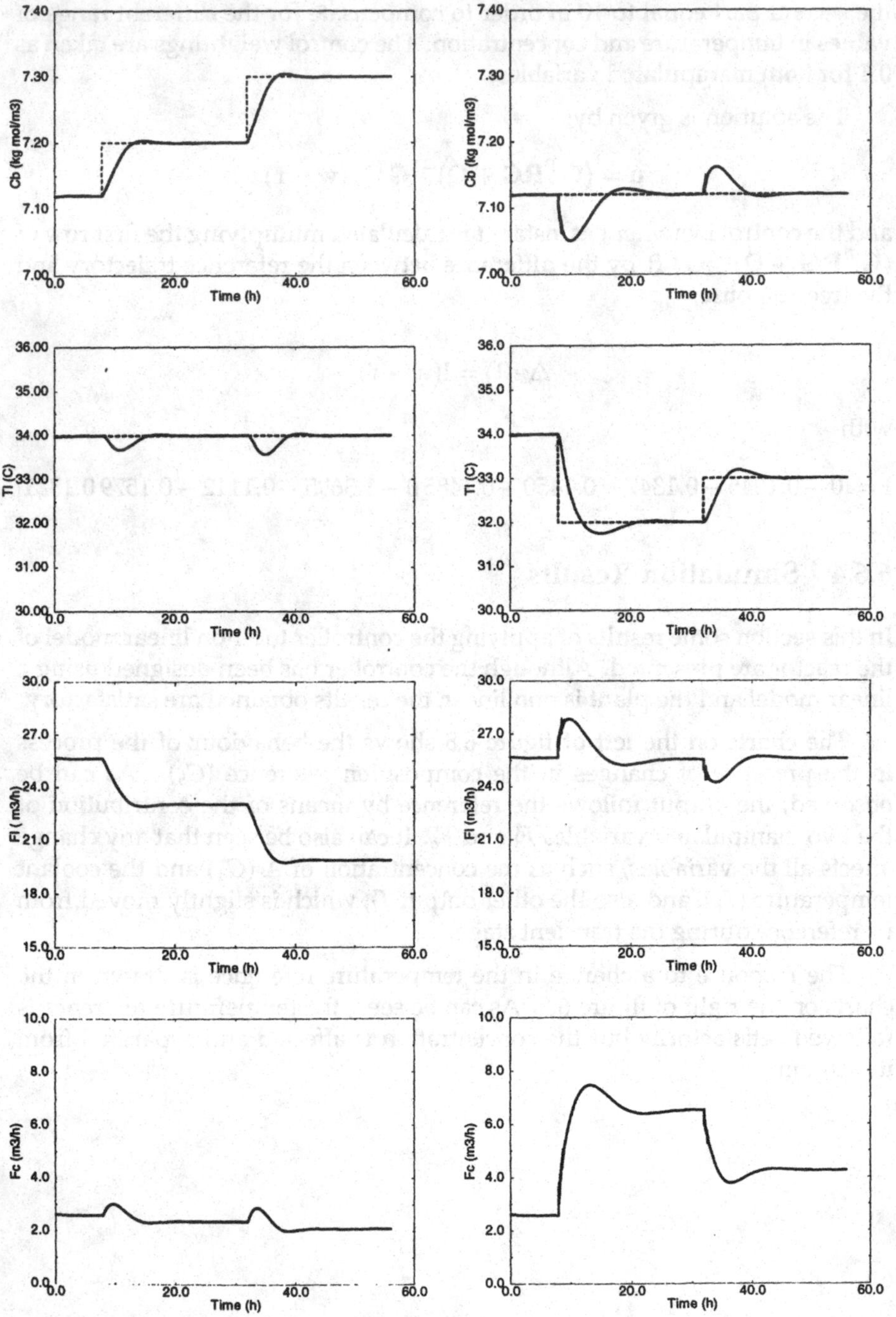

Figure 6.8: Controller response to changes in concentration reference (left) and the liquid temperature reference (right)

Chapter 7

Constrained MPC

The control problem has been formulated in the previous chapters considering all signals to possess an unlimited range. This is not very realistic because in practice all processes are subject to constraints. Actuators have a limited range of action and a limited slew rate, as is the case of control valves which are limited by a fully closed and fully open position and a maximum slew rate. Constructive and/or safety reasons, as well as sensor range, cause bounds in process variables, as in the case of levels in tanks, flows in pipes and pressures in deposits. Furthermore, in practice, the operating points of plants are determined to satisfy economic goals and lie at the intersection of certain constraints. The control system normally operates close to the limits and constraint violations are likely to occur. The control system, especially for long-range predictive control, has to anticipate constraint violations and correct them in an appropriate way. Although input and output constraints are basically treated in the same way, as is shown in this chapter, the implications of the constraints differ. Output constraints are mainly due to safety reasons, and must be controlled in advance because output variables are affected by process dynamics. Input (or manipulated) variables can always be kept in bound by the controller by clipping the control action to a value satisfying amplitude and slew rate constraints.

This chapter concentrates on how to implement generalized predictive controllers for processes with constrained input (amplitude and/or slew rate) and output signals.

7.1 Constraints and MPC

Recall that the MPC control actions were calculated by computing vector u of future control increments that minimizes a quadratic objective function given by:

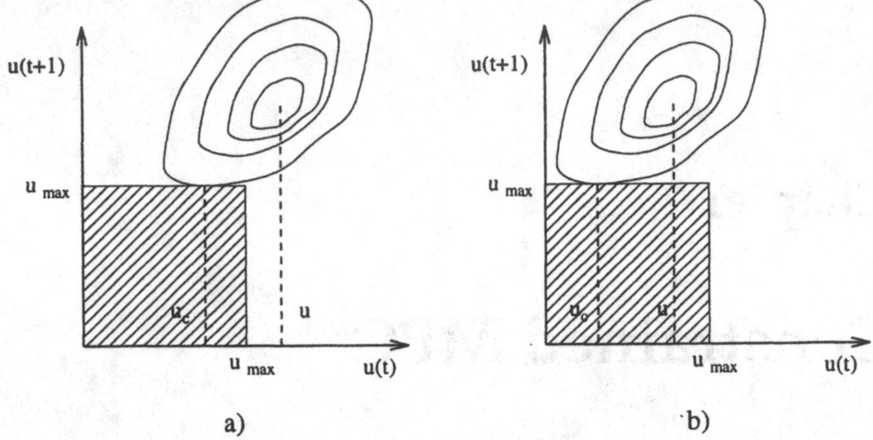

Figure 7.1: Constraints on the control signal

$$J(\mathbf{u}) = \frac{1}{2}\mathbf{u}^T\mathbf{H}\mathbf{u} + \mathbf{b}\mathbf{u} + \mathbf{f}_0 \tag{7.1}$$

The optimal solution of this problem is found by solving the linear equation:

$$\mathbf{H}\mathbf{u} = -\mathbf{b}$$

In practice, the normal way of using a MPC is to compute $u(t)$ as previously described and apply it to the process. If $u(t)$ violates the constraint it is saturated to its limits, either by the control program or by the actuator. The case of $u(t+1), \cdots, u(t+N)$ violating the constraints is not even considered as in most cases these signals are not even computed.

This way of operating does not guarantee that the optimum will be obtained when constraints are violated. The main purpose of GPC, which is to apply the *best* possible control signal by minimizing expression (7.1), will not be achieved.

In order to illustrate this point, consider the cases of constraint violations shown in figure 7.1 of a GPC problem with a control horizon of two. Figure 7.1a shows the case where $u(t) > u_{max}$. In this case the normal way of operating would be to apply u_{max} to the process instead of u_c where the minimum of J is reached when restrictions are considered. In the case shown in figure 7.1b, $u(t)$ does not violate the constraints, and it would be applied to the system instead of the optimum signal u_c that should be applied when constraints are taken into account.

Not considering constraints on manipulated variables to their full extent, may result in higher values of the objective function and thus in a poorer performance of the control system. However, manipulated variables can always be kept to their limits either by the control program or by the actuator and this is not the main reason for treating constraints in an appropriate way. Violating the limits on the controlled variables may be more costly and dangerous as it could cause damage to equipment and losses in production. For example, in most batch reactors the quality of the production requires some of the variables to be kept within specified limits; violating these limits may create a bad quality product and in some cases the loss of the whole batch. When the limits have been set because of safety reasons, the violation of these limits could cause damage to equipment, spillage, or in most cases the activation of the emergency system which will normally produce an emergency stop of the process, losing and/or delaying production, and a normally costly start up procedure.

Constraint violations on the output variables are not contemplated when the only way of handling constraints is by clipping the manipulated variables. One of the main advantages of MPC, its prediction capabilities, is not used to its full potential by this way of operating. Control systems, especially long-range predictive control, should anticipate constraint violations and correct them in an appropriate way.

The constraints acting on a process can originate from amplitude limits in the control signal, slew rate limits of the actuator and limits on the output signals, and can be described respectively by:

$$\begin{array}{rcccl} \underline{U} & \leq & u(t) & \leq & \overline{U} \quad \forall t \\ \underline{u} & \leq & u(t)-u(t-1) & \leq & \overline{u} \quad \forall t \\ \underline{y} & \leq & y(t) & \leq & \overline{y} \quad \forall t \end{array}$$

For a m-input n-output process with constraints acting over a receding horizon N, these constraints can be expressed as:

$$\begin{array}{rcccl} \mathbf{1}\,\underline{U} & \leq & T\mathbf{u} + u(t-1)\,\mathbf{1} & \leq & \mathbf{1}\,\overline{U} \\ \mathbf{1}\,\underline{u} & \leq & \mathbf{u} & \leq & \mathbf{1}\,\overline{u} \\ \mathbf{1}\,\underline{y} & \leq & \mathbf{G}\mathbf{u} + \mathbf{f} & \leq & \mathbf{1}\,\overline{y} \end{array}$$

where $\mathbf{1}$ is an $(N \times n) \times m$ matrix formed by N $m \times m$ identity matrices and T is a lower triangular block matrix whose non null block entries are $m \times m$ identity matrices. The constraints can be expressed in condensed form as:

$$\mathbf{R}\,\mathbf{u} \leq \mathbf{c}$$

with:

$$R = \begin{bmatrix} I_{N\times N} \\ -I_{N\times N} \\ T \\ -T \\ \mathbf{G} \\ -\mathbf{G} \end{bmatrix} \quad \mathbf{c} = \begin{bmatrix} \mathbf{1}\overline{u} \\ -\mathbf{1}\,\underline{u} \\ \mathbf{1}\overline{U} - \mathbf{1}u(t-1) \\ -\mathbf{1}\,\underline{U} + \mathbf{1}u(t-1) \\ \mathbf{1}\overline{y} - \mathbf{f} \\ -\mathbf{1}\,\underline{y} + \mathbf{f} \end{bmatrix}$$

The constraints on the output variables of the type $\underline{y} \leq y(t) \leq \overline{y}$ are normally imposed because of safety reasons. Other types of constraint can be set on the process controlled variables to force the response of the process to have certain characteristics, as shown by Kuznetsov and Clarke [60], and can also be expressed in a similar manner:

Band Constraints

Sometimes one wishes the controlled variables to follow a trajectory within a band. In the food industry, for example, it is very usual for some operations to require a temperature profile that has to be followed with a specified tolerance.

This type of requirement can be introduced in the control system by forcing the output of the system to be included in the band formed by the specified trajectory plus-minus the tolerance. That is:

$$\underline{y}(t) \leq y(t) \leq \overline{y}(t)$$

These constraints can be expressed in terms of the increments of the manipulated variables as follows:

$$\mathbf{G}\mathbf{u} \leq \overline{\mathbf{y}} - \mathbf{f}$$
$$\mathbf{G}\mathbf{u} \geq \underline{\mathbf{y}} - \mathbf{f}$$

Overshoot Constraints

In some processes overshoots are not desirable because of different reasons. In the case of manipulators, for example, an overshoot may produce a collision with the workplace or with the piece it is trying to grasp.

Overshoot constraints have been treated by Kuznetsov and Clarke [60] and are very easy to implement. Every time a change is produced in the setpoint, which is considered to be kept constant for a sufficiently long period, the following constraints are added to the control system:

$$y(t+j) \leq w(t) \text{ for } j = N_{o1} \cdots N_{o2}$$

where N_{o1} and N_{o2} define the horizon where the overshoot may occur (N_{o1} and N_{o2} can always be made equal to 1 and N if this is not known). These constraints can be expressed in terms of the increments of the manipulated variables as follows:

$$\mathbf{G}\mathbf{u} \leq \mathbf{1}w(t) - \mathbf{f}$$

Monotonic Behaviour

Some control systems tend to exhibit oscillations, known as kick back, on the controlled variables before they have gone over the setpoints. These oscillations are not desirable in general because, amongst other reasons, they may cause perturbations in other processes. Constraints can be added to the control system to avoid this type of behaviour by imposing a monotonic behaviour on the output variables. Each time a setpoint changes, and is again considered to be kept constant for a sufficiently long period, new constraints with the following form are added to the control system:

$$\begin{array}{lll} y(t+j) \leq y(t+j+1) & \text{if} & y(t) < w(t) \\ y(t+j) \geq y(t+j+1) & \text{if} & y(t) > w(t) \end{array}$$

These type of constraints can be expressed in terms of the manipulated variables as follows:

$$\mathbf{G}\mathbf{u} + \mathbf{f} \leq \left[\begin{array}{c} \mathbf{0}^T \\ \hline \mathbf{G}' \end{array}\right] \mathbf{u} + \left[\begin{array}{c} y(t) \\ \hline \mathbf{f}' \end{array}\right]$$

where $\mathbf{G}'$ and $\mathbf{f}'$ result from clipping the last n rows (n is the number of output variables) of $\mathbf{G}$ and $\mathbf{f}$. These constraints can be expressed as

$$\left[\begin{array}{cccc} G_0 & 0 & \cdots & 0 \\ G_1 - G_0 & G_0 & \cdots & 0 \\ \vdots & \vdots & \ddots & \vdots \\ G_{N-1} - G_{N-2} & G_{N-2} - G_{N-3} & \cdots & G_0 \end{array}\right] \mathbf{u} \leq \left[\begin{array}{c} y(t) - \mathbf{f}_1 \\ \mathbf{f}_1 - \mathbf{f}_2 \\ \vdots \\ \mathbf{f}_{N-1} - \mathbf{f}_N \end{array}\right]$$

Non-minimum Phase Behaviour

Some processes exhibit a type of non-minimum phase behaviour. That is, when the process is excited by a step in its input the output variable tends to first move in the opposite direction prior to moving to the final position. This kind of behaviour may not be desirable in many cases.

Constraints can be added to the control system to avoid this type of behaviour. The constraints take the form

$$\begin{array}{lll} y(t+j) \geq y(t) & \text{if} & y(t) < w(t) \\ y(t+j) \leq y(t) & \text{if} & y(t) > w(t) \end{array}$$

These constraints can be expressed in terms of the increments of the manipulated variables as follows:

$$\mathbf{G}\mathbf{u} \geq \mathbf{1}y(t) - \mathbf{f}$$

Actuator Nonlinearities

Most actuators in industry exhibit dead zones and other type of nonlinearities. Controllers are normally designed without taking into account actuator nonlinearities. Because of the predictive nature of MPC, actuator nonlinearities can be dealt with as suggested by Chow and Clarke [28].

Dead zones can be treated by imposing constraints on the controller in order to generate control signals outside the dead zone, say $(\underline{u}_d, \overline{u}_d)$ for dead zone on the slew rate of actuators and $(\underline{U}_d, \overline{U}_d)$ for the dead zone on the amplitude of actuators. That is:

$$\begin{array}{ccccc} \mathbf{1}\,\underline{U}_d & \geq & T\mathbf{u} + u(t-1)\,\mathbf{1} & \geq & \mathbf{1}\,\overline{U}_d \\ \mathbf{1}\,\underline{u}_d & \geq & \mathbf{u} & \geq & \mathbf{1}\,\overline{u}_d \end{array}$$

The feasible region generated by this type of constraints is non-convex and the optimization problem is difficult to solve as pointed out in [28].

Terminal State Constraints

These types of constraints appear when applying CRHPC [30] where the predicted output of the process is forced to follow the predicted reference during a number of sampling periods m after the costing horizon N_y. The terminal state constraints can be expressed as a set of equality constraints on the future control increments by using the prediction equation for $\mathbf{y}_m = [y(t+N_y+1)^T \cdots y(t+N_y+m)^T]^T$:

$$\mathbf{y}_m = \mathbf{G}_m\mathbf{u} + \mathbf{f}_m$$

If the predicted response is forced to follow the future reference setpoint $\mathbf{w}_m$, the following equality constraint can be established:

$$\mathbf{G}_m\mathbf{u} = \mathbf{w}_m - \mathbf{f}_m$$

It will be seen later in this chapter that the introduction of this type of constraints simplifies the problem reducing the amount of computation required.

All constraints treated so far can be expressed as $\mathbf{R}\mathbf{u} \leq \mathbf{c}$ and $\mathbf{A}\mathbf{u} = \mathbf{a}$. The GPC problem when constraints are taken into account consists of minimizing expression (7.1) subject to a set of linear constraints. That is, the optimization of a quadratic function with linear constraints, what is usually known as a quadratic programming problem (QP).

7.1.1 Illustrative Examples

Constraints can be included in Generalized Predictive Control to improve performance as demonstrated by Kuznetsov and Clarke [60]. Ordys and Grimble [92] have indicated how to analyze the influence of constraints on the stochastic characteristics of signals for a system controlled by a GPC algorithm. In order to illustrate how constraints can be used to improve the performance of different types of processes some simple illustrative examples are going to be presented.

Input Constraints

In order to show the influence of constraints on the slew rate and on the amplitude of the manipulated variable, consider the reactor described in the previous chapter given by the following left fraction matrix description:

$$\mathbf{A}(z^{-1}) = \begin{bmatrix} 1-1.8629z^{-1}+0.8669z^{-2} & 0 \\ 0 & 1-1.8695z^{-1}+0.8737z^{-2} \end{bmatrix}$$
$$\mathbf{B}(z^{-1}) = \begin{bmatrix} 0.0420-0.0380z^{-1} & 0.4758-0.4559z^{-1} \\ 0.0582-0.0540z^{-1} & 0.1445-0.1361z^{-1} \end{bmatrix}$$

The constraints considered are: maximum slew rate for the manipulated variables of 0.2 units per sampling time and maximum value of -0.3 and 0.3.

The results obtained are shown in 7.2. If we compare the results with the ones obtained in chapter 5 (unconstrained manipulated variables) we can observe how the introduction of the constraints on the manipulated variables has produced a slower closed loop response as was to be expected.

Overshoot Constraints

The system considered for this example corresponds to a discretized version of an oscillatory system $G(s) = 50/(s^2+25)$, taken from [60]. The discrete transfer function for a sampling time of 0.1 seconds is:

$$G(z^{-1}) = \frac{0.244835(1+z^{-1})}{1-1.75516z^{-1}}$$

The results obtained when applying an unconstrained GPC with a prediction horizon and a control horizon of 11 and a weighting factor of 50, are shown in figure 7.3. As can be seen, the output shows a noticeable overshoot. The response obtained, when overshoot constraints is taken into account is shown in the same figure. As can be seen the overshoot has been eliminated.

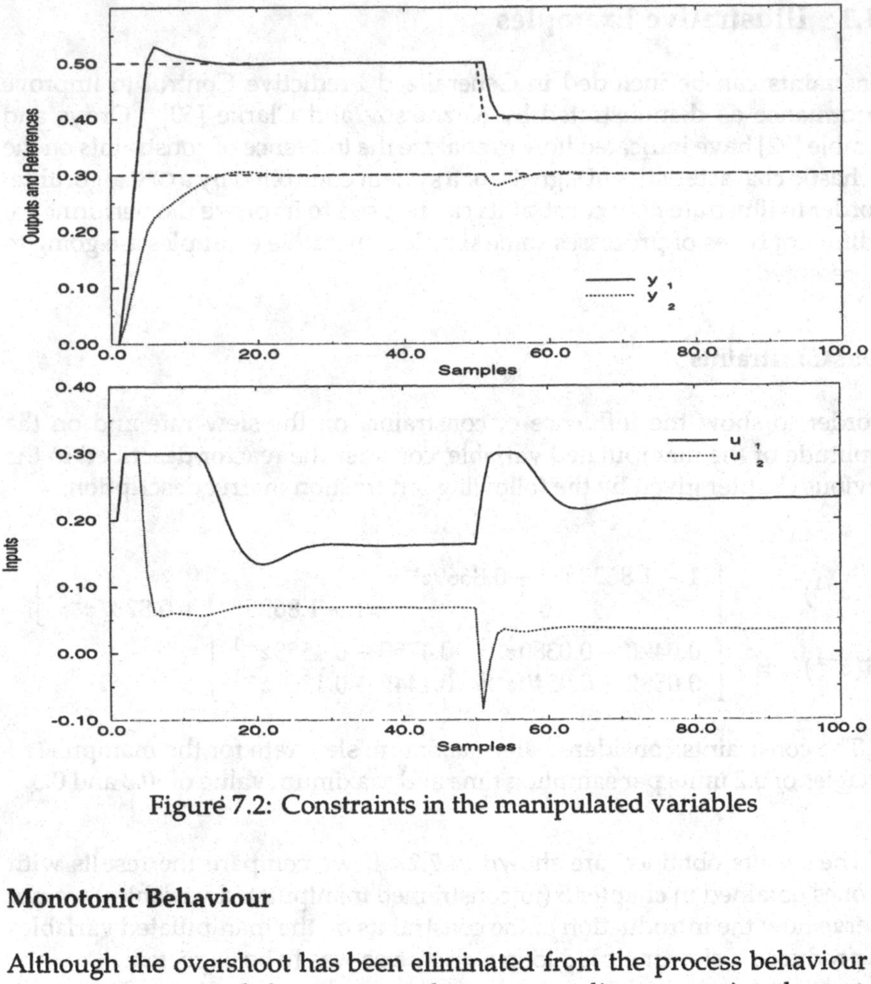

Figure 7.2: Constraints in the manipulated variables

Monotonic Behaviour

Although the overshoot has been eliminated from the process behaviour in the previous example by imposing the corresponding constraint, the system exhibits oscillations prior to reaching the setpoints (kick-back). In order to avoid this type of behaviour, monotonic behaviour constraints were imposed. The results obtained for a prediction horizon and a control horizon of 11 and a weighting factor of 50, are shown in figure 7.4. As can be seen, the oscillations have practically been eliminated. Notice that the prediction and control horizon used are quite large. The reason for this is that, the oscillatory mode of the open loop system has to be cancelled and a large number of control moves has to be considered in order to obtain a feasible solution.

Non-minimum Phase Process

In order to illustrate how constraints may be used to shape the closed loop behaviour consider the non-minimum phase system given by the following

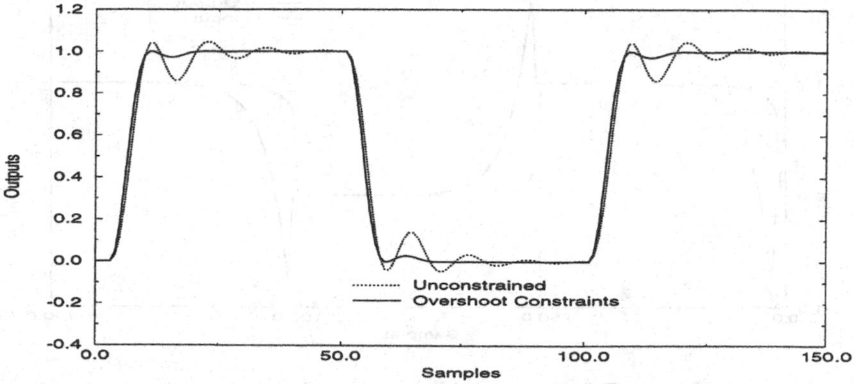

Figure 7.3: Overshoot constraints

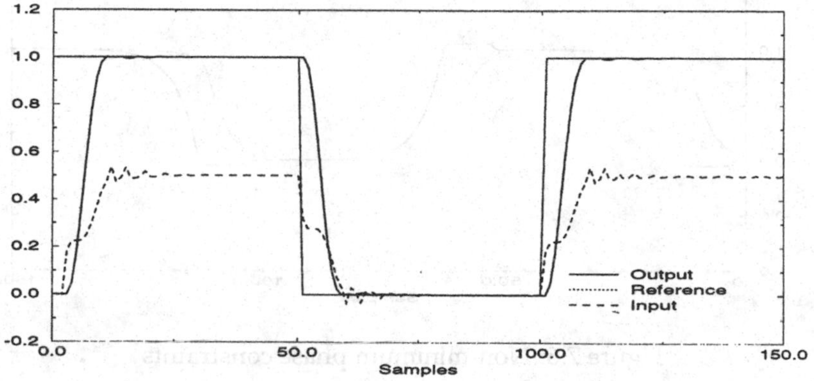

Figure 7.4: Monotonic behaviour constraints

transfer function:

$$G(s) = \frac{1-s}{1+s}$$

If the system is sampled at 0.3 seconds, the discrete transfer function is given by

$$G(z^{-1}) = \frac{-1+1.2592z^{-1}}{1-0.7408z^{-1}}$$

The response obtained for step changes in the reference when a GPC with a prediction horizon of 30, a control horizon of 10 and a weighting factor of 0.1 is applied to the system is shown in figure 7.5. As can be seen, the closed loop behaviour exhibits the typical non-minimum phase behaviour with an initial peak in the opposite direction to the setpoint change. The responses obtained

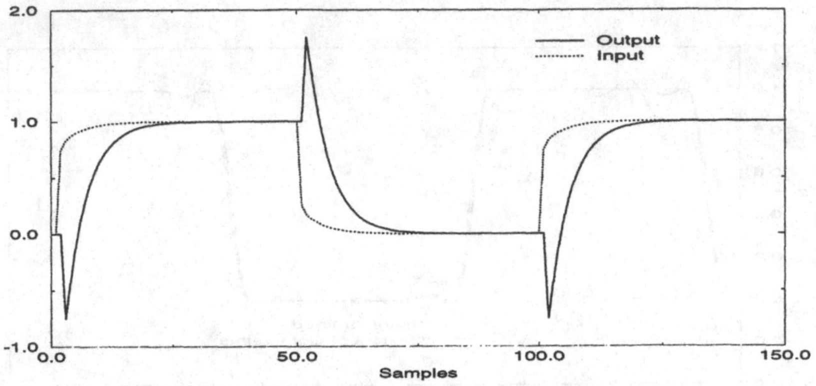

Figure 7.5: Non-minimum phase behaviour

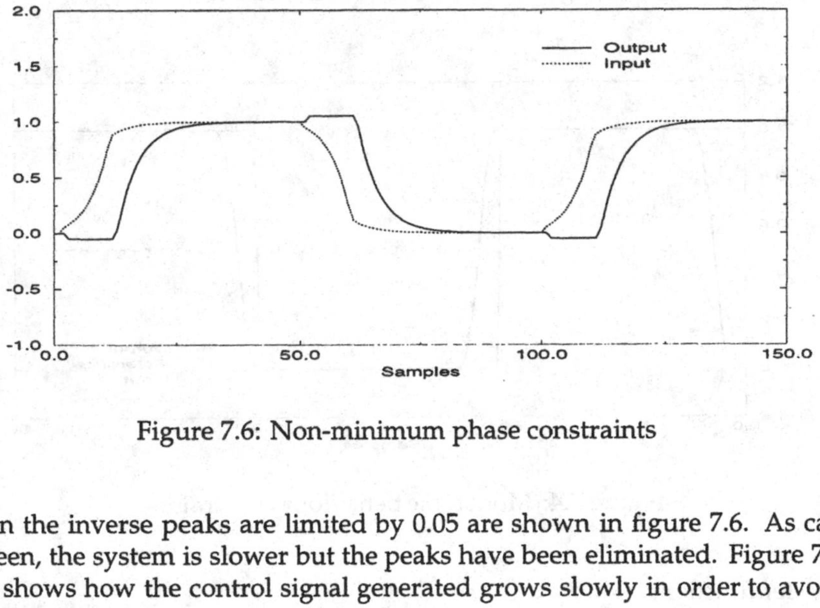

Figure 7.6: Non-minimum phase constraints

when the inverse peaks are limited by 0.05 are shown in figure 7.6. As can be seen, the system is slower but the peaks have been eliminated. Figure 7.6 also shows how the control signal generated grows slowly in order to avoid the inverse peaks.

7.2 Constraints and optimization

The implementation of MPC in industry is not a trivial matter and is certainly a more difficult task than the commissioning of classical control schemes based on PID controllers. An MPC application is more costly, time consuming and requires personnel with a better training in control than when implementing classical control schemes. For a start, a model of the process has to be found, and this requires a significant amount of plant tests, which in most cases implies taking the plant away from its nominal operating conditions. The control equipment needs, in some cases, a more powerful computer and better instrumentation, and commercial MPC packages are expensive. Furthermore, control personnel need appropriate training for commissioning and using MPC.

In spite of these difficulties MPC has proved itself to be economically profitable by reducing operating costs or increasing production and is one of the most successful advanced control techniques used in industry. The reasons for this success depend on the particular application, but are related to the abilities of MPC to optimize cost functions and treat constraints. The following reasons can be mentioned:

- Optimization of operating conditions: MPC optimizes a cost function which can be formulated in order to minimize operating costs or any objective with economic implications.

- Optimization of transitions: The MPC objective function can be formulated in order to optimize a function which measures the cost of taking the process from one operating point to another, with faster process start-ups or commissioning times.

- Minimization of error variance: An MPC can be formulated in order to minimize the variance of the output error. A smaller variance will produce economical benefits for the following reasons:

  - A smaller variance may increase the quality of the product as well as its uniformity.
  - A smaller variance will allow the process to operate closer to the optimal operating conditions: As most processes are constrained, the optimal operating points usually lie at the intersection of some constraints. Processes operating with smaller error variance can operate closer to the optimal operating points. Let us, as an example, consider a process where the production (flow) is related to the operating pressure as illustrated in figure 7.7. Because of operating constraints, the pressure is limited to p_{max} which is the optimal operating point considering production. It is obvious that the process cannot operate at its limit because, due to perturbations, the process would be continuously violating the pressure limits and in most cases emergency systems would shut down the process. If the control system is able to keep the variance small, the set point can be established much closer to the optimal operating point as shown in figure 7.7.

- The explicit handling of constraints may allow the process to operate closer to constraints, and optimal operating conditions.

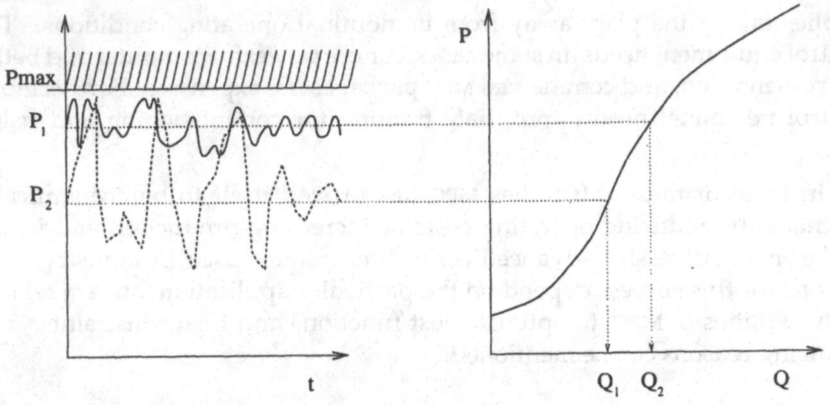

Figure 7.7: Optimal operating point and constraints

- The explicit considerations of constraints may reduce the number of constraint violations reducing the number of costly emergency shut downs.

7.3 Revision of Main Quadratic Programming Algorithms

As was indicated in the previous section, the implementation of Generalized Predictive Controllers for processes with bounded signals requires the solution of a quadratic programming (QP) problem. That is, an optimization problem with a quadratic objective function and linear constraints. This section is dedicated to revising some of the main QP techniques. It is not intended to be an exhaustive description of all QP methods. There are other techniques, such as the ones based on decreasing ellipsoid volume methods that have been used in the GPC context [52] which are not described here.

7.3.1 The Active Set Methods

Equality Constraints

Although a set of inequality constraints is obtained when the GPC control problem is formulated, the first part of the section is dedicated to equality constraints because some of the methods for treating inequality constraints reduce the problem to an equality constraint problem and because in some cases, as in CRHPC [84], some equality constraints appear when the prediction of the future process output is forced to follow exactly the future reference.

The equality constrained QP problem can be stated as:

$$\text{minimize } J(\mathbf{u}) = \frac{1}{2}\mathbf{u}^T\mathbf{H}\mathbf{u} + \mathbf{b}^T\mathbf{u} + \mathbf{f}_0$$
$$\text{subject to } \mathbf{A}\mathbf{u} = \mathbf{a}$$

where $\mathbf{A}$ is an $m \times n$ matrix and $\mathbf{a}$ is an m vector. It is assumed that $m < n$ and that $rank(\mathbf{A}) = m$.

A direct way of solving the problem is to use the constraints to express m of the $\mathbf{u}$ variables as a function of the remaining $n - m$ variables and then substitute them in the objective function. The problem is reduced to minimizing a quadratic function of $n - m$ variables without constraints.

Usually a generalized elimination method is used instead of a direct elimination procedure. The idea is to express $\mathbf{u}$ as a function of a reduced set of $n - m$ variables: $\mathbf{u} = \mathbf{Y}\mathbf{a} + \mathbf{Z}\mathbf{v}$. Where $\mathbf{Y}$ and $\mathbf{Z}$ are $n \times m$ and $n \times (n - m)$ matrices such that $\mathbf{AY} = I, \mathbf{AZ} = 0$ and the matrix $[\mathbf{Y}\ \mathbf{Z}]$ has full rank. Notice that matrix $\mathbf{Y}$ can be interpreted as a generalized left inverse of $\mathbf{A}^T$ and that $\mathbf{Zv}$ is the null column space of $\mathbf{A}^T$.

If this substitution is made, the equality constraints hold and the objective function:

$$\begin{aligned} J(\mathbf{v}) &= \frac{1}{2}[\mathbf{Ya} + \mathbf{Zv}]^T\mathbf{H}[\mathbf{Ya} + \mathbf{Zv}] + \mathbf{b}^T[\mathbf{Ya} + \mathbf{Zv}] + \mathbf{f}_0 = \\ &= \frac{1}{2}\mathbf{v}^T\mathbf{Z}^T\mathbf{HZv} + [\mathbf{b}^T + \mathbf{a}^T\mathbf{Y}^T\mathbf{H}]\mathbf{Zv} + [\frac{1}{2}\mathbf{a}^T\mathbf{Y}^T\mathbf{H} + \mathbf{b}^T]\mathbf{Ya} + \mathbf{f}_0 \end{aligned}$$

That is an unconstrained QP problem of $n - m$ variables. If the matrix $\mathbf{Z}^T\mathbf{HZ}$ is positive definite, there is only one global optimum point that can be found solving the linear set of equations:

$$\mathbf{Z}^T\mathbf{HZv} = -\mathbf{Z}^T(\mathbf{b} + \mathbf{HYa})$$

Notice that if $\mathbf{u}^k$ is a point satisfying the constraints $\mathbf{Au}^k = \mathbf{a}$, any other point $\mathbf{u}$ satisfying the constraints can be expressed as $\mathbf{u} = \mathbf{u}^k + \mathbf{Zv}$. Thus the vector $\mathbf{Ya}$ can be made equal to any point satisfying the constraints. Vector $\mathbf{v}$ can be expressed as the solution of the following linear equation:

$$\mathbf{Z}^T\mathbf{HZv} = -\mathbf{Z}^T\mathbf{g}(\mathbf{u}^k)$$

where $\mathbf{g}(\mathbf{u}^k) = \mathbf{Hu}^k + \mathbf{b}$ is the gradient of $J(\mathbf{u})$ at $\mathbf{u}^k$.

A general way of obtaining appropriate $\mathbf{Y}$ and $\mathbf{Z}$ matrices is by choosing a $(n - m) \times n$ matrix $\mathbf{W}$ such that the matrix $\begin{bmatrix} \mathbf{A} \\ \mathbf{W} \end{bmatrix}$ is non singular. The inverse can then be expressed as:

$$\begin{bmatrix} \mathbf{A} \\ \mathbf{W} \end{bmatrix}^{-1} = [\mathbf{Y}\ \mathbf{Z}]$$

It then follows that $\mathbf{AY} = \mathbf{I}$ and $\mathbf{AZ} = \mathbf{0}$.

If matrix $\mathbf{W}$ is chosen as $[\mathbf{0}\ \mathbf{I}]$, the method coincides with the direct elimination method. Another way of choosing $\mathbf{W}$ is related to the active set method that will be described later. The idea is to use inactive constraints ($\mathbf{a}_i$) as the rows of $\mathbf{W}$. If an inactive constraint, present in $\mathbf{W}$, becomes active (the rows of $\mathbf{R}$ where $\mathbf{r}_i\mathbf{u} = c_i$), the corresponding row of $\mathbf{W}$ is transferred to $\mathbf{A}$. When an active constraint becomes inactive, the corresponding row of $\mathbf{A}$ is transferred to $\mathbf{W}$. By doing this, the inverse of matrix $\begin{bmatrix} \mathbf{A} \\ \mathbf{W} \end{bmatrix}$ need not be recomputed in order to calculate $\mathbf{Y}$ and $\mathbf{Z}$.

Inequality Constraints

As shown at the beginning of the chapter, the GPC of processes with bounded signals results in a QP problem with linear inequality constraints.

The main idea of the active set method is to reduce the inequality constraints QP problem to a sequence of equality constraints QP problems that can be solved by the techniques described previously.

Consider a feasible point $\mathbf{u}^0$, that is, $\mathbf{Ru}^0 \leq \mathbf{c}$ and the set of active constraints (all the equality constraints and the rows of $\mathbf{R}$ where $\mathbf{r}_i\mathbf{u} = c_i$). Form matrix $\mathbf{A}$ and vector $\mathbf{a}$ by adding these rows ($\mathbf{r}_i$) and corresponding limits (c_i) and the equality constraints.

The problem can now be solved with the method described previously. Suppose that $\mathbf{u}^1$ is the solution to the equally constrained QP problem. If $\mathbf{u}^1$ is feasible with respect to the inactive constraints, a test for optimality has to be performed in order to check if the global optimum has been found. This can be accomplished by verifying that the Lagrange multipliers for all equality constraints $\lambda_i \geq 0$. If this is not the case, the constraint with the most negative Lagrange multiplier is dropped from the active constraint set and the previous steps are repeated.

If point $\mathbf{u}^1$ is not feasible with respect to the inactive constraints, the nearest intersection from $\mathbf{u}^0$ of the line joining points $\mathbf{u}^0$ and $\mathbf{u}^1$ and the inactive constraints is computed. The corresponding constraint is added to the active set and the previous steps are repeated.

Notice that the method requires an initial feasible point. Procedures to find a feasible point will be described later in the chapter.

7.3.2 Feasible Directions Methods

The key idea of feasible directions methods is to improve the objective function by moving from a feasible point to an improved feasible point until the optimum is reached. Given a feasible point $\mathbf{u}^k$, an improving feasible direction $\mathbf{d}_k$ is determined such that by taking a sufficiently small step along $\mathbf{d}_k$ the new point will be feasible and will have a smaller value for the objective function.

There are various ways of generating feasible directions, one of the most popular ones in terms of simplicity is the gradient projection method of Rosen which is based on the following:

Definition: An $n \times n$ matrix $\mathbf{P}$ is called a projection matrix if $\mathbf{P} = \mathbf{P}^T$ and $\mathbf{PP} = \mathbf{P}$.

Consider the problem of minimizing $J(\mathbf{u})$ subject to $\mathbf{Au} \leq \mathbf{a}$ and a feasible point $\mathbf{u}^k$ such that $\mathbf{A}_1\mathbf{u}^k = \mathbf{a}_1$ and $\mathbf{A}_2\mathbf{u}^k < \mathbf{a}_2$; where the matrices $\mathbf{A}_1$ and $\mathbf{A}_2$ and vectors $\mathbf{a}_1$ and $\mathbf{a}_2$ correspond to the active constraint and inactive constraint sets respectively.

Lemma 6.1 [10]: A non zero direction $\mathbf{d}$ is an improving feasible direction if and only if $\mathbf{A}_1\mathbf{d} \leq 0$ and $\nabla J(\mathbf{u})^T\mathbf{d} < 0$.

Lemma 6.2 [10]: If $\mathbf{P}$ is a projection matrix such that $\mathbf{P}\nabla J(\mathbf{u}^k) \neq 0$ then $\mathbf{d} = -\mathbf{P}\nabla J(\mathbf{u}^k)$ is an improving direction of J at $\mathbf{u}^k$. Furthermore, if $\mathbf{A}_1$ has full rank and if $\mathbf{P}$ is of the form $\mathbf{P} = \mathbf{I} - \mathbf{A}_1^T(\mathbf{A}_1\mathbf{A}_1^T)^{-1}\mathbf{A}_1$, then $\mathbf{P}$ is an improving feasible direction.

The proof is straightforward:

$$\nabla J(\mathbf{u})^T\mathbf{d} = -\nabla J(\mathbf{u})^T\mathbf{P}\nabla J(\mathbf{u}) = -\nabla J(\mathbf{u})^T\mathbf{P}^T\mathbf{P}\nabla J(\mathbf{u}) = -\left|\mathbf{P}\nabla J(\mathbf{u})\right|^2 < 0$$

That is, $\mathbf{d}$ is an improving direction. Moreover, $\mathbf{A}_1\mathbf{d} = -\mathbf{A}_1\mathbf{P}\nabla J(\mathbf{u}^k) = -\mathbf{A}_1(\mathbf{I} - \mathbf{A}_1^T(\mathbf{A}_1\mathbf{A}_1^T)^{-1}\mathbf{A}_1)\nabla J(\mathbf{u}^k) = 0$ that is, $\mathbf{A}_1\mathbf{d} = 0$ showing that $\mathbf{d}$ is a feasible direction.

Once a non null improving feasible direction $\mathbf{d}$ has been found, function $J(\mathbf{u})$ is minimized along $\mathbf{d}$. This can be done by making:

$$\mathbf{u}^{k+1} = \mathbf{u}^k + \lambda_k\mathbf{d}$$

The value of λ_k is computed as $\lambda_k = \min(\lambda_{opt}, \lambda_{max})$, where λ_{opt} is the value of λ which minimizes $J(\mathbf{u}^k + \lambda\mathbf{d})$ and λ_{max} is the maximum value of λ such that $\mathbf{A}_2(\mathbf{u}^k + \lambda\mathbf{d}) \leq \mathbf{a}_2$.

Because $J(\mathbf{u})$ is a quadratic function, these values can easily be computed:

$$\begin{aligned} J(\mathbf{u}^k + \lambda\mathbf{d}) &= \frac{1}{2}(\mathbf{u}^k + \lambda\mathbf{d})^T\mathbf{H}(\mathbf{u}^k + \lambda\mathbf{d}) + \mathbf{b}^T(\mathbf{u}^k + \lambda\mathbf{d}) + \mathbf{f}_0 = \\ &= \frac{1}{2}\mathbf{d}^T\mathbf{H}\mathbf{d}\lambda^2 + (\mathbf{d}^T\mathbf{H}\mathbf{u}^k + \mathbf{b}^T\mathbf{d})\lambda + J(\mathbf{u}^k) \end{aligned}$$

The optimum can be found for:

$$\lambda_{opt} = -\frac{\mathbf{d}^T\mathbf{H}\mathbf{u}^k + \mathbf{b}^T\mathbf{d}}{\mathbf{d}^T\mathbf{H}\mathbf{d}} \tag{7.2}$$

The value of λ_{max} can be found as the minimum value of $\frac{\mathbf{c}_j - \mathbf{a}_j^T\mathbf{u}^k}{\mathbf{a}_j^T\mathbf{d}}$ for all j such that $\mathbf{a}_j^T$ and $\mathbf{c}_j$ are the rows of the inactive constraint set and respective bound, and such that $\mathbf{a}_j^T\mathbf{d} > 0$.

The algorithm can be summarized as follows:

1. If the active constraint set is empty then let $\mathbf{P} = \mathbf{I}$ otherwise let $\mathbf{P} = \mathbf{I} - \mathbf{A}_1^T(\mathbf{A}_1\mathbf{A}_1^T)^{-1}\mathbf{A}_1$, where $\mathbf{A}_1$ corresponds to the matrix formed by the rows of $\mathbf{A}$ corresponding to active constraints.

2. Let $\mathbf{d}_k = -\mathbf{P}(\mathbf{H}\mathbf{u}^k + \mathbf{b})$

3. If $\mathbf{d}_k \neq 0$ go to step 4 else:

   3.1 If the active constraint set is empty then STOP else:

   3.1.1 Let $\mathbf{w} = -(\mathbf{A}_1\mathbf{A}_1^T)^{-1}\mathbf{A}_1(\mathbf{H}\mathbf{u}^k + \mathbf{b})$

   3.1.2 if $\mathbf{w} \geq 0$ STOP, otherwise choose a negative component of $\mathbf{w}$, say w_j and remove the corresponding constraint from the active set. That is, remove row j from $\mathbf{A}_1$. Go to step 1.

4. Let $\lambda_k = \min(\lambda_{opt}, \lambda max)$ and $\mathbf{u}^{k+1} = \mathbf{u}^k + \lambda_k\mathbf{d}_k$. Replace k by $k+1$ and go to step 1.

7.3.3 Initial Feasible Point

Some of the QP algorithms discussed above start from a feasible point. If the bounds on the process only affect the control signals, a feasible point can very easily be obtained by making the control signal $u(k+j) = u(k-1)$, (supposing that $u(k-1)$ is in the feasible region and that $\underline{u} \leq 0 \leq \overline{u}$). This however may not be a good starting point and may reflect in the efficiency of the optimization algorithm. A better way of obtaining a starting solution could be by using the feasible solution found in the previous iteration shifted one sampling time and adding a last term equal to zero. That is, if $\mathbf{u}^{k-1} = [\Delta u(k-1), \Delta u(k), \cdots, \Delta u(k+n-2), \Delta u(k+n-1)]$, the initial solution is made equal to $[\Delta u(k), \Delta u(k+1), \cdots, \Delta u(k+n-1), 0]$.

If the reference has changed at instant k, this may not be a good starting point and a better solution may be found by computing the unconstrained solution and clipping it to fit the constraints.

If more complex constraints are present, such as output constraints, the problem of finding an initial feasible solution cannot be solved as previously described because just clipping the input signal may not work and a procedure to find an interior point of a polytope has to be used. One of the simplest way of finding an initial solution is by using the following algorithm:

1. Fix any initial point $\mathbf{u}^0$.
2. Let $\mathbf{r} = \mathbf{R}\mathbf{u}^0 - \mathbf{c}$
3. If $\mathbf{r} \leq 0$ STOP. ($\mathbf{u}^0$ is feasible)
4. $r_{max}=max(\mathbf{r})$
5. Solve the following augmented optimization problem using an active set:

$$\min_{\mathbf{u}'} J'(\mathbf{u}') = \min_{\mathbf{u}'}[0\,0 \;\cdots\; 0\,1]\mathbf{u}'$$

$$\mathbf{R}'\mathbf{u}' \leq \mathbf{c}$$

with:

$$\mathbf{u}' = \begin{bmatrix} \mathbf{u} \\ z \end{bmatrix}, \qquad \mathbf{R}' = [\mathbf{R} \;\; -1],$$

and the starting point

$$\mathbf{u}' = \begin{bmatrix} \mathbf{u}^0 \\ r_{max} \end{bmatrix}.$$

Notice that this starting point is feasible for the augmented problem.

6. If $J(\mathbf{u}'_{opt}) \leq 0$ a feasible solution has been found for the original problem, otherwise the original problem is unfeasible.

7.3.4 Pivoting Methods

Pivoting methods such as the Simplex have been widely used in linear programming because these algorithms are simple to program and also because they finish in a finite number of steps finding the optimum or indicating that no feasible solution exists. The minimization of a quadratic function subject to a set of linear constraints can be solved by pivoting methods and they can be applied to MPC as shown by Camacho [16]. One of the most popular pivoting algorithms is based on reducing the QP problem to a linear complementary problem.

The Linear Complementary Problem

Let $\mathbf{q}$ and $\mathbf{M}$ be a given m vector and a given $m \times m$ matrix respectively. The linear complementary problem (LCP) consists of finding two m vectors $\mathbf{s}$ and $\mathbf{z}$ so that:

$$\mathbf{s} - \mathbf{M}\mathbf{z} = \mathbf{q}, \qquad \mathbf{s}, \mathbf{z} \geq 0, \qquad < \mathbf{s}, \mathbf{z} >= 0 \tag{7.3}$$

A solution $(\mathbf{s}, \mathbf{z})$ to the above system is called a complementary basic feasible solution if for each pair of the complementary variables (s_i, z_i) one of them is basic for $i = 1, \cdots, m$, where s_i and z_i are the i-entries of vectors $\mathbf{s}$ and $\mathbf{z}$ respectively.

If $\mathbf{q}$ is non negative, a complementary feasible basic solution can be found by making $\mathbf{s} = \mathbf{q}$ and $\mathbf{z} = \mathbf{0}$. If this is not the case Lemke's algorithm [65] can be used. In this algorithm an artificial variable z_0 is introduced leading to:

$$\mathbf{s} - \mathbf{M}\mathbf{z} - \mathbf{1}\, z_0 = \mathbf{q}, \qquad \mathbf{s}, \mathbf{z}, z_0 \geq 0, \qquad < \mathbf{s}, \mathbf{z} >= 0 \tag{7.4}$$

We obtain a starting solution to the above system by making $z_0 = max(-q_i)$, $\mathbf{z} = 0$ and $\mathbf{s} = \mathbf{q} + \mathbf{1}\, z_0$. If by a sequence of pivoting, compatible with the above system, the artificial variable z_0 is driven to zero a solution to the linear complementary problem is obtained. An efficient way of finding a sequence of pivoting that converges to a solution in a finite number of steps under some mild assumptions on matrix $\mathbf{M}$ is by using Lemke's algorithm from the following tableau:

	$\mathbf{s}$	$\mathbf{z}$	z_0	
$\mathbf{s}$	$\mathbf{I}_{m \times m}$	$-\mathbf{M}$	$-\mathbf{1}$	$\mathbf{q}$

Other advantages of using Lemke's algorithm are that the solution can be traced out as the parameter $z_0 \downarrow 0$, no special techniques are required to resolve degeneracy and when applied to LCP generated by QP problems, the unconstrained solution of the QP problem can be used as the starting point, as will be shown later.

Transforming the GPC into an LCP

The constrained GPC problem can be transformed into an LCP problem as follows:

Make $\mathbf{u} = \mathbf{1}\,\underline{u} + \mathbf{x}$. Constraints can then be expressed in condensed form as:

$$\mathbf{x} \geq 0 \qquad (7.5)$$
$$R\,\mathbf{x} \leq \mathbf{c}$$

with:

$$R = \begin{bmatrix} I_{N\times N} \\ T \\ -T \\ \mathbf{G} \\ -\mathbf{G} \end{bmatrix} \qquad \mathbf{c} = \begin{bmatrix} \mathbf{l}\,(\overline{u} - \underline{u}) \\ \mathbf{l}\,\overline{u} - T\,\mathbf{l}\,\underline{u} - u(t-1)\mathbf{l} \\ -\mathbf{l}\,\underline{u} + T\,\mathbf{l}\,\underline{u} + u(t-1)\mathbf{l} \\ \mathbf{l}\,\overline{y} - \mathbf{f} - \mathbf{G}\,\mathbf{l}\,\underline{u} \\ -\mathbf{l}\,\underline{y} + \mathbf{f} + \mathbf{G}\,\mathbf{l}\,\underline{u} \end{bmatrix} \qquad (7.6)$$

Equation (7.1) can be rewritten as:

$$J \;=\; \frac{1}{2}\mathbf{x}^T\,\mathbf{H}\,\mathbf{x} + \mathbf{a}\,\mathbf{x} + \mathbf{f}_1 \qquad (7.7)$$

$$(7.8)$$

where: $\mathbf{a} = \mathbf{b} + \underline{u}\,\mathbf{l}^T\,\mathbf{H}$ and $\mathbf{f}_1 = \mathbf{f}_0 + \underline{u}^2\,\mathbf{l}^T\,\mathbf{H}\,\mathbf{l} + \mathbf{b}\,\underline{u}$

Denoting the Lagragian multiplier vectors of the constraints $\mathbf{x} \geq 0$ and $R\mathbf{x} \leq \mathbf{c}$ by $\mathbf{v}$ and $\mathbf{v}_1$, respectively, and denoting the vector of slack variables by $\mathbf{v}_2$, the Kuhn-Tucker conditions [10] can be written as

$$\begin{aligned} R\,\mathbf{x} + \mathbf{v}_2 &= \mathbf{c} \\ -\mathbf{H}\,\mathbf{x} - R^T\mathbf{v} + \mathbf{v}_1 &= \mathbf{a} \\ \mathbf{x}^T\mathbf{v}_1 = 0,\ \mathbf{v}^T\mathbf{v}_2 &= 0 \\ \mathbf{x}, \mathbf{v}, \mathbf{v}_1, \mathbf{v}_2 &\geq 0 \end{aligned} \qquad (7.9)$$

These expressions can be rewritten as:

$$\begin{bmatrix} \mathbf{I}_{m\times m} & \mathbf{0}_{m\times N} & \mathbf{0}_{m\times m} & R \\ \mathbf{0}_{N\times m} & \mathbf{I}_{N\times N} & -R^T & -\mathbf{H} \end{bmatrix} \begin{bmatrix} \mathbf{v}_2 \\ \mathbf{v}_1 \\ \mathbf{v} \\ \mathbf{x} \end{bmatrix} = \begin{bmatrix} \mathbf{c} \\ \mathbf{a} \end{bmatrix} \qquad (7.10)$$

The Kuhn-Tucker conditions can be expressed as a linear complementary problem $\mathbf{s} - \mathbf{M}\mathbf{z} = q$, $\mathbf{s}^T\mathbf{z} = 0$, $\mathbf{s}, \mathbf{z} \geq 0$ with:

$$\mathbf{M} = \begin{bmatrix} 0 & -R \\ R^T & \mathbf{H} \end{bmatrix}, \quad \mathbf{q} = \begin{bmatrix} \mathbf{c} \\ \mathbf{a} \end{bmatrix}, \quad \mathbf{s} = \begin{bmatrix} \mathbf{v}_2 \\ \mathbf{v}_1 \end{bmatrix}, \quad \mathbf{z} = \begin{bmatrix} \mathbf{v} \\ \mathbf{x} \end{bmatrix} \qquad (7.11)$$

This problem can be solved using Lemke's algorithm by forming the following tableau

	$\mathbf{v}_2$	$\mathbf{v}_1$	$\mathbf{v}$	$\mathbf{x}$	z_0	
$\mathbf{v}_2$	$\mathbf{I}_{m\times m}$	$\mathbf{0}_{m\times N}$	$\mathbf{0}_{m\times m}$	R	$-\mathbf{l}$	$\mathbf{c}$
$\mathbf{v}_1$	$\mathbf{0}_{N\times m}$	$\mathbf{I}_{N\times N}$	$-R^T$	$-\mathbf{H}$	$-\mathbf{l}$	$\mathbf{a}$

Although the algorithm will converge to the optimum solution in a finite number of steps as matrix $\mathbf{H}$ is positive definite [10], it needs a substantial amount of computation. One of the reasons for this is that the $\mathbf{x}$ variables in the starting solution of Lemke's algorithm are not part of the basis. That is, the algorithm starts from the solution $\mathbf{x} = 0$ which may be far away from the optimum solution. The efficiency of the algorithm can be increased by finding a better starting point.

If equation (7.10) is multiplied by: $\begin{bmatrix} \mathbf{I}_{m\times m} & R\,\mathbf{H}^{-1} \\ \mathbf{0}_{N\times m} & -\mathbf{H}^{-1} \end{bmatrix}$ we have:

$$\begin{bmatrix} \mathbf{I}_{m\times m} & R\mathbf{H}^{-1} & -R\mathbf{H}^{-1}R^T & \mathbf{0}_{m\times N} \\ \mathbf{0}_{N\times m} & -\mathbf{H}^{-1} & \mathbf{H}^{-1}R^T & \mathbf{I}_{N\times N} \end{bmatrix} \begin{bmatrix} \mathbf{v}_2 \\ \mathbf{v}_1 \\ \mathbf{v} \\ \mathbf{x} \end{bmatrix} = \begin{bmatrix} \mathbf{c} + R\mathbf{H}^{-1}\mathbf{a} \\ -\mathbf{H}^{-1}\mathbf{a} \end{bmatrix} \tag{7.12}$$

The vector on the right hand side of equation (7.12) corresponds to the vector of slack variables for the unconstrained solution and to the unconstrained solution respectively. Furthermore, equation (7.12) shows that if Lemke's algorithm is started from this point, all the $\mathbf{x}$ variables are in the basis. In most cases only a few constraints will be violated for the unconstrained solution of the GPC problem. Thus, the constrained solution will be close to the initial condition and the number of iterations required should decrease.

The algorithm can be described as follows:

1. Compute the unconstrained solution $\mathbf{x}_{min} = -\mathbf{H}^{-1}\mathbf{a}$.
2. Compute $\mathbf{v}_{2min} = \mathbf{c} - R\mathbf{x}_{min}$. If $\mathbf{x}_{min}$ and $\mathbf{v}_{2min}$ are non negative then stop with $u(t) = x_1 + \underline{u} + u(t-1)$.
3. Start Lemke's algorithm with $\mathbf{x}$ and $\mathbf{v}_2$ in the basis with the following tableau:

	$\mathbf{v}_2$	$\mathbf{x}$	$\mathbf{v}$	$\mathbf{v}_1$	z_0	
$\mathbf{v}_2$	$\mathbf{I}_{m\times m}$	$\mathbf{0}_{m\times N}$	$R\mathbf{H}^{-1}R^T$	$R\mathbf{H}^{-1}$	-1	$\mathbf{v}_{2min}$
$\mathbf{x}$	$\mathbf{0}_{N\times m}$	$\mathbf{I}_{N\times N}$	$\mathbf{H}^{-1}R^T$	$-\mathbf{H}^{-1}$	-1	$\mathbf{x}_{min}$

4. If x_1 is not in the first column of the tableau, make it zero. Otherwise give it the corresponding value.
5. $u(t) = u(t-1) + \underline{u} + x_1$.

7.4 Constraints Handling

In some cases, depending on the type of constraints acting on the process, some advantages can be obtained from the particular structure of the constraint matrix **R**. This section deals with the way in which this special type of structure can be used to improve the efficiency of the QP algorithms.

7.4.1 Slew Rate Constraints

When only slew rate constraints are taken into account, the constraints can be expressed as:

$$\begin{bmatrix} \mathbf{I} \\ -\mathbf{I} \end{bmatrix} \mathbf{u} \leq \begin{bmatrix} \overline{\mathbf{u}} \\ -\underline{\mathbf{u}} \end{bmatrix}$$

Active Set Methods

It can be seen that the active constraint matrix **A** can be expressed, after appropriate permutations in order to keep the active bounds on the m first variables, as:

$$\mathbf{A}^T = \mathbf{Y} = \begin{bmatrix} \mathbf{I} \\ \mathbf{0} \end{bmatrix} \begin{matrix} m \\ n-m \end{matrix} \qquad \mathbf{Z} = \mathbf{V} = \begin{bmatrix} \mathbf{0} \\ \mathbf{I} \end{bmatrix} \begin{matrix} m \\ n-m \end{matrix}$$

Matrix **H** can be partitioned, after reordering, as:

$$\mathbf{H} \begin{bmatrix} \mathbf{H}_{11}\mathbf{H}_{12} \\ \mathbf{H}_{21}\mathbf{H}_{22} \end{bmatrix}$$

where $\mathbf{H}_{11}$ is an $m \times m$ matrix. The linear system to be solved:

$$\mathbf{Z}^T\mathbf{H}\mathbf{Z}\mathbf{v} = -\mathbf{Z}^T(\mathbf{b} + \mathbf{H}\mathbf{Y}\mathbf{a}) = \mathbf{H}_{22}\mathbf{v} = -\mathbf{b}_2 - \mathbf{H}_{22}\mathbf{a}_2$$

That is, no calculations are needed for the generalized elimination method other than reordering the variables and corresponding matrices.

Rosen's Gradient Projection Method

The active constraint matrix $\mathbf{A}_1$ will have m rows, corresponding to the m values of $\Delta u(k+j)$ which are bounded. Each of the rows of $\mathbf{A}_1$ will have all its elements equal to zero except element j which will be equal to 1 if the bound corresponds to the upper limit or -1 if it is bounded by the lower limit. The product $\mathbf{A}_1\mathbf{A}_1^T$ is then:

$$(\mathbf{A}_1\mathbf{A}_1^T)_{ij} = \sum_{l=1}^{N} a_{il}a_{jl} = \begin{cases} = 1 \text{ if } i = j \text{ and constraint } j \text{ is active} \\ = 0 \text{ otherwise} \end{cases}$$

That is, an $m \times m$ identity matrix. The projection matrix $\mathbf{P}$ is then:

$$\mathbf{P} = \mathbf{I} - \mathbf{A}_1^T \mathbf{A}_1$$

It can then easily be seen [115] that the projection matrix can now be expressed by:

$$p_{ij} = \begin{cases} = 0 \text{ when } i \neq j \text{ or one of the bounds on variable } \triangle u(k+j) \text{ is active} \\ = 1 \text{ when } i = j \text{ and neither bound on variable } \triangle u(k+j) \text{ is active} \end{cases}$$

The search direction is given by:

$$d_i = \begin{cases} = 0 \text{ when one of the bounds on variable } \triangle u(k+j) \text{ is active} \\ = -g_i \text{ when } i = j \text{ and neither bound on variable } \triangle u(k+j) \text{ is active} \end{cases}$$

The value of λ_{max} can easily be found by:

$$\lambda_{max} = min \left[\min_j \left(\frac{\overline{u} - \Delta u(k+j)}{d_j} \middle| d_j > 0 \right), \min_j \left(\frac{\underline{u} - \Delta u(k+j)}{d_j} \middle| d_j < 0 \right) \right]$$

The computation of vector $\mathbf{w}$, which is necessary in order to check the Kuhn-Tucker condition, can be written as $\mathbf{w} = -\mathbf{A}_1\mathbf{g}$. The stopping criterium is also considerably simplified and it can be stated as: for all active constraint j check that $g_j \leq 0$ if j correspond to an upper bound otherwise check that $g_j \geq 0$.

7.4.2 Amplitude Constraints

When the only constraints present are the maximum and minimum value of the control signals $u(k+j)$. The constraints can be expressed as:

$$\mathbf{1}\underline{u} \leq \mathbf{T}\mathbf{u} - \mathbf{1}u(k-1)) \leq \mathbf{1}\overline{u}$$

or

$$\mathbf{1}(\underline{u} - u(k-1)) \leq \mathbf{T}\mathbf{u} \leq \mathbf{1}(\overline{u} - u(k-1))$$

where matrix $\mathbf{T}$ is a lower triangular matrix whose entries are ones and $\mathbf{1}$ is a vector composed of ones the constraint matrix takes the form:

$$\mathbf{R} = \begin{bmatrix} \mathbf{T} \\ -\mathbf{T} \end{bmatrix}$$

Although some advantages can be gained from the particular shape of the constraint matrix, the GPC can be reformulated in order to reduce the case to the much simpler one seen in the previous section.

Recall from chapter 4 that the optimal predictions for the process output can be expressed as: $\mathbf{y} = \mathbf{G}\mathbf{u} + \mathbf{f}$. The vector of future control increments is given by:

$$\mathbf{u} = \begin{bmatrix} u(k) - u(k-1) \\ u(k+1) - u(k) \\ \vdots \\ u(k+N) - u(k+N-1) \end{bmatrix} =$$

$$= \begin{bmatrix} 1 & 0 & 0 & \cdots & 0 \\ -1 & 1 & 0 & \cdots & 0 \\ 0 & -1 & 1 & \cdots & 0 \\ \vdots & & & \ddots & \vdots \\ 0 & 0 & 0 & \cdots & 1 \end{bmatrix} \begin{bmatrix} u(k) \\ u(k+1) \\ \vdots \\ u(k+N) \end{bmatrix} - \begin{bmatrix} u(k-1) \\ 0 \\ \vdots \\ 0 \end{bmatrix} = \mathbf{D}\mathbf{U} - \mathbf{f}_1$$

If this substitution is made in the equation of future predictions we get:

$$\mathbf{y} = \mathbf{G}(\mathbf{D}\mathbf{U} - \mathbf{f}_1) + \mathbf{f}_0 = \mathbf{G}'\mathbf{U} + \mathbf{f}_2$$

where $\mathbf{G}'$ is a lower triangular matrix with all its diagonal elements equal to g_0 and its secondary diagonal elements are given by $g_i - g_{i-1}$. Vector $\mathbf{f}_2$ can be expressed as $(\mathbf{f}_2)_i = (\mathbf{f}_0)_i - g_i u(k-1)$.

The objective function can now be expressed as a function of the future control actions $\mathbf{U}$:

$$J(\mathbf{U}) = (\mathbf{G}'\mathbf{U} + \mathbf{f}_2 - \mathbf{w})^T(\mathbf{G}'\mathbf{U} + \mathbf{f}_2 - \mathbf{w}) + \lambda(\mathbf{D}\mathbf{U} - \mathbf{f}_1)^T(\mathbf{D}\mathbf{U} - \mathbf{f}_1) =$$
$$= \mathbf{U}^T(\mathbf{G}'^T\mathbf{G} + \mathbf{D}^T\mathbf{D})\mathbf{U} + 2[(\mathbf{f}_2 - \mathbf{w})^T\mathbf{G}' - \mathbf{f}_1^T\mathbf{D}]\mathbf{U} + (\mathbf{f}_2 - \mathbf{w})^T(\mathbf{f}_2 - \mathbf{w}) + \mathbf{f}_1^T\mathbf{f}_1$$

That is a quadratic form:

$$J(\mathbf{U}) = \frac{1}{2}\mathbf{U}^T\mathbf{H}'\mathbf{U} + \mathbf{b}' + \mathbf{f}'$$

where

$$\mathbf{H}' = 2(\mathbf{G}'^T\mathbf{G} + \mathbf{D}^T\mathbf{D}),\ \mathbf{b}' = 2[(\mathbf{f}_2 - \mathbf{w})^T\mathbf{G}' - \mathbf{f}_1^T\mathbf{D}],\ \mathbf{f}' = (\mathbf{f}_2 - \mathbf{w})^T(\mathbf{f}_2 - \mathbf{w}) + \mathbf{f}_1^T\mathbf{f}_1$$

Notice that $\mathbf{f}_1^T\mathbf{f}_1 = u(k-1)^2$ and that $\mathbf{D}^T\mathbf{D}$ is a tridiagonal matrix with the elements of the main diagonal equal to 2 and the elements of the other two accompanying subdiagonals equal to -1.

The problem has been reduced to optimizing a quadratic form with the constraint matrix $\mathbf{R} = [\mathbf{I}\ -\mathbf{I}]^T$ and the efficiency of the optimization procedure can be increased as shown in the previous section.

7.4.3 Output Constraints

When the only constraints present are the maximum and minimum value of the output signals $y(k+j)$. The constraints can be expressed as:

$$\mathbf{y}_{min} \leq \mathbf{Gu} + \mathbf{f} \leq \mathbf{y}_{max}$$

Which can also be expressed as:

$$\begin{bmatrix} \mathbf{G} \\ -\mathbf{G} \end{bmatrix} \mathbf{u} \leq \begin{bmatrix} \mathbf{y}_{max} - \mathbf{f} \\ \mathbf{y}_{min} - \mathbf{f} \end{bmatrix}$$

Notice that all the blocks of the constraint matrix are lower triangular blocks and some advantages may be gained from that as will be shown in the following.

7.4.4 Constraints Reduction

The computational requirements of the QP algorithms depend heavily on the number of constraints considered. Only those constraints which limit the feasible region of the space need to be taken into account. The efficiency of the algorithms can be increased if the superfluous constraints, that is, those constraints not limiting the feasible region, are eliminated. There are a number of algorithms for determining the minimum set of limiting constraints, or what is the same, for determining the convex hull, or polytope, corresponding to the feasible region of space. Although the elimination of all superfluous constraints may reduce the amount of computation needed, the procedure itself requires a substantial amount of computation. The fact that in this case the constraint matrices are lower triangular can be used to detect non limiting constraints. Some constraints can easily be eliminated as follows.

By making $\mathbf{u} = \mathbf{1}\underline{u} + \mathbf{x}$, constraints on the slew rate and amplitude of actuators and on the process output signals can be expressed as:

$$\begin{aligned} 0 \leq \mathbf{x} &\leq c_1 \\ c_3 \leq T\mathbf{x} &\leq c_2 \\ \frac{c_5}{g_0} \leq \frac{\mathbf{G}}{g_0}\mathbf{x} &\leq \frac{c_4}{g_0} \end{aligned} \tag{7.13}$$

with $\mathbf{c}^T = [c_1^T \, c_2^T \, c_3^T \, c_4^T \, c_5^T]$

Now consider the first row for each of the constraints (7.13), that is, the constraint affecting only x_1:

$$\begin{aligned} x_1 \leq c_{11}, \quad x_1 \leq c_{21}, \quad x_1 \leq \frac{c_{41}}{g_0} \\ x_1 \geq 0, \quad x_1 \geq c_{31}, \quad x_1 \geq \frac{c_{51}}{g_0} \end{aligned} \tag{7.14}$$

where c_{ij} is the j entry of vector c_i.

Notice that of the first three constraints (7.14) only that with a smaller right hand side has to be kept whilst the other two can be eliminated because they do not limit the feasible region. The same applies to the last three constraints. Thus four constraints can be eliminated in this first step. The x_1 variable will be bounded by $l_1 \leq x_1 \leq r_1$, where r_1 is the smallest of all right hand side terms of the first row of constraints (7.14) and l_1 is the biggest of all right hand side terms of the second row.

Let us now consider the constraints (7.13) limiting x_1 and x_2; that is, the second row of each constraint block (7.13). These constraints can be written as:

$$\begin{array}{lll} x_2 \leq c_{12}, & x_2 \leq c_{22} - x_1, & x_2 \leq \dfrac{c_{42}}{g_0} - \dfrac{g_1}{g_0} x_1 \\ x_2 \geq 0, & x_2 \geq c_{32} - x_1, & x_2 \geq \dfrac{c_{52}}{g_0} - \dfrac{g_1}{g_0} x_1 \end{array} \tag{7.15}$$

The right hand sides of the above constraints depend, in general, on x_1. As x_1 is bounded by $l_1 \leq x_1 \leq r_1$, the right hand side of each of the constraints (7.15) will be bounded by two limits. Consider for example the second of the first row of constraints (7.15). The minimum of the right hand side of this inequality is given by $min(c_{22} - x_1) = c_{22} - max(x_1) = c_{22} - r_1$. The same considerations can be applied to the maximum. The right hand side will therefore be limited by $r_{22min} = c_{22} - r_1$ and $r_{22max} = c_{22} - l_1$. Where the first subindex in r_{kij} refers to variable x_k, the second refers to constraint i and the last subindex indicates whether it is the minimum or maximum limit. Notice that r_{23min} and r_{23max} can be computed in a similar manner and that $r_{k1min} = r_{k1max}$.

A right boundary for x_2 can be defined by $r_2 = min(r_{2jmax})$ for $j = 1, 2, 3$. Notice that variable x_2 must always be smaller than r_2, thus any constraint j having $r_2 < r_{2jmin}$ can be eliminated as it will not be limiting the feasible region.

A left boundary l_2 can be obtained for variable x_2 from the last three constraints (7.15). For each of these constraints a minimum l_{2jmin} and maximum l_{2jmax} limit can be found in the same way. A left boundary for variable x_2 can now be given by $l_2 = max(l_{2jmin})$ for $j = 1, 2, 3$. Constraint j can now be eliminated if $l_2 > l_{2jmax}$.

After this step, variables x_1 and x_2 will be bound by (l_1, r_1) and (l_2, r_2) respectively. As the constraint matrices are lower triangular, the same procedure can be applied to obtain the boundaries (and eliminate the superfluous restrictions) for x_3 and then, recursively for the remaining variables.

Notice that constraints of the type $x_k \geq 0$ do not appear in the constraint matrix R in the algorithm described above. As the algorithm considers all variables to be positive, these constraints are implicitly taken into account. If any left bound l_i is positive, the constraint $x_i \geq 0$ can be eliminated. In order to do this the following substitution can be made: $x_j = l(x_1, x_2, \cdots, x_{j-1}) + z_j$, where $l(x_1, x_2, \cdots, x_{j-1})$ is the right hand side of one of the remaining constraints of type $x_i \geq l(x_1, x_2, \cdots, x_{j-1})$. This constraint can now be substituted by $z_j \geq 0$. The constraint matrices have to be changed accordingly.

Notice that the procedure described does not guarantee a minimum number of constraints, further reductions could be achieved but it would require more computation and a more complex algorithm.

7.5 1-norm

Although quadratic programming algorithms are very efficient, the MPC problem can be solved by a much more efficient linear programming method if a 1-norm type of function is used.

The objective function is now

$$J(\mathbf{u}) = \sum_{j=N_1}^{N_2} \sum_{i=1}^{n} |y_i(t+j) - w_i(t+j)| + \lambda \sum_{j=1}^{N_u} \sum_{i=1}^{m} |\Delta u_i(t+j-1)| \quad (7.16)$$

where N_1 and N_2 define the costing horizon and N_u defines the control horizon. The absolute values of the output tracking error and the absolute values of the control increments are taken instead of the square of them, as is usual in GPC.

If a series of $\mu_i \geq 0$ and $\beta_i \geq 0$ such that

$$\begin{array}{rcccll}
-\mu_i & \leq & (y_i(t+j) - w_i(t+j)) & \leq & \mu_i & i = 1, \ldots n \quad j = 1, \ldots N \\
-\beta_i & \leq & \Delta u_i(t+j-1) & \leq & \beta_i & i = 1, \ldots m \quad j = 1, \ldots N_u \\
0 & \leq & \sum_{i=1}^{n \times N} \mu_i + \lambda \sum_{i=1}^{m \times N_u} \beta_i & \leq & \gamma &
\end{array}$$

then γ is an upper bound of $J(\mathbf{u})$. The problem is now reduced to minimizing the upper bound γ.

When constraints on the output variables $(\underline{\mathbf{y}}, \overline{\mathbf{y}})$, the manipulated variables $(\underline{\mathbf{U}}, \overline{\mathbf{U}})$ and on the slew rate of the manipulated variables $(\underline{\mathbf{u}}, \overline{\mathbf{u}})$ are taken into account, the problem can be interpreted as an LP problem with:

$$
\begin{array}{rcl}
\multicolumn{3}{l}{\min_{\gamma,\mu,\beta,\mathbf{u}} \gamma} \\
\multicolumn{3}{l}{\text{subject to}} \\
\mu & \geq & G\mathbf{u}+\mathbf{f}-\mathbf{w} \\
\mu & \geq & -G\mathbf{u}-\mathbf{f}+\mathbf{w} \\
\overline{\mathbf{y}} & \geq & G\mathbf{u}+\mathbf{f} \\
-\underline{\mathbf{y}} & \geq & -G\mathbf{u}-\mathbf{f} \\
\beta & \geq & \mathbf{u} \\
\beta & \geq & -\mathbf{u} \\
\overline{\mathbf{u}} & \geq & \mathbf{u} \\
-\underline{\mathbf{u}} & \geq & \mathbf{u} \\
\overline{\mathbf{U}} & \geq & T\mathbf{u}+\mathbf{1}u(t-1) \\
-\underline{\mathbf{U}} & \geq & -T\mathbf{u}-\mathbf{1}u(t-1) \\
\gamma & \geq & \mathbf{1}^T\mu+\lambda\mathbf{1}\beta
\end{array}
$$

The problem can be transformed into the form:

$$\min_{\mathbf{x}} \mathbf{c}^T\mathbf{x}, \quad \text{subject to} \quad A\mathbf{x} \leq \mathbf{b}, \quad \mathbf{x} \geq 0$$

with :

$$
\mathbf{x} = \left[\begin{array}{c} \mathbf{u}-\underline{\mathbf{u}} \\ \hline \mu \\ \hline \beta \\ \hline \gamma \end{array}\right] \qquad
\mathbf{c} = \left[\begin{array}{c} \mathbf{0} \\ \hline \mathbf{0} \\ \hline \mathbf{0} \\ \hline \mathbf{1} \end{array}\right]
$$

$$
A = \left[\begin{array}{c|c|c|c}
\mathbf{G} & -I & \mathbf{0} & \mathbf{0} \\
-\mathbf{G} & -I & \mathbf{0} & \mathbf{0} \\ \hline
\mathbf{G} & \mathbf{0} & \mathbf{0} & \mathbf{0} \\
-\mathbf{G} & \mathbf{0} & \mathbf{0} & \mathbf{0} \\ \hline
I & \mathbf{0} & -I & \mathbf{0} \\
-I & \mathbf{0} & -I & \mathbf{0} \\ \hline
I & \mathbf{0} & \mathbf{0} & \mathbf{0} \\ \hline
T & \mathbf{0} & \mathbf{0} & \mathbf{0} \\
-T & \mathbf{0} & \mathbf{0} & \mathbf{0} \\ \hline
\mathbf{0} & \mathbf{1}^T & \mathbf{1}^T\lambda & -\mathbf{1}
\end{array}\right] \qquad
\mathbf{b}_i = \left[\begin{array}{c}
-\mathbf{G}\underline{\mathbf{u}}-\mathbf{f}+\mathbf{w} \\
\mathbf{G}\underline{\mathbf{u}}+\mathbf{f}-\mathbf{w} \\ \hline
\overline{\mathbf{y}}-\mathbf{G}\underline{\mathbf{u}}-\mathbf{f} \\
-\underline{\mathbf{y}}+\mathbf{G}\underline{\mathbf{u}}+\mathbf{f} \\ \hline
-\underline{\mathbf{u}} \\
\underline{\mathbf{u}} \\ \hline
\overline{\mathbf{u}}-\underline{\mathbf{u}} \\ \hline
\overline{\mathbf{U}}-T\underline{\mathbf{u}}-\mathbf{1}u(t-1) \\
-\underline{\mathbf{U}}+T\underline{\mathbf{u}}+\mathbf{1}u(t-1) \\ \hline
\mathbf{0}
\end{array}\right]
$$

The number of variables involved in the linear programming problem is $2 \times m \times N_u + n \times N + 1$, while the number of constraints is $4 \times n \times N + 5 \times m \times N_u + 1$. For a process with 5 input and 5 output variables with control horizon $N_u = 10$ and costing horizon $N = 30$, the number of variables for the LP problem is 251 while the number of constraints would be 851, which can be solved by any LP algorithm. As the number of constraints is higher than the number of decision variables, solving the dual LP problem should also be less computationally expensive. The number of constraints can be reduced because of the special form of the constraint matrix A by applying the constraint reduction algorithm.

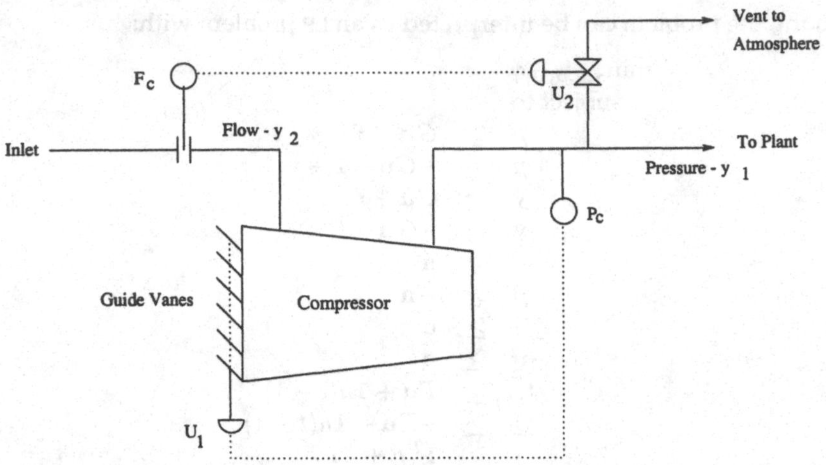

Figure 7.8: Compressor

7.6 Case study : a Compressor

Compressed air is used in most industrial plants for different purposes. Air compressors supplying compressed air to the different processes of a plant can frequently be found in industry. The example studied in this section corresponds to a large air compressor, see figure 7.8, supplying air to a plant. The outlet pressure is controlled by manipulating the guide vanes of the compressor. A blow-off valve is installed to prevent a surge. When the blow-off valve is closed, the compressor is a single input single output process that can be controlled appropriately by using standard control techniques. When the blow off valve opens, the compressor is a multivariable process with two inputs and two outputs. The manipulated variables are the guide vane angle (u_1) and the position of the valve (u_2) and the controlled variables are the air pressure (y_1) and the air flow rate (y_2). The process model is given by the following transfer matrix [24]:

$$\begin{bmatrix} Y_1(s) \\ Y_2(s) \end{bmatrix} = \begin{bmatrix} \dfrac{0.1133e^{-0.715s}}{1+4.48s+1.783s^2} & \dfrac{0.9222}{1+2.071s} \\ \dfrac{0.3378e^{-0.299s}}{1+1.09s+0.361s^2} & \dfrac{-0.321e^{-0.94s}}{1+2.463s+0.104s^2} \end{bmatrix} \begin{bmatrix} U_1(s) \\ U_2(s) \end{bmatrix}$$

The compressor can be controlled as shown in [24], by decoupling the process at zero frequency and using a PI controller for the first loop and a proportional controller for the second. These controllers were obtained with the help of the Inverse Nyquist Array (INA). The simulated closed loop responses of the process to successive step changes in both references are

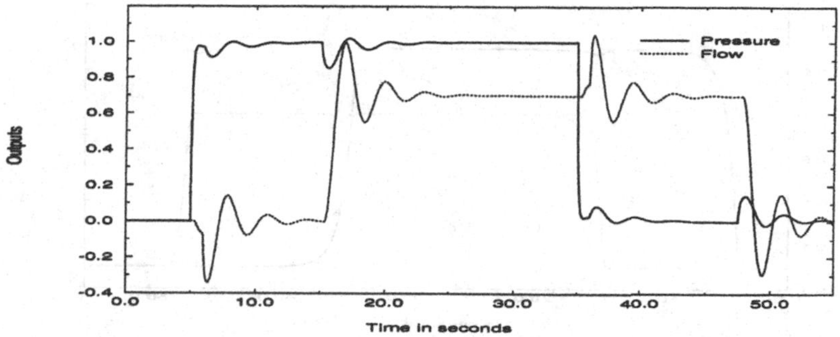

(a)

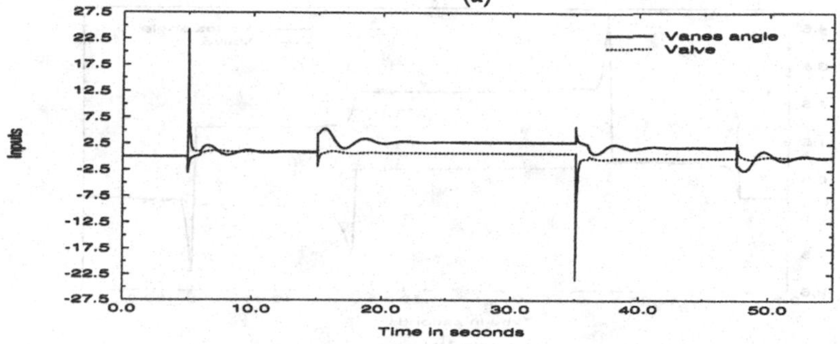

(b)

Figure 7.9: Closed loop responses of compressor: INA controller

shown in figure 7.9a. It can be seen that the responses are quite oscillatory. The evolution of the manipulated variables can be seen in figure 7.9b. As can be seen, the valve position exhibits high peaks for each change in the pressure setpoint.

A sampling time of 0.05 is chosen. The process can be approximated by the following discrete transfer matrix:

$$G(z^{-1}) = \begin{bmatrix} \frac{10^{-4}(0.7619z^{-1}+0.7307z^{-2})}{1-1.8806z^{-1}+0.8819z^{-2}}z^{-14} & \frac{0.022z^{-1}}{1-0.9761z^{-1}} \\ \frac{10^{-2}(0.1112z^{-1}+0.1057z^{-2})}{1-1.8534z^{-1}+0.8598z^{-2}}z^{-6} & \frac{10^{-2}(-0.2692z^{-1}-0.1821z^{-2})}{1-1.2919z^{-1}+0.306z^{-2}}z^{-19} \end{bmatrix}$$

The process can be controlled with a multivariable GPC with the following design parameters: $N_1 = 20$, $N_2 = 23$, $N_3 = 3$ and $\lambda = 0.8$. The beginning of the costing horizon has been chosen as the maximum of the deadtimes. The behaviour of the process can be seen in figure 7.10a. As can be seen, both controlled variables reach their set point rapidly and without oscillations. The perturbations caused in each of the controlled variables by a step change in the reference of the other variable are very small.

The evolution of the manipulated variables are shown in figure 7.10b. As

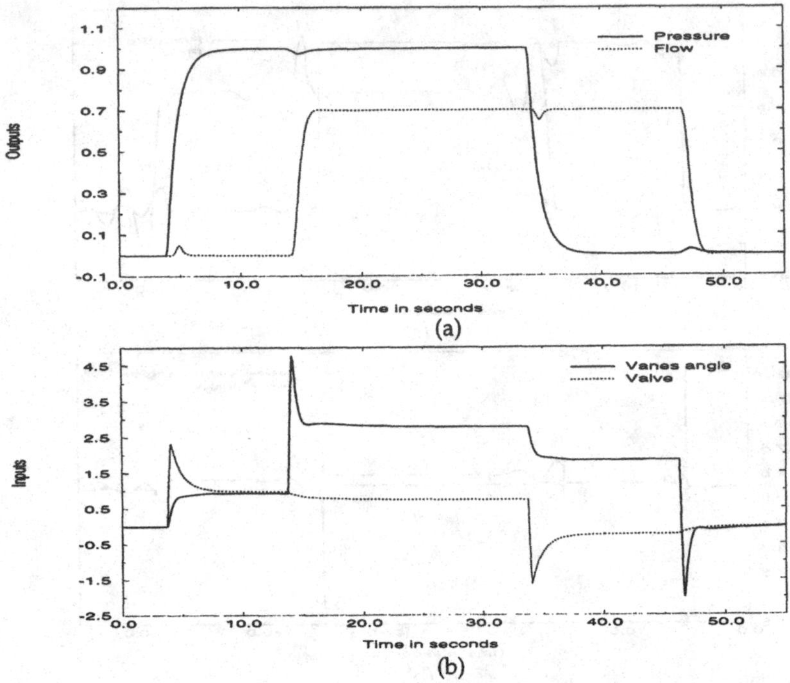

Figure 7.10: Closed loop responses of compressor: unconstrained GPC

can be seen, a high peak in the valve position is observed (although much smaller than the peaks observed when controlling the compressor with the INA controller).

To reduce the manipulated variable peak, a constrained GPC can be used. The manipulated variables are restricted to being in the interval $[-2.75, 2.75]$. The evolution of the controlled variables are shown in figure 7.11a. As can be seen, the response of the pressure to a step change in the reference is a bit slower than in the unconstrained case, but the manipulated variable is kept within the desired limits as shown in figure 7.11b.

7.7 Constraint Management

7.7.1 Feasibility

Sometimes, during the optimization stage, the region defined in the decision variables by the set of constraints is empty. In these conditions, the optimization algorithm cannot find any solution and the optimization problem is said to be infeasible. Unobtainable control objectives and/or perturbations that take the process away from the operating point may cause infeasibility. An

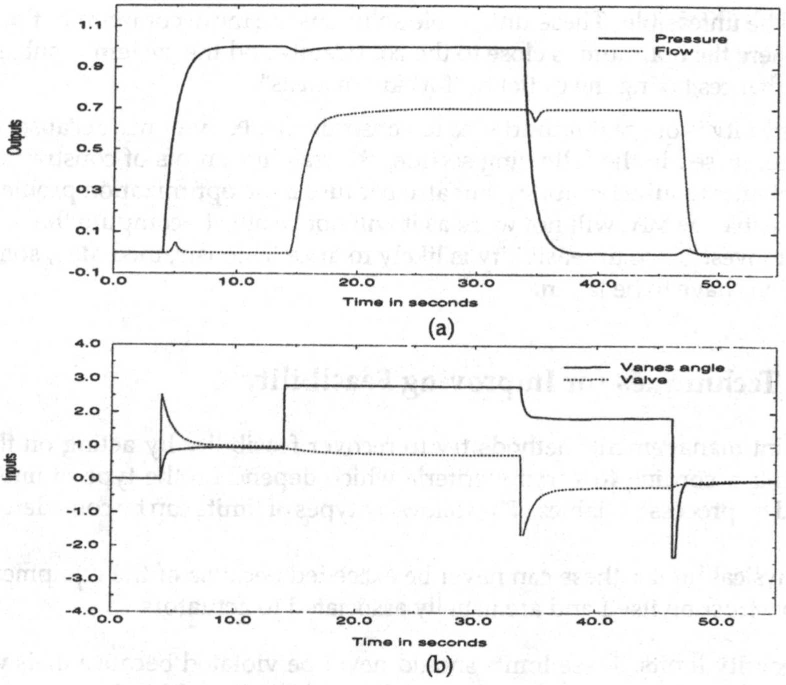

Figure 7.11: Closed loop responses of compressor: constrained GPC

optimization problem is feasible when the objective function is bounded and there are points in the space of decision variables that satisfy all constraints.

Infeasibility may appear in the steady state regime and during transients. The infeasibility problems of steady state regime are usually caused by unobtainable control objectives. This occurs, for example when the setpoints cannot be reached because the manipulated variables are constrained. In general, if the manipulated variables are constrained to be in a hipercube, the reachable set points are in a polytope in the controlled variable space whose vertices are defined by multiplying the vertex of the hipercube by the process DC gain matrix. These unfeasibilities can be easily handled during the design phase by eliminating these types of objectives.

Infeasibility can appear in the transitory regime, even when the imposed constraints seem to be reasonable. Constraints that do not cause problems in normal operation can cause problems under certain circumstances. A perturbation or large reference change may force a variable outside its limits so that it is impossible to introduce it into its permitted zone again using limited energy control signals. In these circumstances the constraints become temporarily incompatible. Infeasibility may also be produced when the operator redefines the operational variable limits while the process is in operation, as mentioned in [4]. If the variables are already outside the new limits, the problem will be unfeasible. These unfeasible solutions are more common in those cases where the optimum is close to the constraints and the system is subject to disturbances,taking the outlet to "forbidden areas".

Feasibility is of great importance to constrained MPC not only because, as will be discussed in the following section, the stability proofs of constrained MPC strategies require feasibility, but also because if the optimization problem is not feasible the MPC will not work as it will not be able to compute the next control moves. Since unfeasibility is likely to arise in constrained MPC, some precautions have to be taken.

7.7.2 Techniques for Improving Feasibility

Constraint management methods try to recover feasibility by acting on the constraints according to varying criteria which depend on the type of limits imposed on process variables. The following types of limits can be considered:

- Physical limits: these can never be exceeded because of the equipment construction itself and are usually associated to actuators.
- Security limits: these limits should never be violated because their violation could be dangerous to process security or could induce a costly shut down of the process by the emergency equipment. These limits are usually associated to process controlled variables.
- Operational limits:these are fixed by the operators as bounds that should not be exceeded by the process variables in order to maintain appropriate operating conditions. They can be exceeded in certain circumstances.
- Real limits:these are the ones used by the control algorithm at each instance. They are provided by the constraints manager who should calculate them in such a way that they never exceed the physical limits.

Possible solutions to this problem can be classified into the following groups:

1. **Disconnection of the controller:** The easiest way of solving this type of problem is to pass the controller to a back up value or to a back up controller when constraint incompatibilities arise and return to automatic operation when the admissibility of the solution is recovered. As can be understood, this method has serious disadvantages. Normally, when constraint incompatibility problems arise it is because the closed loop system is at a critical stage where the operator usually has very little experience in operating. Furthermore, if the constraints have to do with safety or economic aspects then any decisions taken when constraint compatibility problems arise are usually critical because in these cases

some of the control objectives cannot be satisfied. This method is usually used when the constraint incompatibility problems are not frequent.

2. **Constraint Elimination:** Feasibility is analysed at each sampling period and thus the elimination of constraints is done temporarily. The feasibility is checked periodically in order to be able to reinsert eliminated constraints. Notice that given a point in the decision variable space, the constraints that are violated can be computed easily, but optimization methods, in general, do not specify which constraints are causing infeasibility. When some of the constraints are dropped, the optimization algorithm has to be run again with the remaining constraints to check for feasibility.

   It is necessary to eliminate a group of constraints in those cases where the complete set of constraints imposed on the system is incompatible. Each time a constraints incompatibility problem arises, a set of inadmissible constraints is formed which is not taken into account in the optimization process. Various types can be distinguished in the constraint elimination methodology.

   - **Indiscriminate elimination:** With this strategy all constraints are eliminated every time a feasible solution arises. This is not the best method to solve the problem but it is the quickest. This method should not be used in cases where the constraints are directly related to safety.
   - **Hierarchical elimination:** During the design stage, a priority is given to each constraint. Every time feasibility problems arise the controller eliminates, in an orderly manner, the constraints of less priority until the feasibility of the solution is reestablished. This is checked at every sampling period in order to reinsert those constraints that were temporarily dropped.

3. **Constraints relaxation:** This method consists in temporarily relaxing the bounds (ie. increasing their values) or changing *hard* constraints ($\mathbf{Ru} < \mathbf{a}$) to *soft* constraints ($\mathbf{Ru} < \mathbf{a} + \epsilon$) adding a term $\epsilon^T \mathbf{T} \epsilon$ to the cost function, so that any violation of the constraint is penalized. In the long run, the penalizing term in the objective function will take the auxiliary variable to zero.

4. **Changing the constraint horizons:** Most of the constraint unfeasibility arises in the first part of the cost horizon, because sudden perturbations may take the process to an infeasible region. The main idea of this method is not to take into account the constraints during the first part of the horizon. Some commercial MPC use the concept of constraint window.

7.8 Constrained MPC and Stability

Infinite horizon optimal controllers, such as the well known Linear Quadratic Gaussian optimal controller (LQG), are easy to implement and guarantee a stable close loop for linear processes under quite general assumptions. However, infinite horizon control problems can only be solved when all process variables are unconstrained. The main difficulty for using infinite horizons with processes with constrained variables comes from the fact that numerical methods, with a necessarily finite number of decision variables, have to be used to solve the optimization problem involved. The stability analysis of finite horizon controllers is a much more difficult task, especially if the variables are constrained giving rise to a non-linear control law. Furthermore, no explicit functional description of the control law can be found making the problem even more difficult. A breakthrough has been made in the last few years in this field. As pointed out by Morari [76], "the recent work has removed this technical and to some extent psychological barrier (people did not even try) and started wide spread efforts to tackle extensions of this basic problem with the new tools".

The basic idea of the new approaches is that infinite horizon cost functions can be shown to be monotonically decreasing, if there is a feasible solution, and thus can be interpreted as a Lyapunov function which guarantees stability. In order to find a numerical solution to the infinite costing horizon control problem, the number of decision variables has to be finite. Two basic approaches have been used for this: in the first one, the objective function is considered to be composed of two parts; one with a finite horizon and constrained, and the other with an infinite horizon and unconstrained. The second approach is essentially equivalent [30] and it consists of imposing terminal state constraints and using a finite control horizon.

The first type of approach has originated the following results obtained by Rawlings and Muske [100], who demonstrated asymptotic stability for processes described by

$$x(t+1) = Mx(t) + Nu(t)$$

with $u(t)$ generated by minimizing:

$$J_j = \sum_{j=t}^{\infty}(x(j)^t Rx(j) + u(j)^t Sx(j)) \text{ with } R, S > 0$$

and subject to:

$$Du(i) \leq d \text{ for } i = t, t+1, \cdots t+Nu$$
$$Hx(i) \leq h \text{ for } i = t, t+1, \cdots \infty$$

for stabilizable pairs (M, N) with r unstable modes and $N_u \geq r$ if the minimization problem stated above is feasible.

The central idea is that if the minimization problem is feasible at sampling time t, then J_t is finite and $J_{t+1} \leq J_t + x(t)^t R x(t) + u(t)^t S u(t)$. The cost function J_t can then be interpreted as a monotonically decreasing Lyapunov function and asymptotic stability is therefore guaranteed. Notice that the problem at $t+1$ is also feasible (for the noise free case without external perturbations). Also note that the infinite and unconstrained part of the objective function can be solved by a Riccati equation and that a cost function depending on the state can be obtained. This cost function is introduced in the finite horizon optimization problem which is solved by numerical methods.

The second type of approach has been developed in the GPC context following Clarke and Scattolini [30] CRHPC. The main idea is to impose state terminal constraints, or, in the input-output context, to force the predicted output to exactly follow the reference during a sufficiently large horizon m after the costing horizon. The problem can be stated as:

$$\min_{\mathbf{u}} \quad \sum_{j=N_1}^{N_2} \|\hat{y}(t+j \mid t) - w(t+j)\|_R^2 + \sum_{j=1}^{N_u} \| \Delta u(t+j-1)\|_Q^2$$

$$\text{subject to} \quad \Delta u(t+N_u+j-1) = 0$$

$$y(t+N_2+m+j) = w(t+N_2+m+1)$$

Stability results for CRHPC have been obtained [30],[36] for the unconstrained case. Scokaert and Clarke [111] have demonstrated the stability property CRHPC in the presence of constraints. The principal idea is that if a feasible solution is found, and the settling horizon N_y is large enough to cover the transient of the output variables, the cost function is monotonically decreasing (if there are no external disturbances and the process is noise free) and can be interpreted as a Lyapunov function which will guarantee stability. It can also be shown that the problem will be feasible in the next iteration.

Stability results for constrained SGPC have also been obtained by Rossiter and Kouvaritakis [58], [109] who found that for any reference $w(t)$ which assumes a constant value w^* after a number (N) of sampling periods, if the constrained SGPC is feasible for sufficiently large values of the horizons (depending on N), the closed loop will be stable and the output will asymptotically go to w^*. The stability has been also demonstrated in [107] without imposing the terminal state conditions, implicitly used in SGPC. The work is based on characterizing all stable predictions which are not necessarily of a finite impulse response type as in standard SGPC. This allows for more degrees of freedom and increases the feasibility of the problem.

All stability results require the feasibility of the control law. If no feasible solution is found, one can always use the unconstrained solution and clip it to the manipulated variable bounds, but this way of operating would not guarantee the stability of the closed loop (for the nominal plant). Note that input constraints can always be satisfied by saturating the control signals, but this is not the case of output or state constraints which are the real cause of infeasibility. Some suggestions have been made in literature to cope with infeasibility.

Rawlings and Muske [100] proposed dropping the state constraints during the initial portion of the infinite horizon in order to make the problem feasible. Zheng and Morari [130] proposed changing the hard constraints on the state ($Hx(i) \leq h$) for soft constraints ($Hx(i) \leq h + \epsilon$ with $\epsilon \geq 0$) to ensure feasibility, adding the term $\epsilon^t Q \epsilon$ to the costing function in order to penalize constraint violation and thus obtain a better performance. They also demonstrated that any stabilizable system can be asymptotically stabilized by the MPC with soft constraints and state feedback if N_u is chosen to be sufficiently large and also that it stabilizes any open loop stable system with output feedback (state vector computed by an observer). Muske *et al* [81] have shown that an infinite horizon MPC with output feedback which does not enforce state constraints during the first stages, produces a stable closed loop when controlling open loop stable systems and that it also stabilizes unstable processes provided that the initial process and observer states are inside the feasible region. Scokaert and Clarke [111] have proposed a way of removing constraints when no feasible solutions are found. Their idea is to increase the lower constraint horizon until a feasible solution has been found. They also suggest that another possible way of removing constraints would be by having them organized in a hierarchical way with the critical ones at one end and the less important one at the other. This ordering may be used to drop the constraints when no feasible solution is found.

An idea is proposed in [46] to ensure the feasibility of constrained SGPC in the presence of bounded disturbances that could take the process away from the constrained region. The idea is to determine the minimum required control power to reject the worst perturbations in the future. To implement this idea, tighter constraints than the physical limits are imposed on the manipulated variables. The difference between the physical limits and the new constraints is the minimum control effort required to maintain the feasibility of the constrained optimization problem in order to guarantee stability.

Model predictive control schemes for nonlinear systems which guarantee stability have also been proposed [72],[73]. The main idea is to solve the constrained MPC in a finite horizon driving the state to zero or inside a region W where control is transferred to a linear stabilizing controller. The main problem with this idea is that region W is very difficult to compute and that, in general, the resulting optimization problem is non convex. This makes the optimization problem much more difficult to solve and the optimality of the solution cannot be assured. Fortunately, stability is guaranteed when the optimization problem is feasible and does not require optimality for this type of controller, although the performance may suffer by using local minima. These ideas have been extended by Chen and Allgower [26] who proposed a quasi infinite constrained MPC. The idea of the terminal region and linear stabilizing controller is used, but only for computing the terminal cost. In quasi-infinite horizon nonlinear MPC the control signal is determined by solving the optimization problem on-line and the control is never transferred to the linear stabilizing controller even when inside the terminal region.

Another way to achieve closed loop stability of nonlinear predictive control proposed by Yang and Polak [124] is by imposing *contraction* constraints on the state. The idea is to impose the following constraint

$$||x(t+1)|| < \alpha||x(t)||$$

with $\alpha \in [0, 1)$. Stability is guaranteed if the optimization problem is feasible. The main advantage of the algorithm is that if it is feasible the closed loop is exponentially stable. Imposing the *contraction* constraints is, however, very restrictive for many control problems and unfeasibility is encountered in many situations.

7.9 Multiobjective MPC

All the MPC strategies, analysed previously, are based on optimizing a single objective cost function, which is usually quadratic, in order to determine the future sequence of control moves that makes the process behave best. However, in many control problems the behaviour of the process cannot be measured by a single objective function but, most of the time, there are different, and sometimes conflicting, control objectives. The reasons for multiple control objectives are varied:

- Processes have to be operated differently when they are at different operating stages. For example at the start up phase of the process, a minimum start-up time may be desired, while once the process has reached the operating regime, a minimum variance of the controlled variables may be the primary control objective.
- Even if the process is working at a particular operating stage, the control objective may depend on the value of the variables. For example the control objective when the process is working at the nominal operating point may be to minimize the weighted sum of the square errors of the controlled variables with respect to their prescribed values. But if the value of one of the variables is too high, because of a sudden perturbation for example, the main control objective may be to reduce the value of this variable as soon as possible.

Furthermore, in many cases, the control objective is not to optimize the sum of the squared errors, but to keep some variables within specified bounds. Notice that this situation is different to the constraint control MPC, as the objective is to keep the variable there, although excursions of the variable outside this region, though not desirable, are permitted. In constrained MPC the variables should be kept within the prescribed region because of physical limitations, plant safety or other considerations. Constraints which cannot be violated are referred to as *hard* constraints, while those which can are known as *soft* constraints. These types of objectives can be expressed by penalizing the amount by which the offending variable violates the limit. Consider, for example, the process with a controlled variable $y(t)$ where the control objective is to keep $y_l \leq y(t) \leq y_h$. The control objective may be formulated as

$$J = p(y(t+j) - y_h) \sum_{j=N_1}^{N_2} (y(t+j) - y_h)^2 + p(y_l - y(t+j)) \sum_{j=N_1}^{N_2} (y(t+j) - y_l)^2$$

where function p is a step function. That is, p takes the value one when its argument is greater than or equal to zero and the value zero when its argument is negative.

Notice that the objective function is no longer a quadratic function and QP algorithms cannot be used. The problem can be transformed into a QP problem by introducing slack variables $\epsilon_h(j)$ and $\epsilon_l(j)$. That is:

$$\begin{aligned} y(t+j) &\leq y_h + \epsilon_h(j) \\ y(t+j) &\geq y_l - \epsilon_l(j) \end{aligned}$$

The manipulated variable sequence is now determined by minimizing:

$$J = \sum_{j=N_1}^{N_2} \epsilon_h(j)^2 + \sum_{j=N_1}^{N_2} \epsilon_l(j)^2$$

subject to $\epsilon_l(j) \geq 0$ and $\epsilon_h(j) \geq 0$ and the rest of the constraints acting on the problem. Notice that the problem has been transformed into a QP problem with more constraints and decision variables.

Sometimes all control objectives can be summarized in a single objective function. Consider, for example, a process with a series of control objectives J_1, J_2, ..., J_m. Some of the control objectives may be to keep some of the controlled variables as close to their references as possible, while other control objectives may be related to keeping some of the variables within specified regions. Consider all objectives to have been transformed into minimizing a quadratic function J_i, subject to a set of linear constraints on the decision variables $\mathbf{R}_i \mathbf{u} \leq \mathbf{a}_i$. The future control sequence can be determined by minimizing the following objective function:

$$J = \sum_{i=1}^{m} \beta_i J_i$$

subject to

$$\mathbf{R}_i\mathbf{u} \leq \mathbf{a}_i \text{ for } i = 1, \cdots, m$$

The importance of each of the objectives can be modulated by appropriate setting of all β_i. This is, however, a nontrivial matter in general as it is very difficult to determine the set of weights which will represent the relative importance of the control objectives. Furthermore, practical control objectives are sometime qualitative, making the task of determining the weights even more difficult.

7.9.1 Priorization of Objectives

In some cases, the relative importance of the control objectives can be established by priorization. That is, the objectives of greater priority, for example objectives related to security, must be accomplished before other objectives of less priority are considered. Objectives can be prioritized by giving much higher values to the corresponding weights. However this is a difficult task which is usually done by a trial and error method.

In [120] a way of introducing multiple prioritized objectives into the MPC framework is given. Consider a process with a series of m prioritized control objectives O_i. Suppose that objective O_i has a higher priority than objective O_{i+1} and that the objectives can be expressed as:

$$\mathbf{R}_i\mathbf{u} \leq \mathbf{a}_i$$

The main idea consists of introducing integer variables L_i which take the value one when the corresponding control objective is met and zero otherwise. Objectives are expressed as:

$$\mathbf{R}_i\mathbf{u} \leq \mathbf{a}_i + K_i(1 - L_i) \tag{7.17}$$

where K_i is a conservative upper bound on $\mathbf{R}_i\mathbf{u} - \mathbf{a}_i$. If objective O_i is satisfied, $L_i = 1$ and the reformulated objective coincides with the original control objective. By introducing K_i , the reformulated objective (constraint) is always satisfied even when the corresponding control objective O_i is not met ($L_1 = 0$).

The priorization of objectives can be established by imposing the following constraints:

$$L_i - L_{i+1} \geq 0 \text{ for } i = 1, \cdots, m-1$$

The problem is to maximize the number of satisfied control objectives:

$$\sum_{i=1}^{m} L_i$$

If the process model is linear, the problem can be solved with a Mixed Integer Linear Programming (MILP) algorithm. The number of integer variables can be reduced as indicated in [120]. The idea is to use the same variable L_i for constraints that cannot be violated at the same time, as is the case of upper and lower bounds on the same control or manipulated variable.

The set of constraints (7.17) can be modified [120], to improve the degree of the constraint satisfaction of objectives that cannot be satisfied. Suppose that not all objectives can be satisfied at a particular instance. Suppose that objective O_f is the first objective that failed. In order to come as close as possible to satisfying this objective, a slack variable α satisfying the following set of constraints is introduced:

$$\mathbf{R}_i\mathbf{u} \leq \mathbf{a}_i + \alpha + K_i\left((i-1) + (1-L_i) - \sum_{j=1}^{i-1} L_j\right)$$

and the objective function to be minimized is:

$$J = -K_\alpha \sum_{i=1}^{m} L_i + f(\alpha) \tag{7.18}$$

where f is a penalty function of the slack variable α (positive and strictly increasing) and K_α is an upper bound on f. The optimization algorithm will try to maximize the number of satisfied objectives ($L_i = 1$) before attempting to reduce $f(\alpha)$ because the overall objective function can be made smaller by increasing the number of non zeros L_i variables than by reducing $f(\alpha)$. As all objectives O_i for $i < f$ are satisfied ($L_i = 1$), constraints (7.9.1) will also be satisfied. As O_f is the first objective that failed:

$$\sum_{i=1}^{i-1} L_i = f - 1 \text{ for } i \geq f$$

That is, the term multiplying K_i of constraint (7.9.1) is zero for $i = f$, while for $i > f$ this term is greater than one. This implies that all constraints (7.9.1) will be satisfied for $i > f$. The only active constraint is:

$$\mathbf{R}_f\mathbf{u} \leq \mathbf{a}_f + \alpha$$

That is, the optimization method will try to optimize the degree of satisfaction of the first objective that failed only after all more prioritized objectives have been satisfied. Notice that $L_i = 0$ does not imply that objective O_i is not satisfied, it only indicates that the corresponding constraint has been relaxed.

If the process is linear and function f is linear, the problem of maximizing (7.18) can be solved by a MILP. If f is a quadratic function, the problem can be solved by a Mixed Integer Quadratic Programming (MIQP) algorithm. Although there are efficient algorithms to solve mixed integer programming problems, the amount of computation required is much higher than that required for LP or QP problems. The number of objectives should be kept small in order to implement the method in real time.

Chapter 8

Robust MPC

Mathematical models of real processes cannot contemplate every aspect of reality. Simplifying assumptions have to be made, especially when the models are going to be used for control purposes, where models with simple structures (linear in most cases) and sufficiently small size have to be used due to available control techniques and real time considerations. Thus, mathematical models, and especially control models, can only describe the dynamics of the process in an approximative way.

Most control design techniques need a control model of the plant with fixed structure and parameters (*nominal model*), which is used throughout the design. If the control model were an exact, rather than an approximate, description of the plant and there were no external disturbances, processes could be controlled by an open-loop controller. Feedback is necessary in process control because of the external perturbations and model inaccuracies in all real processes. The objective of robust control is to design controllers which preserve stability and performance in spite of the models inaccuracies or uncertainties. Although the use of feedback contemplates the inaccuracies of the model implicitly, the term of robust control is used in literature for control systems that explicitly consider the discrepancies between the model and the real process [70].

There are different approaches for modelling uncertainties depending mainly on the type of technique used for designing the controllers. The most extended techniques are frequency response uncertainties and transfer function parametric uncertainties. Most of the approaches assume that there is a family of models and that the plant can be exactly described by one of the models belonging to the family. That is, if the family of models is composed of linear models, the plant is also linear. The approach considered here is the one relevant to the key feature of MPC which is to predict future values of the output variables. The uncertainties can be defined about the prediction capability of the model. It will be shown that no assumptions have to be made regarding the linearity of the plant in spite of using a family of linear models for control purposes.

8.1 Process Models and Uncertainties

Two basic approaches are extensively used in literature to describe modelling uncertainties: frequency response uncertainties and transfer function parametric uncertainties. Frequency uncertainties are usually described by a band around the nominal model frequency response. The plant frequency response is presumed to be included in the band. In the case of parametric uncertainties, each of the coefficients of the transfer function is presumed to be bounded by the uncertainties limit. The plant is presumed to have a transfer function, with parameters within the uncertainty set.

Both ways of modelling uncertainties consider that the exact model of the plant belongs to the family of models described by the uncertainty bounds. That is, that the plant is linear with a frequency response within the uncertainty band for the first case and that the plant is linear and of the same order as that of the family of models for the case of parametric uncertainties.

Control models in MPC are used to predict what is going to happen: future trajectories. The appropriate way of describing uncertainties in this context seems to be by a model, or a family of models that, instead of generating a future trajectory, may generate a band of trajectories in which the process trajectory will be included when the same input is applied, in spite of the uncertainties. One should expect this band to be narrow when a good model of the process is available and the uncertainty level is low and to be wide otherwise.

The most general way of posing the problem in MPC is as follows: consider a process whose behaviour is dictated by the following equation:

$$y(t+1) = f(y(t), \cdots, y(t-n_y), u(t), \cdots, u(t-n_u), z(t), \cdots, z(t-n_z), \psi) \quad (8.1)$$

where $y(t) \in \mathbf{Y}$ and $u(t) \in \mathbf{U}$ are n and m vectors of outputs and inputs, $\psi \in \mathbf{\Psi}$ is a vector of parameters possibly unknown and $z(t) \in \mathbf{Z}$ is a vector of possibly random variables.

Now consider the model, or the family of models, for the process described by:

$$\hat{y}(t+1) = \hat{f}(y(t), \cdots, y(t-n_{n_a}), u(t), \cdots, u(t-n_{n_b}), \theta) \quad (8.2)$$

where $\hat{y}(t+1)$ is the prediction of output vector for instant $t+1$ generated by the model, $\hat{f}$ is a vector function, usually a simplification of f, n_{n_a} and n_{n_b} are the number of past outputs and inputs considered by the model and $\theta \in \mathbf{\Theta}$ is a vector of uncertainties about the plant. Variables that although influencing plant dynamics are not considered in the model because of the necessary simplifications or for other reasons are represented by $z(t)$.

The dynamics of the plant (8.1) are completely described by the family of models (8.2) if for any $y(t), \cdots, y(t-n_y) \in \mathbf{Y}$, $u(t), \cdots, u(t-n_u) \in \mathbf{U}$, $z(t), \cdots, z(t-n_z) \in \mathbf{Z}$ and $\psi \in \Psi$, there is a vector of parameters $\theta_i \in \Theta$ such that

$$f(y(t), \cdots, y(t-n_y), u(t), \cdots, u(t-n_u), z(t), \cdots, z(t-n_z), \psi) =$$
$$= \hat{f}(y(t), \cdots, y(t-n_{n_a}), u(t), \cdots, u(t-n_{n_b}), \theta_i)$$

The way in which the uncertainties parameter θ and its domain Θ are defined mainly depends on the structures of f and $\hat{f}$ and on the degree of certainty about the model. In the following the most used model structures in MPC will be considered.

8.1.1 Truncated Impulse Response Uncertainties

For an m-input n-output MIMO stable plant the truncated impulse response is given by N real matrices $(n \times m)$ H_t. The (i,j) entry of H_t corresponds to the ith-output of the plant when an impulse has been applied to input variable u_j.

The natural way of considering uncertainties is by supposing that the coefficients of the truncated impulse response, which can be measured experimentally, are not known exactly and are a function of the uncertainty parameters. Different type of functions can be used. The most general way will be by considering that the impulse response may be within a set defined by $(\underline{H}_t)_{ij} \leq (H_t)_{ij} \leq (\overline{H}_t)_{ij}$. That is $(H_t)_{ij}(\Theta) = (Hm_t)_{ij} + \Theta_{t_{ij}}$, with Θ defined by $(Hm_t)_{ij} - (\overline{H}_t)_{ij} \leq \Theta_{t_{ij}} \leq (\underline{H}_t)_{ij} - (Hm_t)_{ij}$ and Hm_t is the *nominal* response. The dimension of the uncertainty parameter vector is $N \times (m \times n)$. For the case of $N = 40$ and a 5 inputs - 5 outputs MIMO plant, the number of uncertainty parameters is 1000, which will normally be too high for the min-max problem involved.

This way of modelling does not take into account the possible structures of the uncertainties. When these are considered, the dimension of the uncertainty parameter set may be considerably reduced.

In [25] and in [93] a linear function of the uncertainty parameters is suggested:

$$H_t = \sum_{j=1}^{q} G_{t_j} \theta_j$$

The idea is that the plant can be described by a linear combination of q known stable linear time invariant plants with unknown weighting θ_j. This approach is suitable in the case where the plant is non-linear and linear models are obtained at different operating regimes. It seems plausible that a linear combination of linearized models can describe the behaviour of the plant over a wider range of conditions than a single model.

As will be seen later, considering the impulse response as a linear function of the uncertainty parameters is of great interest for solving the robust MPC problem. Furthermore, note that the more general description of uncertainties $(H_t)_{ij}(\Theta) = (\underline{H}_t)_{ij} + \Theta_{t_{ij}}$ can also be expressed this way by considering:

$$H_t(\theta) = \sum_{i=1}^{n}\sum_{j=1}^{m}(\underline{H}_t)_{ij} + \Theta_{t_{ij}} H_{ij}$$

where H_{ij} is a matrix having entry (i,j) equal to one and the remaining entries to zero.

The predicted output can be computed as

$$y(t+j) = \sum_{i=1}^{N}(Hm_i + \theta_i)u(t+j-i)$$

while the predicted nominal response is

$$ym(t+j) = \sum_{i=1}^{N} Hm_i u(t+j-i)$$

The prediction band around the nominal response is then limited by:

$$\min_{\theta\in\Theta}\sum_{i=1}^{N}\theta_i u(t+j-i) \text{ and } \max_{\theta\in\Theta}\sum_{i=1}^{N}\theta_i u(t+j-i)$$

8.1.2 Matrix Fraction Description Uncertainties

Let us consider the following n-output m-input multivariable discrete-time model:

$$\mathbf{A}(z^{-1})y(t) = \mathbf{B}(z^{-1})\,u(t-1) \tag{8.3}$$

where $A(z^{-1})$ and $B(z^{-1})$ are polynomial matrices of appropriate dimensions.

Parametric uncertainties about the plant can be described by $(\underline{A}_k)_{ij} \leq (A_k)_{ij} \leq (\overline{A}_k)_{ij}$ and $(\underline{B}_k)_{ij} \leq (B_k)_{ij} \leq (\overline{B}_k)_{ij}$. That is, $(A_k)_{ij} = (\underline{A}_k)_{ij} + \Theta_{a_{k_{ij}}}$ $(B_k)_{ij} = (\underline{B}_k)_{ij} + \Theta_{b_{k_{ij}}}$.

The number of uncertainty parameters for this description is $n_a \times n \times n + (n_b+1) \times n \times m$. Note that uncertainties about actuators and dead times will mainly reflect on coefficients of the polynomial matrix $B(z^{-1})$ while uncertainties about the time constants will mainly affect the polynomial matrix $A(z^{-1})$. Note that if the parameters of the polynomial matrices $\mathbf{A}(z^{-1})$ and

$\mathbf{B}(z^{-1})$ have been obtained via identification, the covariance matrix indicates how big the uncertainty band for the coefficients is.

The most frequent case in industry is that each of the entries of the transfer matrix has been characterized by its static gain, time constant and equivalent dead time. Bounds on the coefficients of matrices $\mathbf{A}(z^{-1})$ and $\mathbf{B}(z^{-1})$ can be obtained from bounds on the gain and time constants. Uncertainties about the dead time are, however, difficult to handle. If the uncertainty band about the dead time is higher than the sampling time used, it will translate into a change in the order of the polynomial or coefficients that can change from zero and to zero. If the uncertainty band about the dead time is smaller than the sampling time, the pure delay time of the discrete time model does not have to be changed. The fractional delay time can be modelled by the first terms of a Padé expansion and the uncertainty bound of these coefficients can be calculated from the uncertainties of the dead time. In any case dead time uncertainty bounds tend to translate into a very high degree of uncertainty about the coefficients of the polynomial matrix $\mathbf{B}(z^{-1})$.

The prediction equations can be expressed in terms of the uncertainty parameters. Unfortunately, for the general case, the resulting expressions are too complicated and of little use because the involved min-max problem would be too difficult to solve in real time. If the uncertainties only affect polynomial matrix $\mathbf{B}(z^{-1})$, the prediction equation is an affine function of the uncertainty parameter and the resulting min-max problem is less computationally expensive as will be shown later in the chapter. Uncertainties on $\mathbf{B}(z^{-1})$ can be given in various ways. The most general way is by considering uncertainties on the matrices ($B_i = Bn_i + \theta_i$). If the plant can be described by a linear combination of q known linear time invariant plants with unknown weighting θ_j, polynomial matrix $\mathbf{B}(z^{-1})$ can be expressed as:

$$\mathbf{B}(z^{-1}) = \sum_{i=1}^{q} \theta_i \mathbf{P}_i(z^{-1})$$

The polynomial matrices $\mathbf{P}_i(z^{-1})$ are a function of the polynomial matrices $\mathbf{B}_i(z^{-1})$ and $\mathbf{A}_i(z^{-1})$ corresponding to the matrix fraction description of each plant. For the case of diagonal $\mathbf{A}_i(z^{-1})$ matrices, the polynomial matrices $\mathbf{P}_i(z^{-1})$ can be expressed as:

$$\mathbf{P}_i(z^{-1}) = \prod_{j=1, i\neq j}^{q} \mathbf{A}_j(z^{-1})\mathbf{B}_i(z^{-1})$$

Note that the general case of uncertainties on the coefficients parameters could have also been expressed this way but with a higher number of uncertainty parameters. Using prediction equation (6.5)

$$\begin{aligned} y(t+j|t) &= \mathbf{F}_j(z^{-1})y(t) + \mathbf{E}_j(z^{-1})\mathbf{B}(z^{-1}) \triangle u(t+j-1) = \\ &= \mathbf{F}_j(z^{-1})y(t) + \mathbf{E}_j(z^{-1})(\textstyle\sum_{i=1}^{q} \theta_i \mathbf{P}_i(z^{-1})) \triangle u(t+j-1) \end{aligned}$$

That is, an affine function in θ_i

8.1.3 Global Uncertainties

The key idea of this way of modelling the uncertainties is to assume that all modelling errors are globalized in a vector of parameters, such that the plant can be described by the following family of models:

$$\hat{y}(t+1) = \hat{f}(y(t), \cdots, y(t-n_{n_a}), u(t), \cdots, u(t-n_{n_b})) + \theta(t)$$

with dim($\theta(t)$)=n.

Notice that global uncertainties can be related to other types of uncertainties. For the impulse response model, the output at instant $t+j$ with parametric and temporal uncertainties description is given by:

$$\begin{aligned}
\hat{y}(t+j) &= \sum_{i=1}^{N}(Hm_i + \theta_i)u(t+j-i) \\
\hat{y}(t+j) &= \sum_{i=1}^{N}(Hm_i)u(t+j-i) + \theta(t+j)
\end{aligned}$$

Therefore $\theta(t+j) = \sum_{i=0}^{N} \theta_i u(t+j-i)$ and the limits for i component ($\underline{\theta}_i$, $\overline{\theta}_i$) of vector $\theta(t+j)$ when $u(t)$ is bounded (in practice always) are given by:

$$\begin{aligned}
\underline{\theta}_i &= \min_{u(\cdot)\in\mathbf{U}, \theta_i\in\Theta} \sum_{i=0}^{N} \theta_{t_i} u(t+j-i) \\
\overline{\theta}_i &= \max_{u(\cdot)\in\mathbf{U}, \theta_i\in\Theta} \sum_{i=0}^{N} \theta_{t_i} u(t+j-i)
\end{aligned}$$

The number of uncertainty parameters is reduced from $N \times (m \times n)$ to n but the approach is more conservative because the limits of the uncertainty parameter domain have to be increased to contemplate the worst global situation. The way of defining the uncertainties is, however, much more intuitive and directly reflects how good the j-step-ahead prediction model is.

For the left matrix fraction description, the uncertainty model is defined by:

$$\tilde{\mathbf{A}}(z^{-1})y(t) = \mathbf{B}(z^{-1}) \triangle u(t-1) + \theta(t) \tag{8.4}$$

with $\theta(t) \in \Theta$ defined by $\underline{e}(t) \leq \theta(t) \leq \overline{e}(t)$.

Notice that with this type of uncertainty one does not have to presume the model to be linear, as is the case of parametric uncertainty or frequency uncertainty modelling. Here it is only assumed that the process can be approximated by a linear model in the sense that all trajectories will be included in bands that depend on $\underline{\theta}(t)$ and $\overline{\theta}(t)$. It may be argued that this type of global uncertainties are more disturbances than uncertainties because they seem to work as external perturbations. However, the only assumption made is that they are bounded, in fact $\theta(t)$ may be a function of past inputs and outputs. If the process variables are bounded, the global uncertainties can also be bounded.

The model given by expression (8.4) is an extension of the integrated error concept used in CARIMA models. Because of this, the uncertainty band will grow with time. In order to illustrate this point consider the system described by the first order difference equation

$$y(t+1) = ay(t) + bu(t) + \theta(t+1) \text{ with } \underline{\theta} \leq \theta(t) \leq \overline{\theta}$$

That is, a model without integrated uncertainties. Let us suppose that the past inputs and outputs, and future inputs are zero, thus producing a zero nominal trajectory. The output of the uncertain system is given by

$$\begin{aligned} y(t+1) &= \theta(t+1) \\ y(t+2) &= a\,\theta(t+1) + \theta(t+2) \\ &\vdots \\ y(t+N) &= \sum_{j=0}^{N-1} a^j \theta(t+1+j) \end{aligned}$$

The upper bound will grow as $|a|^{(j-1)}\overline{\theta}$ and the lower bound as $|a|^{(j-1)}\underline{\theta}$. The band will stabilize for stable systems to a maximum value of $\overline{\theta}/(1-|a|)$ and $\underline{\theta}/(1-|a|)$. This type of model will not incorporate the possible drift in the process caused by external perturbations.

For the case of integrated uncertainties defined by the following model:

$$y(t+1) = ay(t) + bu(t) + \frac{\theta(t)}{\Delta}$$

the increment of the output is given by:

$$\Delta y(t+k) = \sum_{j=0}^{k-1} a^j \theta(t+j+1)$$

and

$$y(t+N) = \sum_{k=1}^{N} \Delta y(t+j) = \sum_{k=1}^{N}\sum_{j=0}^{k-1} a^j \theta(t+j+1)$$

indicating that the uncertainty band will grow continuously. The rate of growth of the uncertainty band stabilizes to $\overline{\theta}/(1-|a|)$ and $\underline{\theta}/(1-|a|)$, after the transient caused by process dynamics.

In order to generate the j-step-ahead prediction for the output vector, let us consider the Bezout identity:

$$I = \mathbf{E}_j(z^{-1})\tilde{\mathbf{A}}(z^{-1}) + \mathbf{F}_j(z^{-1})z^{-j} \tag{8.5}$$

Using equations (8.4) and (8.5) we get

$$y(t+j) = \mathbf{F}_j(z^{-1})y(t) + \mathbf{E}_j(z^{-1})\mathbf{B}(z^{-1})\,\Delta u(t+j-1) + \mathbf{E}_j(z^{-1})\theta(t+j) \tag{8.6}$$

Notice that the prediction will be included in a band around the nominal prediction $ym(t+j) = \mathbf{F}_j(z^{-1})y(t) + \mathbf{E}_j(z^{-1})\mathbf{B}(z^{-1})\,\Delta u(t+j-1)$ delimited by

$$ym(t+j) + \min_{\theta(\cdot)\in\mathbf{\Theta}} \mathbf{E}_j(z^{-1})\theta(t+j) \leq y(t+j) \leq ym(t+j) + \max_{\theta(\cdot)\in\mathbf{\Theta}} \mathbf{E}_j(z^{-1})\theta(t+j)$$

Because of the recursive way in which polynomial $\mathbf{E}_j(z^{-1})$ can be obtained, when $\overline{e}(t)$ and $\underline{e}(t)$ are independent of t, the band can also be obtained recursively by increasing the limits obtained for $y(t+j-1)$ by:

$$\max_{\theta(t+j)\in\mathbf{\Theta}} E_{j,j-1}\theta(t+1) \text{ and } \min_{\theta(t+j)\in\mathbf{\Theta}} E_{j,j-1}\theta(t+1)$$

where $\mathbf{E}_j(z^{-1}) = \mathbf{E}_{j-1}(z^{-1}) + E_{j,j-1}z^{-(j-1)}$.

Consider the set of j-ahead optimal predictions $\mathbf{y}$ for $j = 1, \cdots, N$, which can be written in condensed form as:

$$\mathbf{y} = \mathbf{G}_u\mathbf{u} + \mathbf{G}_\theta\theta + \mathbf{f} \tag{8.7}$$

where $\mathbf{y} = [y(t+1), y(t+2), \cdots, y(t+N)]^t$, $\mathbf{u} = [u(t), u(t+1), \cdots, u(t+N-1)]^t$, $\theta = [\theta(t+1), \theta(t+2), \cdots, \theta(t+N)]^t$ and $\mathbf{f}$ is the free response, that is the response due to past outputs (up to time t) and past inputs. Vectors $\mathbf{u}$ and θ correspond to the present and future values of the control signal and uncertainties.

8.2 Objective Functions

The objective of predictive control is to compute the future control sequence $u(t)$, $u(t+1)$, ..., $u(t+N_u)$ in such a way that the optimal j-step ahead predictions $y(t+j \mid t)$ are driven close to $w(t+j)$ for the prediction horizon. The way in which the system will approach the desired trajectories will be indicated by a function J which depends on the present and future control signals and uncertainties. The usual way of operating, when considering a stochastic type of uncertainty, is to minimize function J for the most expected situation. That is, supposing that the future trajectories are going to be the future expected trajectories. When bounded uncertainties are considered explicitly, bounds on the predictive trajectories can be calculated and it would seem that a more robust control would be obtained if the controller tried to minimize the objective function for the worst situation. That is, by solving:

$$\min_{\mathbf{u}\in\mathbf{U}} \max_{\theta\in\Theta} J(\mathbf{u},\theta)$$

The function to be minimized is the maximum of a norm that measures how well the process output follows the reference trajectories. Different type of norms can be used for this purpose.

8.2.1 Quadratic Norm

Let us consider a finite horizon quadratic criterion

$$J(N_1,N_2,N_u) = \sum_{j=N_1}^{N_2} [\hat{y}(t+j \mid t) - w(t+j)]^2 + \sum_{j=1}^{N_u} \lambda[\Delta u(t+j-1)]^2 \quad (8.8)$$

If the prediction equation (8.7) is used, equation (8.8) can now be written as:

$$\begin{aligned} J = (\mathbf{G}_u\mathbf{u} + \mathbf{G}_\theta\theta + \mathbf{f} - \mathbf{w})^T(\mathbf{G}_u\mathbf{u} + \mathbf{G}_\theta\theta + \mathbf{f} - \mathbf{w}) + \lambda\mathbf{u}^T\mathbf{u} = \\ \mathbf{u}^T M_{uu}\mathbf{u} + M_u\mathbf{u} + M + M_\theta\theta + \theta^t M_{\theta\theta}\theta + \theta^t M_{eu}\mathbf{u} \end{aligned} \quad (8.9)$$

where $\mathbf{w}$ is a vector containing the future reference sequences $\mathbf{w} = [w(t+N_1),\cdots,w(t+N_2)]^t$

The function $J(\mathbf{u},\theta)$ can be expressed as a quadratic function of θ for each value of $\mathbf{u}$:

$$J(\mathbf{u},\theta) = \theta^t M_{\theta\theta}\theta + M'_e(\mathbf{u})\theta + M'\mathbf{u} \quad (8.10)$$

with $M'_\theta = M_\theta + \mathbf{u}^t M_\theta$ and $M' = M + M_u\mathbf{u} + \mathbf{u}^t M_{uu}\mathbf{u}$.

Let us define:

$$Jm(\mathbf{u}) = \max_{\theta\in\Theta} J(\mathbf{u},\theta)$$

Matrix $M_{\theta\theta} = \mathbf{G}_\theta^t\mathbf{G}_\theta$ is a positive definite matrix because G_θ is a lower triangular matrix having all the elements on the principal diagonal equal to one. Since matrix $M_{\theta\theta}$ is positive definite, the function is strictly convex, ([10] theorem 3.3.8) and the maximum of J will be reached in one of the vertex of the polytope Θ ([10] theorem 3.4.6).

For a given $\mathbf{u}$ the maximization problem is solved by determining which of the $2^{(N\times n)}$ vertices of the polytope Θ produces the maximum value of $J(\mathbf{u},\theta)$

It can easily be seen that function $Jm(\mathbf{u})$ is a piece-wise quadratic function of $\mathbf{u}$. Let us divide the $\mathbf{u}$ domain $\mathbf{U}$ in different regions $\mathbf{U}_p$ such that $\mathbf{u}\in\mathbf{U}_p$ if the maximum of $J(\mathbf{u},\theta)$ is attained for the polytope vertex θ_p. For the region $\mathbf{U}_p$ the function $J_m(\mathbf{u})$ is defined by:

$$Jm(\mathbf{u}) = \mathbf{u}^t M_{uu}\mathbf{u} + M_u^*(\theta_p)\mathbf{u} + M^*\theta_p$$

with $M_u^* = M_u + \theta_p^t M_u$ and $M^* = M + M_\theta\theta_p + \theta_p^t M_{\theta\theta}\theta_p$.

Matrix M_{uu}, which is the Hessian matrix of function $Jm(\mathbf{u})$, can be assured to be positive definite by choosing a value of $\lambda > 0$. This implies that the function is convex ([10] theorem 3.3.8) and that there are no local optimal solutions different to the global optimal solution ([10] theorem 3.4.2).

One of the main problems of non-linear programming algorithms, local minima, are avoided and any non-linear programming method can be used to minimize function $Jm(\mathbf{u})$. However, and because the evaluation of $Jm(\mathbf{u})$ implies finding the minimum at one of the vertex of the polytope Θ, the computation time can be prohibitive for real time application with long costing and control horizons. The problem gets even more complex when the uncertainties on the parameters of the transfer function are considered. The amount of computation required can be reduced considerably if other type of objective functions are used, as will be shown in the following sections.

8.2.2 $\infty - \infty$ norm

Campo and Morari [25], showed that by using an $\infty - \infty$ type of norm the min-max problem involved can be reduced to a linear programming problem which requires less computation and can be solved with standard algorithms. Although the algorithm proposed by Campo and Morari was developed for processes described by the truncated impulse response, it can easily be extended to the left matrix fraction descriptions used throughout the text.

The objective function is now described as:

$$J(\mathbf{u},\theta) = \max_{j=1\cdots N} \|\hat{y}(t+j|t) - w(t)\|_\infty = \max_{j=1\cdots N}\max_{i=1\cdots n} |\hat{y}_i(t+j|t) - w_i(t)| \quad (8.11)$$

Note that this objective function will result in an MPC which minimizes the maximum error between any of the process outputs and the reference trajectory for the worst situation of the uncertainties, the control effort required to do so is not taken into account.

By making use of the prediction equation $\mathbf{y} = G_u\mathbf{u} + G_\theta\theta + \mathbf{f}$, and defining $g(\mathbf{u},\theta) = (\mathbf{y} - \mathbf{w})$, the control problem can now be expressed as:

$$\min_{\mathbf{u}\in\mathbf{U}} \max_{\theta\in\Theta} \max_{i=1\cdots n\times N} |g_i(\mathbf{u},\theta)|$$

Define $\mu^*(\mathbf{u})$ as:

$$\mu^*(\mathbf{u}) = \max_{\theta\in\Theta} \max_{i=1\cdots n\times N} |g_i(\mathbf{u},\theta)|$$

If there is any positive real value μ satisfying $-\mu \leq g_i(\mathbf{u},\theta) \leq \mu, \forall\theta \in \boldsymbol{\Theta}$ and for $i = 1 \cdots n \times N$ it is clear that μ is an upper bound of $\mu^*(\mathbf{u})$. The problem can now be transformed into finding the smallest upper bound μ and some $\mathbf{u} \in \mathbf{U}$ for all $\theta \in \boldsymbol{\Theta}$. When constraints on the controlled variables $(\underline{\mathbf{y}}, \overline{\mathbf{y}})$ are taken into account, the problem can be expressed as:

$$\min_{\mu,\mathbf{u}} \mu$$

subject to

$$\left.\begin{array}{r} -\mu \leq g_i(\mathbf{u},\theta) \leq \mu \\ \underline{\mathbf{y}}_i - \mathbf{w}_i \leq g_i(\mathbf{u},\theta) \leq \overline{\mathbf{y}}_i - \mathbf{w}_i \end{array}\right\} \quad \begin{array}{l} \text{for } i = 1, \cdots, n \times N \\ \forall\theta \in \boldsymbol{\Theta} \end{array}$$

The control problem has been transformed into an optimization problem with an objective function which is linear in the decision variables (μ, $\mathbf{u}$) and with an infinite (continuous) number of constraints. If $g(\mathbf{u},\theta)$ is an affine function of θ, $\forall\mathbf{u} \in \mathbf{U}$, the maximum and minimum of $g(\mathbf{u},\theta)$ can be obtained at one of the extreme points of $\boldsymbol{\Theta}$ [25]. Let us call $\mathcal{E}$ the set formed by the $2^{n\times N}$ vertices of $\boldsymbol{\Theta}$. If constraints are satisfied for every point of $\mathcal{E}$ they will also be satisfied for every point of $\boldsymbol{\Theta}$. Thus the infinite, and continuous, constraints can be replaced by a finite number of constraints.

When the global uncertainty model is used and constraints on the manipulated variables $(\underline{\mathbf{U}}, \overline{\mathbf{U}})$ and on the slew rate of the manipulated variables $(\underline{\mathbf{u}}, \overline{\mathbf{u}})$ are also taken into account, the problem can be stated as:

$$\min_{\mu,\mathbf{u}} \mu$$

subject to

$$\left.\begin{array}{rcr} \mathbf{1}\mu & \geq & G_u\mathbf{u} + G_\theta\theta + \mathbf{f} - \mathbf{w} \\ \mathbf{1}\mu & \geq & -G_u\mathbf{u} - G_\theta\theta - \mathbf{f} + \mathbf{w} \\ \overline{\mathbf{y}} & \geq & G_u\mathbf{u} + G_\theta\theta + \mathbf{f} \\ -\underline{\mathbf{y}} & \geq & -G_u\mathbf{u} - G_\theta\theta - \mathbf{f} \end{array}\right\} \forall\theta \in \mathcal{E}$$

$$\begin{array}{rcr} \overline{\mathbf{u}} & \geq & \mathbf{u} \\ -\underline{\mathbf{u}} & \geq & -\mathbf{u} \\ \overline{\mathbf{U}} & \geq & T\mathbf{u} + \mathbf{1}u(t-1) \\ -\underline{\mathbf{U}} & \geq & -T\mathbf{u} - \mathbf{1}u(t-1) \end{array}$$

where $\mathbf{1}$ is an $(N \times n) \times m$ matrix formed by N $m \times m$ identity matrices and T is a lower triangular block matrix whose non null block entries are $m \times m$ identity matrices. The problem can be transformed into the usual form

$$\min_{\mathbf{x}} \mathbf{c}^t\mathbf{x} \quad \text{subject to} \quad A\mathbf{x} \leq \mathbf{b}, \quad \mathbf{x} \geq 0$$

with :

$$\mathbf{x} = \begin{bmatrix} \mathbf{u} - \underline{\mathbf{u}} \\ \mu \end{bmatrix} \quad \mathbf{c}^t = [\overbrace{0, \cdots, 0}^{m\times N_u}, 1] \quad A^t = [A_1^t, \cdots, A_{2^N}^t, A_u^t] \quad \mathbf{b}^t = [\mathbf{b}_1^t, \cdots, \mathbf{b}_{2^N}^t, \mathbf{b}_u^t]$$

The block matrices have the form:

$$A_i = \begin{bmatrix} \mathbf{G}_u & -1 \\ -\mathbf{G}_u & -1 \\ \mathbf{G}_u & 0 \\ -\mathbf{G}_u & 0 \end{bmatrix} \quad A_u = \begin{bmatrix} I & 0 \\ T & 0 \\ -T & 0 \end{bmatrix} \quad \mathbf{b}_i = \begin{bmatrix} -\mathbf{G}_u\underline{\mathbf{u}} - \mathbf{G}_\theta\theta_i - \mathbf{f} + \mathbf{w} \\ \mathbf{G}_u\underline{\mathbf{u}} + \mathbf{G}_\theta\theta_i + \mathbf{f} - \mathbf{w} \\ \overline{\mathbf{y}} - \mathbf{G}_u\underline{\mathbf{u}} - \mathbf{G}_\theta\theta_i - \mathbf{f} \\ -\underline{\mathbf{y}} + \mathbf{G}_u\underline{\mathbf{u}} + \mathbf{G}_\theta\theta_i - \mathbf{f} \end{bmatrix}$$

$$\mathbf{b}_u = \begin{bmatrix} \overline{\mathbf{u}} - \underline{\mathbf{u}} \\ \overline{\mathbf{U}} + T\mathbf{u} + \mathbf{1}u(t-1) \\ -\underline{\mathbf{U}} - T\mathbf{u} - \mathbf{1}u(t-1) \end{bmatrix}$$

where θ_i is the ith-vertex of $\mathcal{E}$.

The number of variables involved in the linear programming problem is $m \times N_u + 1$ while the number of constraints is $4 \times n \times N \times 2^{n \times N} + 3m \times N_u$. As the number of constraints is much higher than the number of decision variables, solving the dual LP problem should be less computationally expensive, as pointed out by Campo and Morari [25].

The number of constraints can, however, be dramatically reduced because of the special form of matrix A. Consider the jth-row for each of the constraint blocks $A_i\mathbf{x} \leq \mathbf{b}_i$. As $A_1 = A_2 = \cdots = A_{2^N}$, the only constraint limiting the feasible region will be the one having the smallest value on the jth element of vector $\mathbf{b}_i$. Therefore, all the other $(2^N - 1)$ constraints can be eliminated and the number of constraints can be reduced to $4 \times n \times N + 3m \times N_u$. Notice that any uncertainty model giving rise to an affine function $g(\mathbf{u}, \theta)$ can be transformed into an LP problem as shown by Campo and Morari [25]. The truncated impulse response uncertainty model or uncertainties in the $\mathbf{B}(z^{-1})$ polynomial matrix produce an affine function $g(\mathbf{u}, \theta)$. However, the constraint reduction mechanism described above cannot be applied and the number of constraints would be very high.

8.2.3 1-norm

Although the type of $(\infty - \infty)$-norm used seems to be appropriate in terms of robustness, the norm is only concerned with the maximum deviation and the rest of the behaviour is not taken explicitly into account. Other types of norms are more appropriate for measuring the performance. Allwright [3] has shown that this method can be extended to the 1-norm for processes described by their truncated impulse response. The derivation for the left matrix representation is also straightforward:

The objective function is now

$$J(\mathbf{u}, \theta) = \sum_{j=N_1}^{N_2} \sum_{i=1}^{n} |y_i(t+j \mid t, \theta) - w_i(t+j)| + \lambda \sum_{j=1}^{N_u} \sum_{i=1}^{m} |\Delta u_i(t+j-1)| \quad (8.12)$$

where N_1 and N_2 define the prediction horizon and N_u defines the control horizon. If a series of $\mu_i \geq 0$ and $\beta_i \geq 0$ such that for all $\theta \in \Theta$,

$$\begin{array}{rcccl} -\mu_i & \leq & (y_i(t+j) - w_i(t+j)) & \leq & \mu_i \\ -\beta_i & \leq & \Delta u_i(t+j-1) & \leq & \beta_i \\ 0 & \leq & \sum_{i=1}^{n\times N} \mu_i + \lambda \sum_{i=1}^{m\times N_u} \beta_i & \leq & \gamma \end{array}$$

then γ is an upper bound of

$$\mu^*(\mathbf{u}) = \max_{\theta\in\mathcal{E}} \sum_{j=1}^{n}\sum_{i=1} |y_i(t+j,\theta) - w_i(t+j)| + \lambda \sum_{j=1}^{N_u}\sum_{i=1}^{m} |\Delta u_i(t+j-1)|$$

The problem is now reduced to minimizing the upper bound γ.

When the global uncertainty model is used and constraints on the output variables, the manipulated variables ($\underline{\mathbf{U}}, \overline{\mathbf{U}}$) and on the slew rate of the manipulated variables ($\underline{\mathbf{u}}, \overline{\mathbf{u}}$) are taken into account, the problem can be interpreted as an LP problem with:

$$\min_{\gamma,\mu,\beta,\mathbf{u}} \gamma$$

subject to

$$\left.\begin{array}{rcl} \mu & \geq & G_u\mathbf{u} + G_\theta\theta + \mathbf{f} + \mathbf{w} \\ \mu & \geq & -G_u\mathbf{u} - G_\theta\theta - \mathbf{f} + \mathbf{w} \\ \overline{\mathbf{y}} & \geq & G_u\mathbf{u} + G_\theta\theta + \mathbf{f} \\ -\underline{\mathbf{y}} & \geq & -G_u\mathbf{u} - G_\theta\theta - \mathbf{f} \end{array}\right\} \forall\theta\in\mathcal{E}$$

$$\begin{array}{rcl} \beta & \geq & \mathbf{u} \\ \beta & \geq & -\mathbf{u} \\ \overline{\mathbf{u}} & \geq & \mathbf{u} \\ -\underline{\mathbf{u}} & \geq & \mathbf{u} \\ \overline{\mathbf{U}} & \geq T & \mathbf{u} + \mathbf{1}u(t-1) \\ -\underline{\mathbf{U}} & \geq & -T\mathbf{u} - \mathbf{1}u(t-1) \\ \gamma & \geq & \mathbf{1}^t\mu + \lambda\mathbf{1}\beta \end{array}$$

The problem can be transformed into the usual form:

$$\min_{\mathbf{x}} \mathbf{c}^t\mathbf{x}, \quad \text{subject to } A\mathbf{x} \leq \mathbf{b}, \ \mathbf{x} \geq 0$$

with :

$$\mathbf{x} = \left[\begin{array}{c} \mathbf{u} - \underline{\mathbf{u}} \\ \hline \mu \\ \hline \beta \\ \hline \gamma \end{array}\right] \qquad \mathbf{c}^t = [\overbrace{0,\cdots,0}^{m\times N_u},\overbrace{0,\cdots,0}^{n\times N},\overbrace{0,\cdots,0}^{m\times N_u},1]$$

$$A^t = [A_1^t, \cdots, A_{2N}^t, A_u^t] \quad \mathbf{b}^t = [\mathbf{b}_1^t, \cdots, \mathbf{b}_{2N}^t, \mathbf{b}_u^t]$$

where the block matrices take the following form:

$$A_i = \left[\begin{array}{c|c|c|c} \mathbf{G}_u & -I & 0 & 0 \\ -\mathbf{G}_u & -I & 0 & 0 \\ \hline \mathbf{G}_u & 0 & 0 & 0 \\ -\mathbf{G}_u & 0 & 0 & 0 \end{array}\right] \qquad \mathbf{b}_i = \left[\begin{array}{c} -\mathbf{G}_u\underline{\mathbf{u}} - \mathbf{G}_\theta\theta_i - \mathbf{f} + \mathbf{w} \\ \mathbf{G}_u\underline{\mathbf{u}} + \mathbf{G}_\theta\theta_i + \mathbf{f} - \mathbf{w} \\ \hline \overline{\mathbf{y}} - \mathbf{G}_u\underline{\mathbf{u}} - \mathbf{G}_\theta\theta_i - \mathbf{f} \\ -\underline{\mathbf{y}} + \mathbf{G}_u\underline{\mathbf{u}} + \mathbf{G}_\theta\theta_i + \mathbf{f} \end{array}\right]$$

$$A_u = \left[\begin{array}{c|c|c|c} I & 0 & 0 & 0 \\ \hline I & 0 & -I & 0 \\ -I & 0 & -I & 0 \\ \hline T & 0 & 0 & \\ -T & 0 & 0 & 0 \\ \hline 0 & \underbrace{1,\cdots,1}_{n\times N} & \underbrace{1,\cdots,1}_{m\times N_u} & -1 \end{array}\right] \qquad \mathbf{b}_u = \left[\begin{array}{c} \overline{\mathbf{u}} - \underline{\mathbf{u}} \\ \hline -\underline{\mathbf{u}} \\ \underline{\mathbf{u}} \\ \hline \overline{\mathbf{U}} - T\underline{\mathbf{u}} - \mathbf{1}u(t-1) \\ -\underline{\mathbf{U}} + T\underline{\mathbf{u}} + \mathbf{1}u(t-1) \\ \hline 0 \end{array}\right]$$

where θ_i is the ith-vertex of $\mathcal{E}$. The number of variables involved in the linear programming problem is $2 \times m \times N_u + n \times N + 1$, while the number of constraints is $4 \times n \times N \times 2^{n\times N} + 5 \times m \times N_u + 1$. As the number of constraints is much higher than the number of decision variables, solving the dual LP problem should also be less computationally expensive than the primal problem.

The number of constraints can be reduced considerably as in the $\infty - \infty$-norm case because of the special form of the constraint matrix A. Consider the jth-row for each of the constraint blocks $A_i\mathbf{x} \leq \mathbf{b}_i$. As $A_1 = A_2 = \cdots = A_{2^N}$, the only constraint limiting the feasible region will be the one having the smallest value on the jth element of vector $\mathbf{b}_i$. Therefore, all the other $(2^N - 1)$ constraints can be eliminated. The number of constraints can be reduced to $4 \times n \times N + 5m \times N_u + 1$.

8.3 Illustrative Examples

8.3.1 Bounds on the Output

The setpoint of many processes in industry is determined by an optimization program in order to satisfy economic objectives. As a result, the optimal setpoint is usually on the intersection of some constraints. This is, for example, the normal situation when maximizing the throughput, which normally results in operating the process at extreme conditions as near as possible to the safety or quality constraints. Consideration of uncertainties may be of great interest for this type of situation. If an MPC that takes into account the constraints is used, the MPC will solve the problem keeping the expected values of the output signals within the feasible region, but, because of external perturbations and/or uncertainties, this does not guarantee that the output is going to be bound. When uncertainties are taken into account, the MPC will

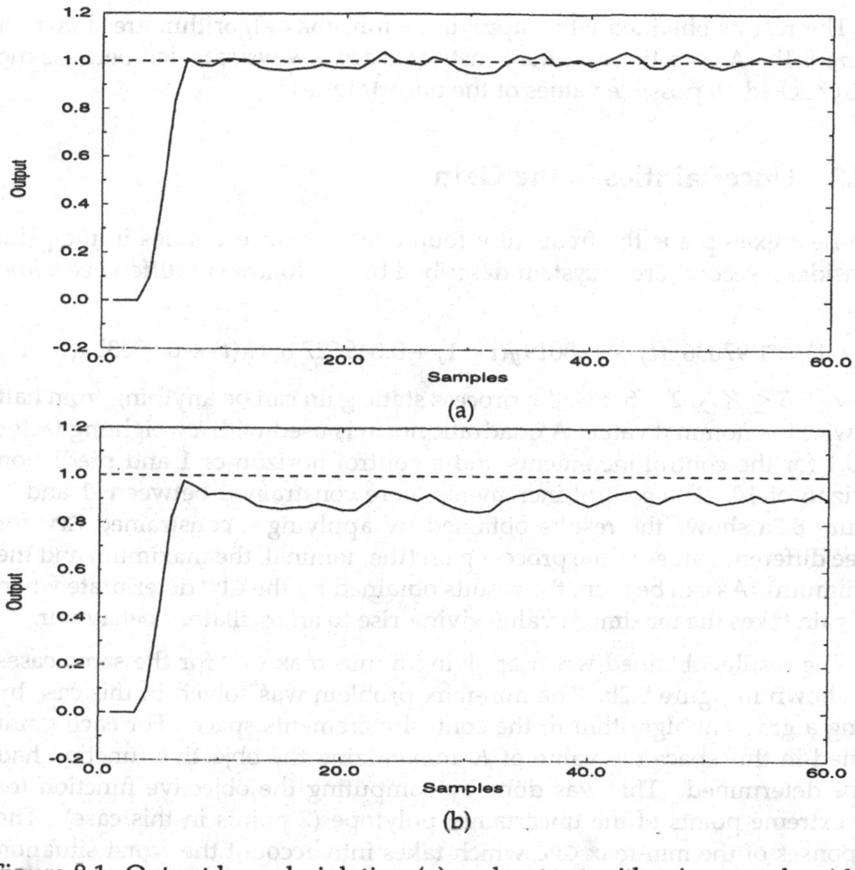

Figure 8.1: Output bound violation (a) and output with min-max algorithm (b)

minimize the objective function for the worst situation and keep the value of the variables within the constraint region for all possible cases of uncertainties.

In order to illustrate this point, consider the process described by the following difference equation:

$$y(t+1) = -1.4y(t) + 0.42y(t-1) + 0.1u(t) + 0.2u(t-1) + \frac{\theta(t+1)}{\Delta}$$

with $-0.03 \leq \theta(t) \leq 0.03$, $y(t) \leq 1$, and $-1 \leq \Delta u(t) \leq 1$. A 1-norm MPC is applied with a weighting factor of 0.2, and predictions and control horizon of 3 and 1 respectively. The setpoint is set at the output constraint value. The uncertainties are randomly generated within the uncertainty set with a uniform distribution. The results obtained are shown in figure 8.1a. Note that the output signal violates the constraint because the MPC only checked the constraints for the expected values.

The results obtained when applying a min-max algorithm are shown in figure 8.1b. As can be seen, the constraints are always satisfied because the MPC checked all possible values of the uncertainties.

8.3.2 Uncertainties in the Gain

The next example is the frequently found case of uncertainties in the gain. Consider a second order system described by the following difference equation:

$$y(t+1) = 1.97036y(t) - 0.98019y(t-1) + 0.049627\,K\,(u(t) + 0.99335u(t-1))$$

where $0.5 \leq K \leq 2$. That is, the process static gain can be anything from half to twice the nominal value. A quadratic norm is used with a weighting factor of 0.1 for the control increments and a control horizon of 1 and prediction horizon of 10. The control increments were constrained between -1 and 1. Figure 8.2a shows the results obtained by applying a constrained GPC for three different values of the process gain (the nominal, the maximum and the minimum). As can be seen, the results obtained by the GPC deteriorate when the gain takes the maximum value giving rise to an oscillatory behaviour.

The results obtained when applying a min-max GPC for the same cases are shown in figure 8.2b. The min-max problem was solved in this case by using a gradient algorithm in the control increments space. For each point visited in this space the value of K maximizing the objective function had to be determined. This was done by computing the objective function for the extreme points of the uncertainty polytope (2 points in this case). The responses of the min-max GPC which takes into account the worst situation are acceptable for all situations as can be seen in figure 8.2b.

A simulation study was carried out with 600 cases varying the process gain uniformly in the parameter uncertainty set from the minimum to maximum value. The bands limiting the output for the constrained GPC and the min-max constrained GPC are shown in figure 8.3. As can be seen, the uncertainty band for the min-max constrained GPC is much smaller than the one obtained for the constrained GPC.

8.4 Robust MPC and Linear Matrix Inequalities

Linear matrix inequalities (LMI) are becoming very popular in control and they have also been used in the MPC context.

A linear matrix inequality is an expression of the form:

$$F(x) = F_0 + \sum_{i=1}^{m} x_i F_i > 0 \tag{8.13}$$

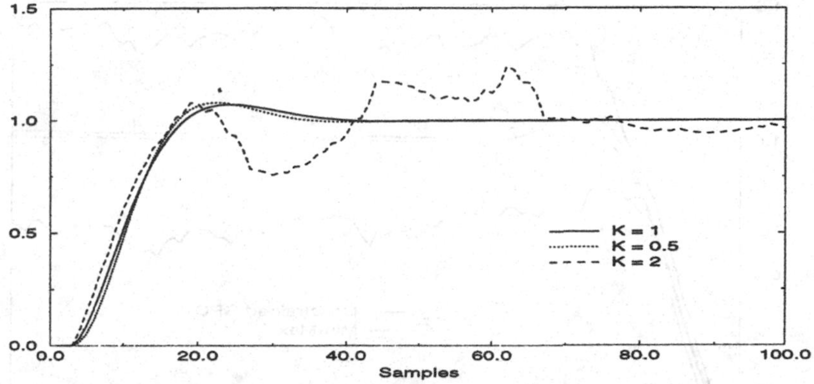

(a)

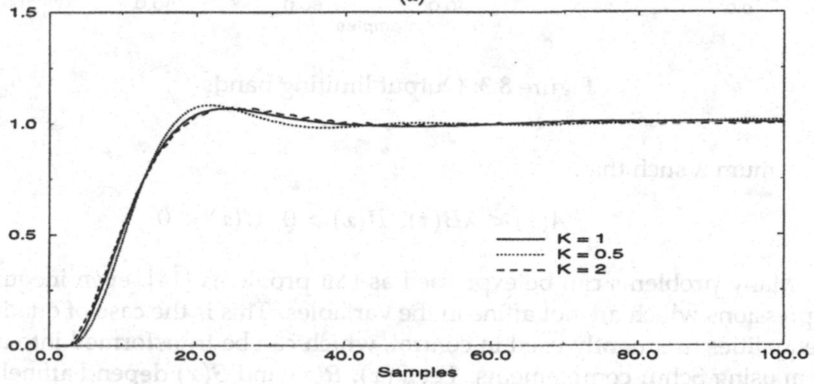

(b)

Figure 8.2: Uncertainty in the gain for constrained GPC (a) and min-max GPC (b)

where F_i are symmetrical real $n \times n$ matrices, the x_i are variables and $F(x) > 0$ means that $F(x)$ is positive definite. The three main LMI problems are:

1. The feasibility problem: determining variables x_1,x_2,...,x_m so that inequality (8.13) holds.

2. The linear programming problem: finding the optimum of:

$$\sum_{i=1}^{m} c_i x_i$$

subject to $F(x) > 0$.

3. The generalized eigenvalue minimization problem: Finding the mini-

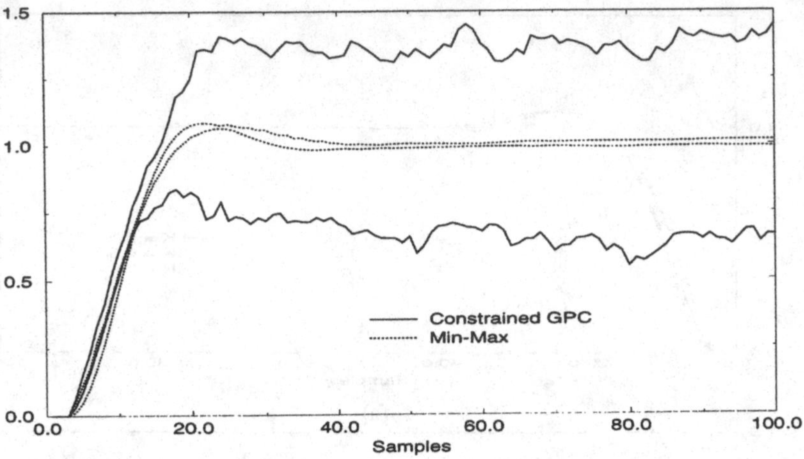

Figure 8.3: Output limiting bands

mum λ such that:

$$A(x) < \lambda B(x), \ B(x) > 0, \ C(x) < 0$$

Many problems can be expressed as LMI problems [14], even inequality expressions which are not affine in the variables. This is the case of quadratic inequalities, frequently used in control, which can be transformed into a LMI form using Schur complements: Let $Q(x)$, $R(x)$ and $S(x)$ depend affinely on x, and are $Q(x)$, $R(x)$ symmetrical. Then the LMI problem:

$$\begin{bmatrix} Q(x) & S(x) \\ S(x)^T & R(x) \end{bmatrix} > 0$$

is equivalent to

$$\begin{array}{cc} R(x) > 0, & Q(x) - S(x)R(x)^{-1}S(x)^T > 0 \\ \text{and} & \\ Q(x) > 0, & R(x) - S(x)^T Q(x)^{-1} S(x) > 0 \end{array}$$

There are efficient algorithms to solve LMI problems which have been applied to solve control problems such as robust stability, robust pole placement, optimal LQG and robust MPC. In this last context, Kothare *et al* [57] proposed a robust constrained model predictive control as follows:

Consider the linear time varying system:

$$x(k+1) \ = \ A(k)x(k) + B(k)x(k)$$

$$\begin{aligned} y(k) &= Cx(k) \\ [A(k)B(k)] &\in \Omega \\ R_x x(k) &\leq a_x \\ R_u u(k) &\leq a_u \end{aligned} \tag{8.14}$$

and the following cost function

$$J(k) = \sum_{i=0}^{\infty} \left(\hat{x}(k+i|k)^T Q_1 \hat{x}(k+i|k) + u(k+i)^T R u(k+i)\right) \tag{8.15}$$

The robust optimization problem is stated as

$$\min_{u(k+i|k), i\geq 0} \quad \max_{[A(k+i), B(k+i)]\in\Omega, i\geq 0} \quad J(k) \tag{8.16}$$

Define a quadratic function $V(x) = x^T P x$ with $P > 0$ such that $V(x)$ is an upper bound on $J(k)$:

$$\max_{[A(k+i), B(k+i)]\in\Omega, i\geq 0} J(k) \leq V(x(k|k)) \tag{8.17}$$

The problem is solved by finding a linear feedback control law $u(k+i|k) = Fx(k+i|k)$ such that $V(x(k|k))$ is minimized. Suppose that there are no constraints in the state and inputs and that the model uncertainties are defined as follows:

$$\Omega = Co\{[A_1, B1], [A_2, B_2], \cdots, [A_L, B_L]\} \tag{8.18}$$

where Co denotes the convex hull defined by vertices $[A_i, B_i]$. That is, any plant $[A, B] \in \Omega$ can be expressed as:

$$[A, B] = \sum_{i=1}^{L} \lambda_i [A_i, B_i]$$

with $\lambda_i \geq 0$ and $\sum_{i=1}^{L} \lambda_i = 1$

In these conditions, the robust MPC can be transformed into the following LMI problem:

$$\min_{\gamma, Q, Y} \gamma \tag{8.19}$$

subject to:

$$\begin{bmatrix} 1 & x(k|k)^T \\ x(k|k) & Q \end{bmatrix} \geq 0 \tag{8.20}$$

$$\begin{bmatrix} Q & QA_j^T + Y^T B_j^T & QQ_1^{1/2} & Y^T R^{1/2} \\ A_j Q + B_j Y & Q & 0 & 0 \\ Q_1^{1/2} Q & 0 & \gamma I & 0 \\ R^{1/2} & 0 & 0 & \gamma I \end{bmatrix} \geq 0, \quad j = 1, \cdots, L \tag{8.21}$$

Once the above LMI problem is solved, the feedback gain can be obtained by:

$$F = YQ^{-1}$$

Kothare *et al* [57] demonstrated that constraints on the state and manipulated variables and other types of uncertainties can also be formulated, and solved, as LMI problems.

The main drawbacks of this method are that:

- Although LMI algorithms are supposed to be numerically efficient, they are not as efficient as specialized LP or QP algorithms.
- The manipulated variables are computed as a linear feedback of the state vector satisfying constraints, but when constraints are present, the optimum does not have to be linear.
- Feasibility problems are more difficult to treat as the physical meaning of constraints is somehow lost when transforming them into the LMI format.

Chapter 9

Applications

This chapter is dedicated to presenting some MPC applications to the control of different real and simulated processes. The first application presented corresponds to a self-tuning and a gain scheduling GPC for a distributed collector field of a solar power plant. In order to illustrate how easily the control scheme shown in chapter 5 can be used in any commercial distributed control system, some applications concerning the control of typical variables such as flows, temperatures and levels of different processes of a pilot plant are presented. Finally the application of a GPC to a diffusion process of a sugar factory is presented.

9.1 Solar Power Plant

This section presents an application of an adaptive long-range predictive controller to a solar power plant and shows how this type of controller, which normally requires a substantial amount of computation (in the adaptive case), can easily be implemented with few computation requirements. The results obtained when applying the controller to the plant are also shown.

The controlled process is the distributed collector field (Acurex) of the Solar Platform of Tabernas (Spain). The distributed collector field consists mainly of a pipe line through which oil is flowing and on to which the solar radiation is concentrated by means of parabolic mirrors, which follow the sun by rotating on one axis, in order to heat the oil. It consists of 480 modules arranged in twenty lines which form ten parallel loops. A simplified diagram of the solar collector field is shown in figure 9.1. The field is also provided with a sun tracking mechanism which causes the mirrors to revolve around an axis parallel to that of the pipe line.

On passing through the field the oil is heated and then introduced into a

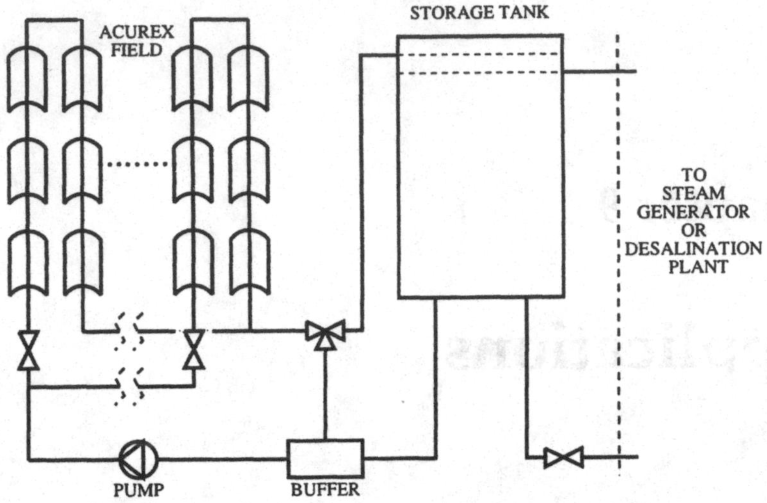

Figure 9.1: Schematic diagram of collectors field

storage tank to be used for the generation of electrical energy. The hot oil can also be used directly for feeding the heat exchanger of a desalination plant. The cold inlet oil to the field is extracted from the bottom of the storage tank.

Each of the loops mentioned above is formed by four twelve-module collectors, suitably connected in series. The loop is 172 metres long, the active part of the loop (exposed to concentrated radiation) measuring 142 metres and the passive part 30 metres.

The system is provided with a three way valve which allows the oil to be recycled in the field until its outlet temperature is adequate to enter into the storage tank. A more detailed description of the field can be found in [51].

A fundamental feature of a solar power plant is that the primary energy source, whilst variable, cannot be manipulated. The intensity of the solar radiation from the sun, in addition to its seasonal and daily cyclic variations, is also dependent on atmospheric conditions such as cloud cover, humidity, and air transparency. It is important to be able to maintain a constant outlet temperature for the fluid as the solar conditions change, and the only means available for achieving this is via the adjustment of the fluid flow.

The objective of the control system is to maintain the outlet oil temperature at a desired level in spite of disturbances such as changes in the solar irradiance level (caused by clouds), mirror reflectivity or inlet oil temperature. This is accomplished by varying the flow of the fluid through the field. The field exhibits a variable delay time that depends on the control variable (flow). The transfer function of the process varies with factors such as irradiance level, mirror reflectance and oil inlet temperature.

The distributed collector field is a nonlinear system which can be approximated by a linear system when considering small disturbances. The maintenance of a constant outlet temperature throughout the day as the solar conditions change requires a wide variation in the operational flow level. This in turn leads to substantial variations in the general dynamic performance and in particular, from the control viewpoint, gives rise to a system time delay which varies significantly. The controller parameters need to be adjusted to suit the operating conditions, and self-tuning control offers one approach which can accommodate such a requirement.

Because of the changing dynamics and strong perturbations, this plant has been used to test different types of controllers [18],[19].

For self-tuning control purposes a simple, linear model is required which relates changes in fluid flow to changes in outlet temperature.

Observations of step responses obtained from the plant indicate that in the continuous time domain behaviour it can be approximated by a first order transfer function with a time delay. Since the time delay τ_d is relatively small compared to the fundamental time constant τ, a suitable discrete model can be constructed by choosing the sample period T to be equal to the lowest value of the time delay τ_d. This corresponds to the operating condition where the flow level is highest. The discrete transfer function model then takes the form:

$$g(z^{-1}) = z^{-k}\frac{bz^{-1}}{1-az^{-1}}$$

and at the high flow level condition k = 1.

9.1.1 Self tuning GPC Control Strategy

A particular feature of the system is the need to include a series feedforward term in the control loop [17]. The plant model upon which the self-tuning controller is based relates changes in outlet temperature to changes in fluid flow only. The outlet temperature of the plant, however, is also influenced by changes in system variables such as solar radiation and fluid inlet temperature. During estimation if either of these variables changes it introduces a change in the system output unrelated to fluid flow which is the control input signal, and in terms of the model would result in unnecessary adjustments of the estimated system parameters.

Since both solar radiation and inlet temperature can be measured, this problem can be eased by introducing a feedforward term in series to the system, calculated from steady state relationships, which makes an adjustment in the fluid flow input, aimed to eliminate the change in outlet temperature caused by the variations in solar radiation and inlet temperature. If the feedforward term perfectly countered the changes in solar radiation and inlet temperature, then the observed outlet temperature changes would only be caused by changes in the control input signal. Although exact elimination obviously cannot be achieved, a feedforward term based on steady state considerations overcomes the major problems inherent in the single input model and permits the successful estimation of model parameters. The basic idea is to compute the necessary oil flow to maintain the desired outlet oil temperature given the inlet oil temperature and the solar radiation. The feedforward signal also provides control benefits when disturbances in solar radiation and fluid inlet temperature occur, but the basic reason for its inclusion is that of preserving the validity of the assumed system model in the self-tuning algorithm. For more details see [23].

In the control scheme, the feedforward term is considered as a part of the plant, using the set point temperature to the feedforward controller as the control signal.

In this section, the precalculated method described in chapter 5 is used. As the deadtime d is equal to 1, the control law is given by:

$$\triangle u(t) = l_{y1}\hat{y}(t+1 \mid t) + l_{y2}y(t) + l_{r1}r(t) \tag{9.1}$$

where the value $\hat{y}(t+1 \mid t)$ is obtained by the use of the predictor:

$$\hat{y}(t+j+1 \mid t) = (1+a)\hat{y}(t+j \mid t) - a\hat{y}(t+j-1 \mid t) + b \triangle u(t+j-1) \tag{9.2}$$

The control scheme proposed is shown in figure 9.2. The plant estimated parameters are used to compute the controller coefficients (l_{y1}, l_{y2}, l_{r1}) via the adaptation mechanism. Notice that in this scheme, the feedforward term is considered as a part of the plant (the control signal is the set point temperature for the feedforward controller instead of the oil flow). This signal is saturated and filtered before its use in the estimation algorithm. The controller also has a saturation to limit the increase of the error signal.

As suggested in chapter 5, a set of GPC parameters were obtained for $\delta(i) = 1$, $\lambda(i) = 5$ and $N = 15$. The pole of the system has been changed with a 0.0005 step from 0.85 to 0.95, which are the values that guarantee the system stability if the parameter set estimation is not accurate enough. Notice that owing to the fact that the closed-loop static gain must equal the value unity, the sum of the three parameters equals zero. The curves shown in figure 9.3 correspond to the controller parameters l_{y1},l_{y2}, l_{r1} for the values of the pole mentioned above.

The adjustment of analytical functions to the calculated values provide:

$$\begin{aligned} l_{y1} &= 0.4338 - 0.6041\, \hat{a} \,/\, (1.11 - \hat{a}) \\ l_{y2} &= -0.4063 + 0.4386\, \hat{a} \,/\, (1.082 - \hat{a}) \\ l_{r1} &= -l_{y1} - l_{y2} \end{aligned} \tag{9.3}$$

These expressions give a very good approximation to the true controller parameters and fit the set of computed data with a maximum error of less

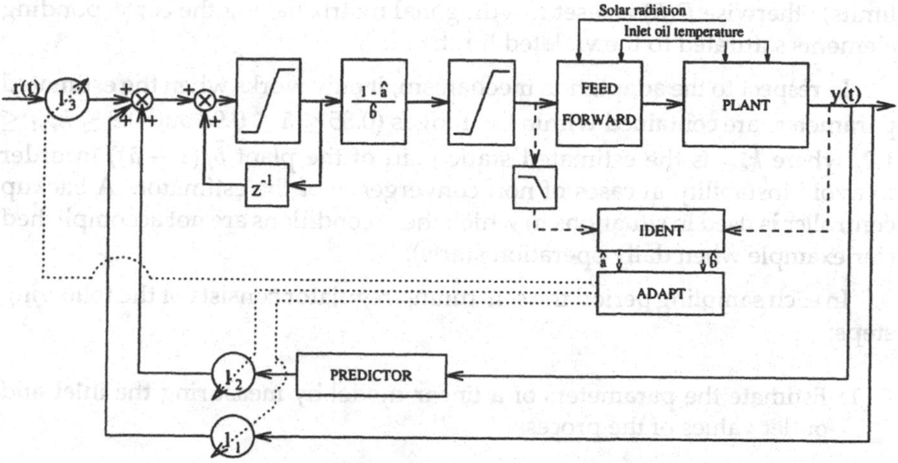

Figure 9.2: Self-tuning control scheme

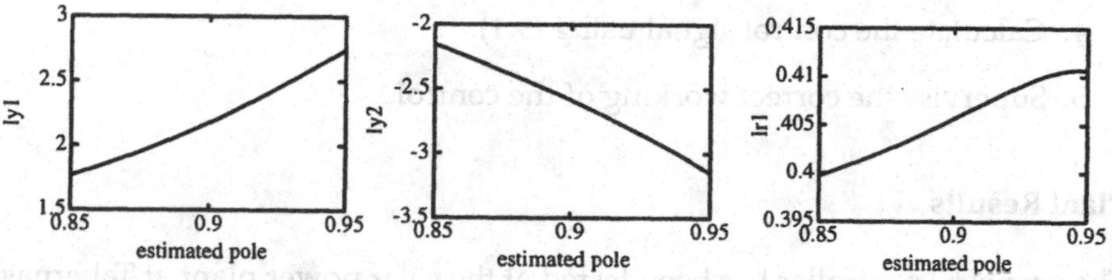

Figure 9.3: Controller parameters for $\lambda = 5$

than 0.6 per cent of the nominal values for the range of interest of the open loop pole. For more details see [18].

The parameters of the system model in the control scheme are determined on-line via recursive least squares estimation. The estimation algorithm incorporates a variable forgetting factor and only works when the input signal contains dynamic information. These considerations can be taken into account by checking the following conditions:

$$| \Delta u | \geq A$$

$$\sum_{k=-N}^{k=0} | \Delta u(k) | \geq B$$

If one of these conditions is true, the identifier is activated. Otherwise, the last set of estimated parameters are used. Typical values of A, B and N chosen from simulation studies are: $A = 9, 7 \leq B \leq 9$ and $N = 5$. The covariance matrix $P(k)$ is also monitored by checking that its diagonal elements are kept within limits; otherwise $P(k)$ is reset to a diagonal matrix having the corresponding elements saturated to the violated limit.

In respect to the adaptation mechanism, it only works when the estimated parameters are contained within the ranges ($0.85 \leq \hat{a} \leq 0.95$ and $0.9 \leq \hat{k}_{est} \leq 1.2$, where $\hat{k}_{est}$ is the estimated static gain of the plant $\hat{b}/(1-\hat{a})$) in order to avoid instability in cases of non convergence of the estimator. A backup controller is used in situations in which these conditions are not accomplished (for example when daily operation starts).

In each sampling period the self-tuning regulator consists of the following steps:

1. Estimate the parameters of a linear model by measuring the inlet and outlet values of the process.

2. Adjust the parameters of the controller using expressions (9.3).

3. Compute $\hat{y}(t+d \mid t)$ using the predictor (9.2).

4. Calculate the control signal using (9.1).

5. Supervise the correct working of the control.

Plant Results

The proposed controller has been tested at the solar power plant at Tabernas, Almería (Spain) under various operating conditions.

Figure 9.4 shows the outlet oil temperature and reference when the proposed self-tuning generalized predictive controller is applied to the plant. The value of the control weighting λ was made equal to 5 and, as can be seen, a fast response to changes in the set point is obtained (the rising time observed is approximately 7 minutes). When smaller overshoots are required, the control weighting factor has to be increased.

The evolution of the irradiation for the test can be seen in the same figure and, as can be seen, it corresponds to a day with small scattered clouds. The oil flow changed from 4.5 l/s to 7 l/s, and the controller could maintain performance in spite of the changes in the dynamics of the process caused by flow changes.

9.1.2 Gain scheduling Generalized Predictive Control

There are many situations in which it is known how the dynamics of a process change with the operating conditions. It is then possible to change the controller parameters taking into account the current operating point of the system. Gain scheduling is a control scheme with open loop adaptation, which

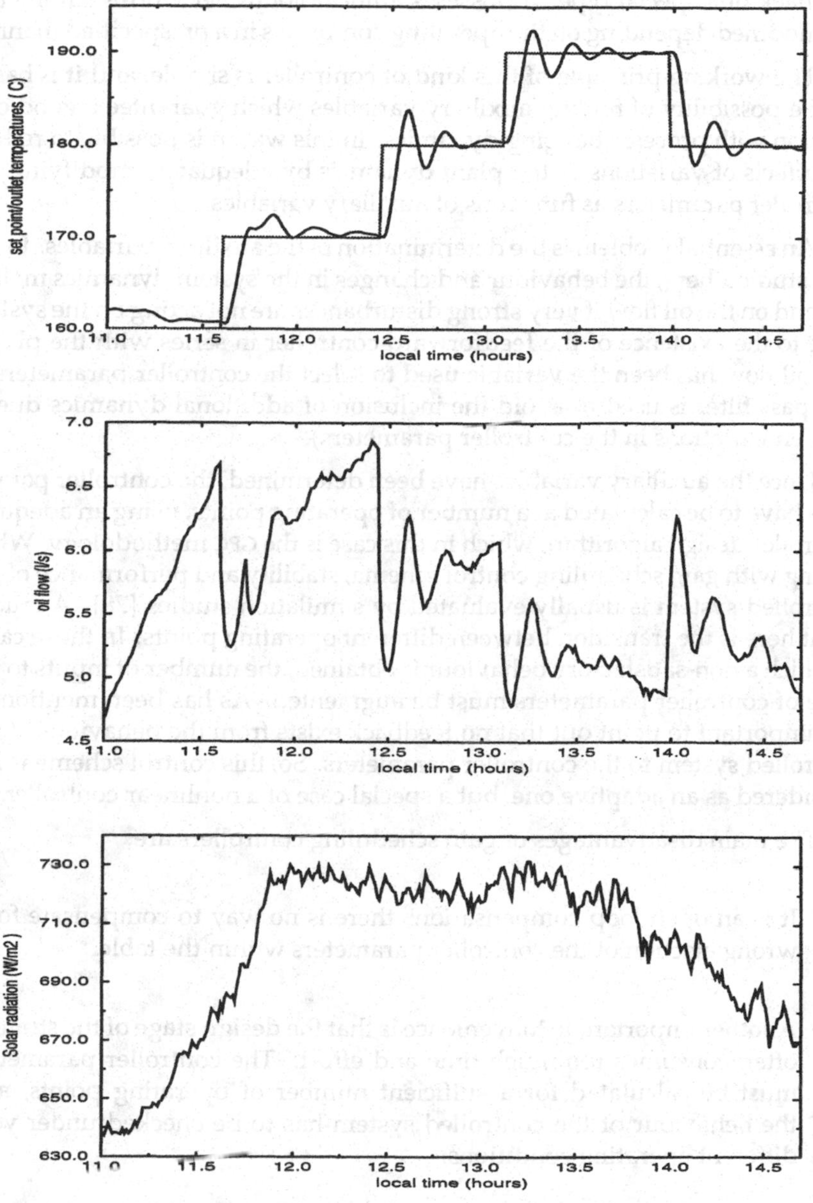

Figure 9.4: Adaptive GPC: Plant outlet oil temperature, flow and solar radiation

can be seen as a feedback control system in which the feedback gains are adjusted by a feedforward compensation. Gain scheduling control is a nonlinear feedback of a special type: it posseses a linear controller whose parameters are modified depending on the operating conditions in a prespecified manner.

The working principle of this kind of controller is simple, and it is based on the possibility of finding auxiliary variables which guarantee a good correlation with process changing dynamics. In this way, it is possible to reduce the effects of variations in the plant dynamics by adequately modifying the controller parameters as functions of auxiliary variables.

An essential problem is the determination of the auxiliary variables. In the case studied here, the behaviour and changes in the system dynamics mainly depend on the oil flow if very strong disturbances are not acting on the system (due to the existence of the feedforward controller in series with the plant). The oil flow has been the variable used to select the controller parameters (a low pass filter is used to avoid the inclusion of additional dynamics due to sudden variations in the controller parameters).

Once the auxiliary variables have been determined, the controller parameters have to be calculated at a number of operating points, using an adequate controller design algorithm, which in this case is the GPC methodology. When coping with gain scheduling control schema, stability and performance of the controlled system is usually evaluated by simulation studies [78]. A crucial point here is the transition between different operating points. In those cases in which a non-satisfactory behaviour is obtained, the number of inputs to the table of controller parameters must be augmented. As has been mentioned, it is important to point out that no feedback exists from the behaviour of the controlled system to the controller parameters. So, this control scheme is not considered as an adaptive one, but a special case of a nonlinear controller.

The main disadvantages of gain scheduling controllers are:

- It is an open loop compensation: there is no way to compensate for a wrong election of the controller parameters within the table.

- Another important inconvenience is that the design stage of the strategy often consumes too much time and effort. The controller parameters must be calculated for a sufficient number of operating points, and the behaviour of the controlled system has to be checked under very different operating conditions.

Its main advantage is the easiness of changing controller parameters in spite of changes in process dynamics. As classical examples of applications of this kind of controller, the following control fields can be mentioned: design of ship steering autopilots, pH control, combustion control, engine control, design of flight autopilots, etc. [9].

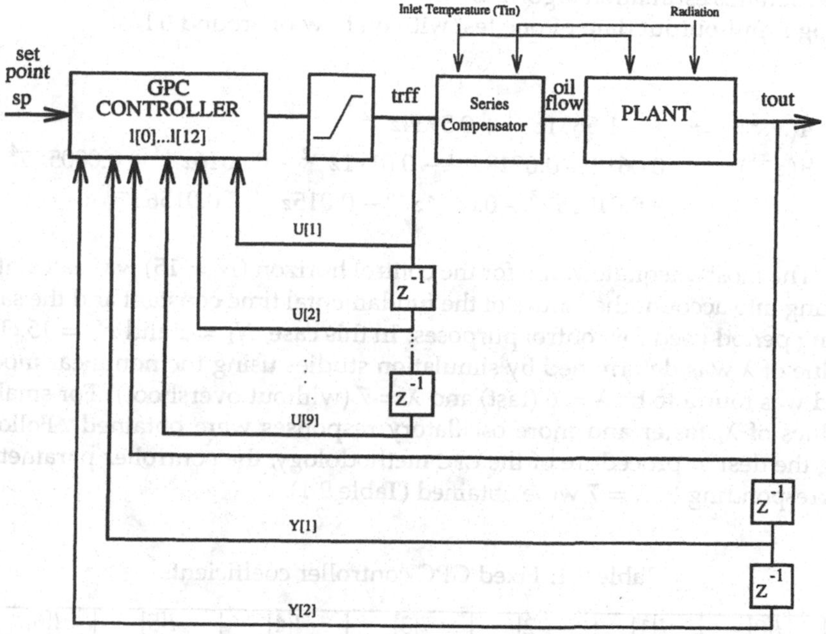

Figure 9.5: Control scheme using high order models

Plant models and fixed parameter controllers

The frequency response of the plant has been obtained by performing PRBS test in different operating conditions, using both the plant and a nonlinear distributed parameter model[1] [11]. In this way, different linear models were obtained from input-output data in different working conditions. These models relate changes in the oil flow to those of the outlet oil temperature, and can take into account the antiresonance characteristics of the plant if they are adequately adjusted. The control structure proposed is shown in Figure 9.5. As can be seen, the output of the generalized predictive controller is the input (t_{rff}) of the series compensation controller, which also uses the solar radiation, inlet oil temperature and reflectivity to compute the value of the oil flow, which is sent to the pump controller.

The controller parameters were obtained from a linear model of the plant. From input output data of the plant, the degrees of the polynomials A and B and the delay (of a CARIMA plant model) that minimizes Akaike's Information Theoretic Criterion (AIC) were found to be $n_a = 2$, $n_b = 8$ and $d = 0$. By a

[1]The software simulation package for the solar distributed collector field with real data from the plant can be obtained by contacting the authors or by accessing the author's home page http://www.esi.us.es/eduardo

least squares estimation algorithm, the following polynomials were obtained using input-output data of one test with oil flow of around 6 l/s:

$$\begin{aligned} A(z^{-1}) &= 1 - 1.5681z^{-1} + 0.5934z^{-2} \\ B(z^{-1}) &= 0.0612 + 0.0018z^{-1} - 0.0171z^{-2} + 0.0046z^{-3} + 0.0005z^{-4} \\ &\quad +0.0101z^{-5} - 0.0064z^{-6} - 0.015z^{-7} - 0.0156z^{-8} \end{aligned}$$

The most adequate value for the control horizon ($N = 15$) was calculated taking into account the values of the fundamental time constant and the sampling period used for control purposes. In this case, $N_1 = 1$ and $N_2 = 15$. The value of λ was determined by simulation studies using the nonlinear model and was found to be $\lambda = 6$ (fast) and $\lambda = 7$ (without overshoot). For smaller values of λ, faster and more oscillatory responses were obtained. Following the design procedure of the GPC methodology, the controller parameters corresponding to $\lambda = 7$ were obtained (Table 9.1).

Table 9.1: Fixed GPC controller coefficients

$l[0]$	$l[1]$	$l[2]$	$l[3]$	$l[4]$	$l[5]$	$l[6]$
−2.4483	6.8216	−4.7091	−0.0644	−0.0526	−0.0084	0.0629
$l[7]$	$l[8]$	$l[9]$	$l[10]$	$l[11]$	$l[12]$	
0.0161	0.0311	−0.0631	0.0231	1.0553	0.3358	

The control law can be written by:

$$\begin{aligned} t_{rff} &= l[2]t_{out} + l[1]y[1] + l[0]y[2] + l[6]u[6] + l[3]u[9] + l[4]u[8] + l[5]u[7] \\ &\quad +l[7]u[5] + l[8]u[4] + l[9]u[3] + l[10]u[2] + l[11]u[1] + l[12]sp \end{aligned} \tag{9.4}$$

where:

t_{rff}: Reference temperature for the feedforward controller.

t_{out}: Outlet temperature of the field.

sp: Set point temperature.

$l[i]$: Controller parameters.

$y[i]$: Outlet temperature of the field at sampling time $(t - i)$.

$u[i]$: Reference temperature for the feedforward controller at sampling time $(t - i)$.

With these values, the behaviour of this fixed parameter controller was analyzed in operation with the distributed solar collector field. The outlet oil

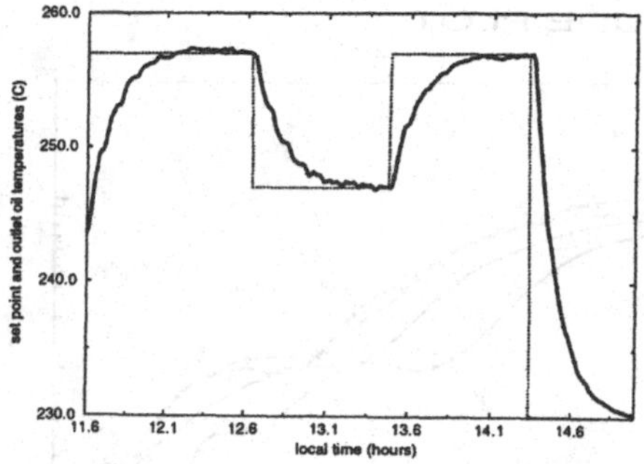

set point and outlet oil temperatures (^{o}C)

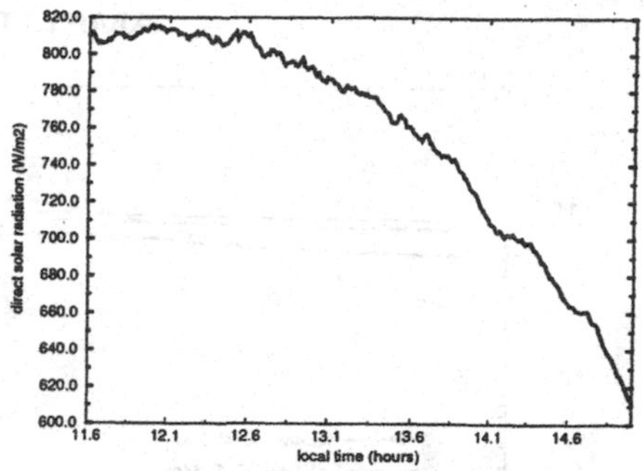

direct solar radiation (W/m^2)

Figure 9.6: Test with the fixed GPC high order controller

temperature of the field evolution and corresponding set point can be seen in Figure 9.6. The evolution of the solar radiation during this test can also be seen in Figure 9.6. Although direct solar radiation goes from 810 W/m^2 to 610 W/m^2, the field was working in midflow conditions because the set point was also changed from 258°C to 230°C. When operating conditions in the field change, the dynamics of the plant also change and the controller should be redesigned to cope with the control objectives.

The dynamics of the field are mainly dictated by the oil flow, which depends on the general operating conditions of the field: solar radiation, reflectivity, oil inlet temperature, ambient temperature and outlet oil temperature set point. These changes in plant dynamics are illustrated in Figure 9.7, where the frequency response of the nonlinear distributed parameter dynamic model of the field can be seen. The curves shown in Figure 9.7 were obtained by a spectral analysis of the input-output signals of the model at different operating points (PRBS signals were used for the input).

As can be seen, the frequency response changes significantly for different operating conditions. The steady state gain changes for different operating points, as well as the location of the antiresonance modes.

Taking into account the frequency response of the plant and the different linear models obtained from it, it is clear that a self-tuning controller based on this type of model is very difficult to implement. The fundamental reason is the fact that the estimation of the model parameters requires a lot of computation when the number of estimated parameters increases and the convergence of the estimation process is seldom accurate and fast enough.

Another way of coping with changing dynamics is to use a gain scheduling controller, making the controller parameters dependent on some variables which indicate the operating conditions.

With the input-output data used to obtain the frequency responses shown

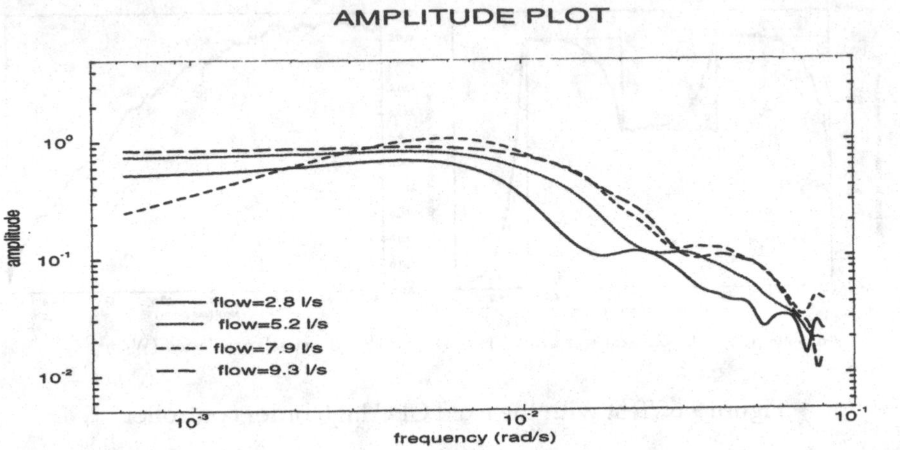

Figure 9.7: Frequency response of the field under different operating conditions

in Figure 9.7 and using the method and type of model previously described for the case of a high order fixed parameter controller, process ($a[i]$ and $b[i]$) and controller ($l[i]$) parameters were obtained for several oil flow conditions (q_1 = 2.8 l/s, q_2 = 5.2 l/s, q_3 = 7.9 l/s and q_4 = 9.3 l/s), using different values of the weighting factor λ. Tables 9.2 and 9.3 contain model and control parameters respectively for a weighting factor $\lambda = 6$. A value of $\lambda = 7$ has also been used to obtain responses without overshoot.

Table 9.2: Coefficients of polynomials $A(z^{-1})$ and $B(z^{-1})$ for different flows

	q_1	q_2	q_3	q_4
$a[1]$	−1.7820	−1.438	−1.414	−1.524
$a[2]$	0.81090	0.5526	0.5074	0.7270
$b[0]$	0.00140	0.0313	0.0687	0.0820
$b[1]$	0.03990	0.0660	0.0767	0.0719
$b[2]$	−0.0182	−.0272	−.0392	−.0474
$b[3]$	−0.0083	0.0071	0.0127	0.0349
$b[4]$	0.00060	0.0118	0.0060	0.0098
$b[5]$	−.00001	0.0138	−.0133	−.0031
$b[6]$	0.00130	0.0098	−.0156	0.0111
$b[7]$	0.00160	0.0027	−.0073	0.0171
$b[8]$	0.00450	−.0054	0.0037	0.0200

The controller parameters which are applied in real operation are obtained by using a linear interpolation with the data given in Table 9.3. It is important

Table 9.3: GPC controller coefficients in several operating points ($\lambda = 6$)

$l[0]$	−7.0481	−1.4224	−1.1840	−1.3603
$l[1]$	16.2223	3.84390	3.48440	3.02280
$l[2]$	−9.5455	−2.7794	−2.6527	−2.0142
$l[3]$	0.03910	−0.0139	0.00860	0.03740
$l[4]$	0.00980	0.00830	−0.0184	0.02730
$l[5]$	0.00560	0.02610	−0.0352	0.01080
$l[6]$	−0.0070	0.03390	−0.0239	−0.0197
$l[7]$	−0.0016	0.02480	0.02630	0.00460
$l[8]$	−0.0793	0.00880	0.03980	0.05070
$l[9]$	−0.1575	−0.0822	−0.0869	−0.1098
$l[10]$	0.36470	0.16410	0.19600	0.12480
$l[11]$	0.82620	0.83010	0.89360	0.87390
$l[12]$	0.37130	0.35800	0.35230	0.35170

to point out that to avoid the injection of disturbances during the controller gain adjustment, it is necessary to use a smoothing mechanism of the transition surfaces of the controller gains. In this case, a linear interpolation in combination with a first order filter has been used, given a modified flow $Q(t) = .95\,Q(t-1) + .05\,q(t)$ (being $q(t)$ the value of the oil flow at instant t and $Q(t)$ the filtered value used for controller parameter adjustment). The linear interpolation has also been successfully applied by [50]. Another kind of gain scheduling approach can be obtained by switching from one set of controller parameters to another depending on the flow conditions, without interpolating between controller parameters. The set of controller parameters c can be obtained by choosing one of the sets c_i in table 9.3, related to flow conditions q_i ($i = 1, 2, 3, 4$):

$$\begin{array}{rll} \text{if } \dfrac{q_{i-1}+q_i}{2} < q \leq \dfrac{q_i+q_{i+1}}{2} & \text{then} & c = c_i,\ i = 2,3 \\ \text{if } q \leq q_1 & \text{then} & c = c_1 \\ \text{if } q \geq q_4 & \text{then} & c = c_4 \end{array}$$

The control structure is similar to the one obtained for the fixed controller previously studied. The optimal realization of the gain scheduling controller consists of calculating the controller parameters under a number of operating conditions, and suppose that the values of the controller coefficients are constant between different operating conditions, generating a control surface based on an optimization criterion which takes into account the tracking error and control effort. It is evident that if the procedure is applied at many working points, an optimum controller will be achieved for those operating

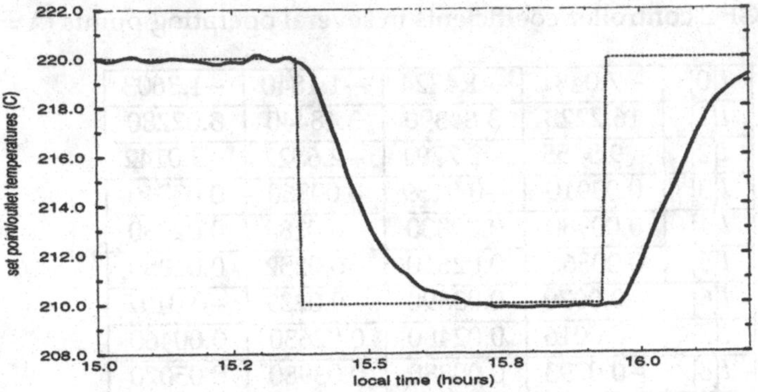

Set point and oil outlet temperatures °C

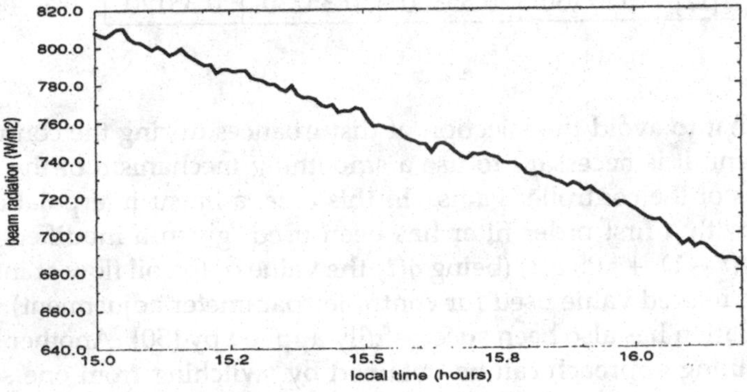

Direct solar radiation (W/m^2)

Figure 9.8: Test with the gain scheduling GPC controller. $\lambda = 7$

conditions if there is a high correlation between the process dynamics and the auxiliary variable. The drawback to this solution is that the design process become tedious. This is one of the main reasons for including a linear interpolation between the controller parameters.

Plant results

In the case of real tests, similar results were obtained and depending on the operating point, disturbances due to passing clouds, inlet oil temperature variations, etc., different performance was achieved.

Figure 9.8 shows the results of one of these tests with the gain scheduling

GPC with $\lambda = 7$. The operating conditions correspond to a clear afternoon with the solar radiation changing from 800 W/m^2 to 660 W/m^2 and oil flow changing from 3.75 l/s to 2 l/s. As can be seen, the effect of the antiresonance modes does not appear in the response, due to the use of an extended high order model which accounts for these system characteristics.

Figure 9.9 shows the result of a test with a weighting factor $\lambda = 7$ corresponding to a day of intermittent scattered clouds which produce large changes in the solar radiation level and the inlet oil temperature changing from 170°C to 207°C. As can be seen, the outlet oil temperature follows the set point in spite of changing operating conditions and the high level of noise in the radiation level produced by clouds.

The results of a test corresponding to a day with sudden changes in the solar radiation caused by clouds can be seen in Figure 9.10. As can be seen the controller (also designed with $\lambda = 7$) is able to handle different operating conditions and the sudden perturbations caused by the clouds. After the tests presented using a weighting factor $\lambda = 7$, two new test campaigns were carried out to test the behaviour of the controller with a weighting factor $\lambda = 6$. In the first campaign, the evaluation of the controller performance was considered. In the second, the behaviour of the controller operating under extreme working conditions was studied.

Figure 9.11 shows the results obtained in the operation on a day with normal levels of solar radiation, but on which a wide range of operating conditions is covered (oil flow changing between 2 l/s and 8.8 l/s) by performing several set point changes. At the start of the operation there is an overshoot of 6°C, due to the irregular conditions of the oil flowing through the pipes because the operation starts with a high temperature level at the bottom of the storage tank. After the initial transient, it can be observed that the controlled system quickly responds to set point changes under the whole range of operating conditions with a negligible overshoot. The rise time is about 6 minutes with a set point change of 15 degrees, as can be seen in Figure 9.11, with smooth changes in the control signal, constituting one of the best controllers implemented at the plant. It is important to note that the controller behaves well even with great set point changes.

9.2 Pilot Plant

In order to show how GPC can be implemented on an industrial SCADA, applications to the control of the most typical magnitudes found in the process industry (flow, level, temperatures) are introduced in this section.

The tests are carried out on a Pilot Plant existing in the *Departamento de Ingeniería de Sistemas y Automática* of the University of Seville. The Pilot Plant is provided with industrial instrumentation and is used as a test bed

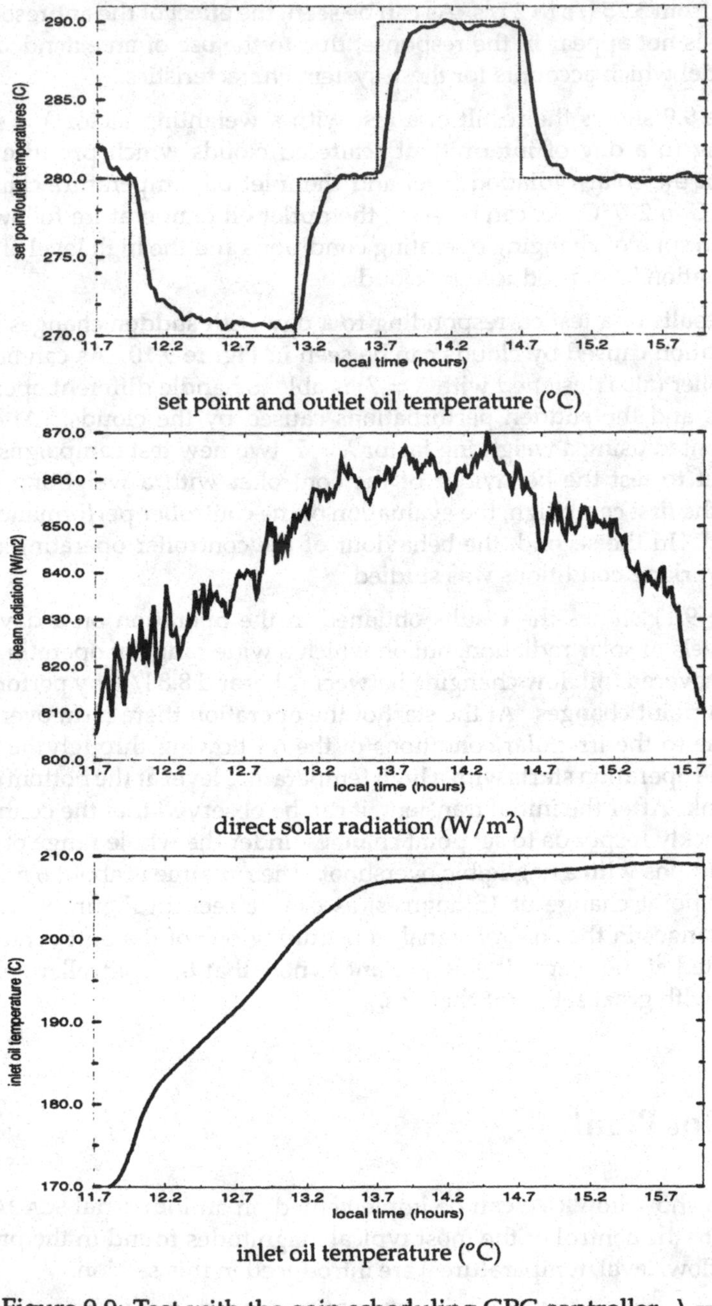

Figure 9.9: Test with the gain scheduling GPC controller. $\lambda = 7$

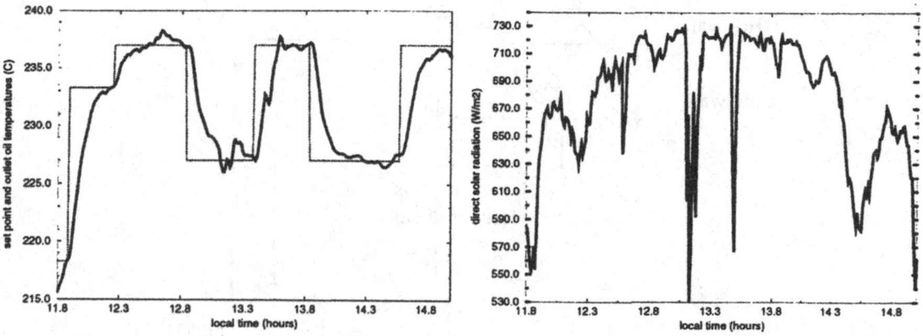

set point and outlet oil temperatures (°C) direct solar radiation (W/m^2)

Figure 9.10: Test with the gain scheduling GPC controller. $\lambda = 7$

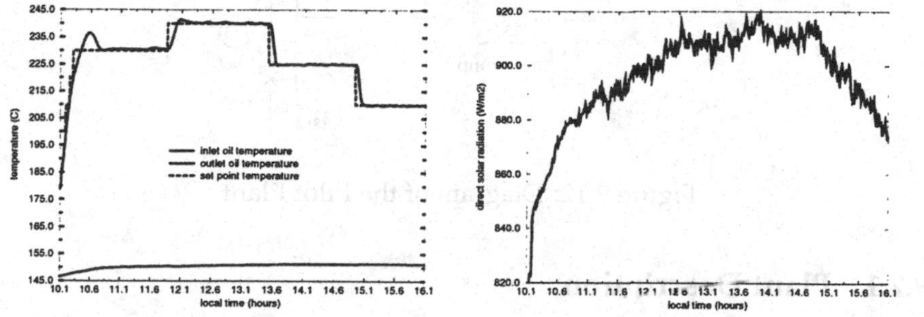

Figure 9.11: Test with the gain scheduling GPC controller. $\lambda = 6$

for new control strategies which can be implemented on an industrial SCADA connected to it. This Plant is basically a system using water as the working fluid in which various thermodynamic processes with interchange of mass and energy can take place. It essentially consists of a tank with internal heating with a series of input-output pipes and recirculation circuit with a heat exchanger.

The design of the Plant allows for various control strategies to be tested in a large number of loops. Depending on the configuration chosen, it is possible to control the types of magnitudes most frequently found in the process industry such as temperature, flow, pressure and level. For this, four actuators are available: three automatic valves and one electric heater that heats the interior of the tank. Later some of the possible loops are chosen (considered as being independent) for implanting the GPC controllers.

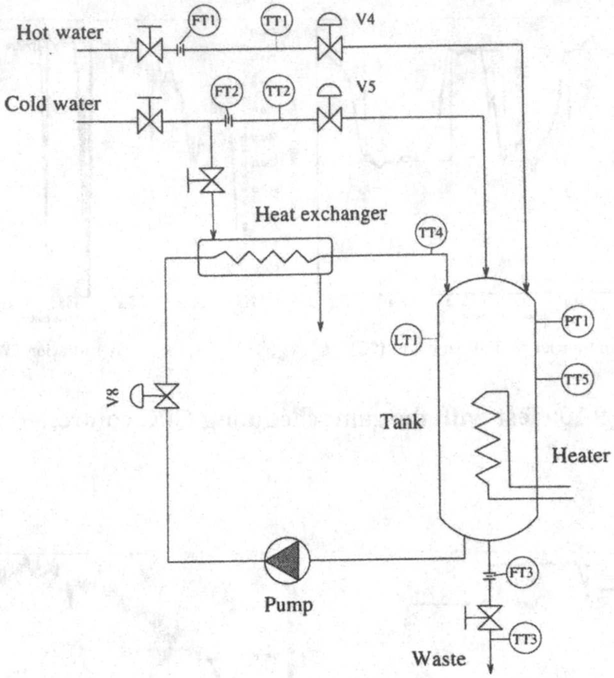

Figure 9.12: Diagram of the Pilot Plant

9.2.1 Plant Description

A diagram of the Plant which shows its main elements as well as the localization of the various instruments is given in figure 9.12.

The main elements are:

- Feed circuit. The plant has two input pipes, a cold water one (at air temperature) and a hot water one (at about 70 $^{\circ}C$) with nominal flow and temperature conditions of 10 l/min and 2 bar for the cold water and 5 l/min and 1 bar for the hot. The temperatures and the flows of the inputs are measured with thermocouples and orifice plates respectively, with motorized valves for regulating the input flows.

- Tank. It has a height of 1 m and an interior diameter of 20 cm, it is thermically insulated, and with an approximate volume of 31 l. It can work pressurized (up to a limit of 4 bar) or at atmospheric pressure, depending on the position of the vent valve. In its interior there is a 15 kW electric resistance for heating, also an overflow, an output pipe and another one for recirculating the water through the exchanger.

- Recirculation circuit. The hot water in the tank can be cooled by entering cold water through the cooling circuit. This circuit is composed of a centrifugal pump that circulates the hot water from the bottom of the tank through a tube bundle heat exchanger returning at a lower temperature at its top.

9.2.2 Plant Control

To control the installation there is an ORSI Integral Cube distributed control system, composed of a controller and a supervisor connected by a local data highway. The former is in charge of carrying out the digital control and analogous routines whilst the latter acts as a programming and communications platform with the operator. On this distributed control system the GPC algorithms seen before will be implemented. This control system constitutes a typical example of an industrial controller, having the most normal characteristics of medium size systems to be found in the market today. As in most control computers the calculation facilities are limited and there is little time available for carrying out the control algorithm because of the attention called for by other operations. It is thus an excellent platform for implanting precalculated GPC in industrial fields.

From all the possible loops that could be controlled the results obtained in certain situations will be shown. These are: control of the cold water flow FT_2 with valve V_5, control of the output temperature of the heat exchanger TT_4 with valve V_8, control of the tank level LT_1 with the cold water flow by valve V_5 and control of the tank temperature TT_5 with the resistance.

9.2.3 Flow Control

The control of the cold water flow has been chosen as an example of regulating a simple loop. Because all the water supplied to the plant comes from only one pressure group, the variations affecting the hot water flow or the cold water flow of the heat exchanger will affect the cold water flow as disturbances. Regulating the cold water flow is not only important as a control loop but also it may be necessary as an auxiliary variable to control the temperature or the level in the tank.

The dynamics of this loop are mainly governed by the regulation valve. This is a motorized valve with a positioner with a total open time of 110 seconds, thus causing slow dynamics in the flow variation. The flow behaviour will approximate that of a first order system with delay.

Firstly the parameters identifying the process are obtained using the reaction curve, and then the coefficients of the GPC are found using the method described in chapter 5. In order to do this, working with the valve 70% open (flow of 3.98 l/min), a step of up to 80% is produced, obtaining after the transition a stationary value of 6.33 l/min. From the data obtained it can be calculated that:

$$K = 0.25 \qquad \tau = 10.5 \text{ seconds} \qquad \tau_d = 10 \text{ seconds}$$

when a sampling time $T = 2$ seconds is used, the parameters for the corresponding discrete model are:

$$a = 0.8265 \qquad b = 0.043 \qquad d = 5$$

The control signal can easily be computed using the expression:

$$u(t) = u(t-1) + (l_{y1}\hat{y}(t+5) + l_{y2}\hat{y}(t+4) + l_{r1}r(t))/K$$

where $u(t)$ is the position of the valve V_5 and $y(t)$ is the value of the flow FT_2. Using the approximation formulas (5.10) with $\lambda = 0.8$, the controller gains result as:

$$\begin{aligned} l_{y1} &= -2.931 \\ l_{y2} &= 1.864 \\ l_{r1} &= 1.067 \end{aligned}$$

The behaviour of the closed loop when set point changes are produced can be seen in figure 9.13. The disturbances appearing in the flow are sometimes produced by changes in other flows of the installation and at other times by electrical noise produced by equipment (mainly robots) located near by. The setpoint was changed from 4 to 6.3 litres per minute. The measured flow follows the step changes quite rapidly (taking into account the slow valve dynamics) with an overshoot of 13%. The manual valve at the cold water inlet was partially closed in order to introduce an external disturbance. As can be seen, the controller rejected the external perturbation quite rapidly.

In order to illustrate how simple it is to implement GPC on an industrial SCADA using the approximation formulas, figure 9.14 shows the code for the Integral Cube distributed control system. The computation of the controller parameters is based on process parameters, and these can be changed on-line by the operator or by an identifier if an adaptive control scheme is used. If the GPC controllers were derived by other procedures, much more complex computation would be required and it would have been virtually impossible to program the GPC using this SCADA programming language (ITER).

9.2.4 Temperature Control at the Exchanger Output

The heat exchanger can be considered to be an independent process within the Plant. The exchanger reduces the temperature of the recirculation water, driven by the pump, using a constant flow of cold water for this. The way of controlling the output temperature is by varying the flow of the recirculation

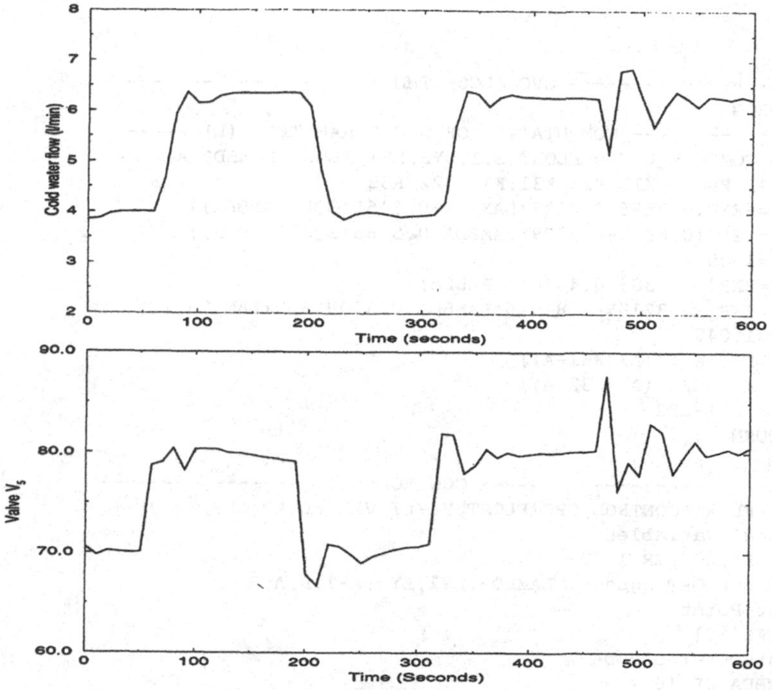

Figure 9.13: Flow control

water with the motorized valve V_8; thus the desired temperature is obtained by variations in the flow. In brief, the heat exchanger is nothing more than a tube bundle with hot water inside that exchanges heat with the exterior cold water. It can thus be considered as being formed of a large number of first order elements that together act as a first order system with pure dead time (see chapter 5). Thus the TT_4-V_8 system will be approached by a transfer function of this type.

Following the procedure used up to now the system parameters which will be used for the control law are calculated. Some of the results obtained are shown in figure 9.15. It should be bore in mind that as the exchanger is not independent from the rest of the plant, its output temperature affects, through the tank, that of the input, producing changes in the operating conditions. In spite of this its behaviour is reasonably good.

The set point was changed from 38 ^oC to 34 ^oC. As can be seen in figure 9.15, the heat exchanger outlet temperature evolved to the new set point quite smoothly without exhibiting oscillations. Two different types of external disturbances were introduced. First the manual valve of the refrigerating cold water was closed for a few seconds. As was to be expected, the outlet temperature of the heat exchanger increased very rapidly because of this strong external perturbation and then it was taken back to the desired value

```
;--------------------- GPC ALGORITHM ---------------------
STARTPROG
;---------------- COMPUTATION OF GPC PARAMETERS (L) -------
DEFSUB COMPUTE_L(VAR FLOAT LY1,LY2,LR1;FLOAT LAMBDA,A)
   LOCAL FLOAT K11,K21,K31,K12,K22,K32
   K11=-EXP(0.3598-0.9127*LAMBDA+0.3165*SQR(LAMBDA))
   K21=-EXP(0.0875-1.2309*LAMBDA+0.5086*SQR(LAMBDA))
   K31=1.05
   K12=EXP(-1.7383-0.40403*LAMBDA)
   K22=EXP(-0.32157-0.81926*LAMBDA+0.3109*SQR(LAMBDA))
   K32=1.045
   LY1=K11+K21*(A/(K31-A))
   LY2=K12+K22*(A/(K32-A))
   LR1=-LY1-LY2
   RETURN
ENDSUB
;--------------------------- CONTROLLER ------------------
DEFSUB FLOAT CONTROL_GPC(FLOAT Y,Yt1;VAR FLOAT CFP,YP,U)
   ;Local Variables
   LOCAL INTEGER I, D
   LOCAL FLOAT Ugpc,G, LAMBDA,LY1,LY2,LR1,R,A,B
   ;Set Point
   R=CFP[10]
   ;Parameter LAMBDA
   LAMBDA=CFP[0]
   ;Output Span
   SPAN_GPC=CFP[13]
   ;Transfer Function Parameters
   A=CFP[6]
   B=CFP[7]
   D=FTOI(CFP[9])
   ;Process Gain
   G=B/(1.0-A)
  ;Predictor
   YP[0]=Y
   YP[1]=(1.0+A)*YP[0]-A*Yt1+B*(U[D]-U[D+1])
   FOR I=2 TO D
   YP[I]=(1.0+A)*YP[I-1]-A*YP[I-2]+B*(U[D-I+1]-U[D-I+2])
   ENDFOR
  ;Controller parameters
   COMPUTE_L(LY1,LY2,LR1,LAMBDA,A)
  ;Control Signal
      IF (D=0) THEN
         Ugpc=U[1]+(LY1*YP[D]+LY2*Yt1+LR1*R)/G
      ELSE
         Ugpc=U[1]+(LY1*YP[D]+LY2*YP[D-1]+LR1*R)/G
      ENDIF
   RETURN(Ugpc)
ENDSUB
ENDPROG
```

Figure 9.14: GPC code in ITER language

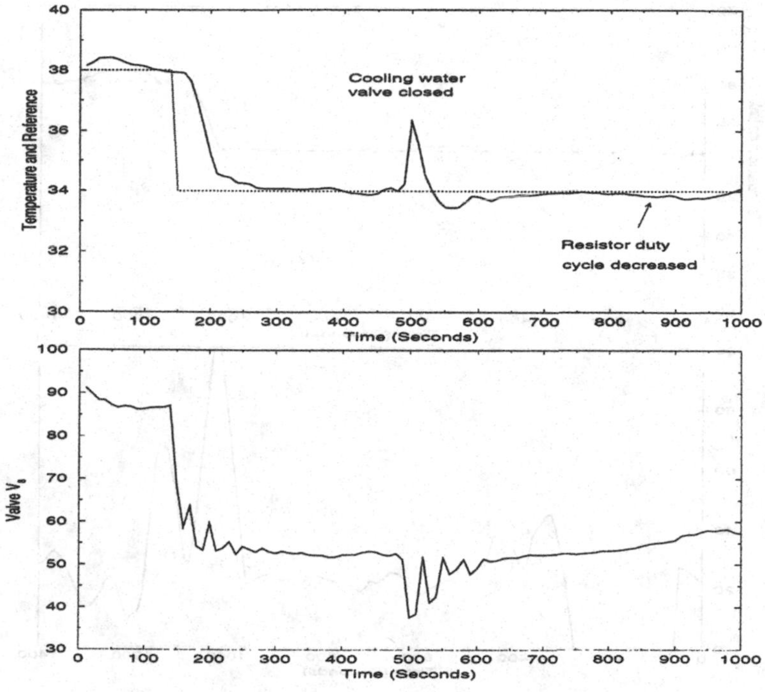

Figure 9.15: Behaviour of the heat exchanger

by the GPC. The second perturbation is caused by decreasing the duty cycle of the resistor in the tank, thus decreasing the inlet hot water temperature and changing the heat exchanger operating point. As can be seen, the GPC rejects almost completely this perturbation, caused by a change in its dynamics.

9.2.5 Temperature Control in the Tank

The next example chosen is also that of a very typical case in the process industry: the temperature of the liquid in a tank. The manipulated variable in this case is the duty cycle of the heating resistor.

The process has integral effect and was identified around the nominal operating conditions (50 ^oC). The following model was obtained:

$$G(s) = \frac{0.41}{s(1+50s)}e^{-50s}$$

The GPC was applied with a sampling time $T = 10$ seconds, $\lambda = 1.2$ and $N = 15$. As in the previous case, the controller parameters were computed by the formulas given in chapter 5 for integrating processes. The results obtained are shown in figure 9.16. A perturbation (simulating a major failure

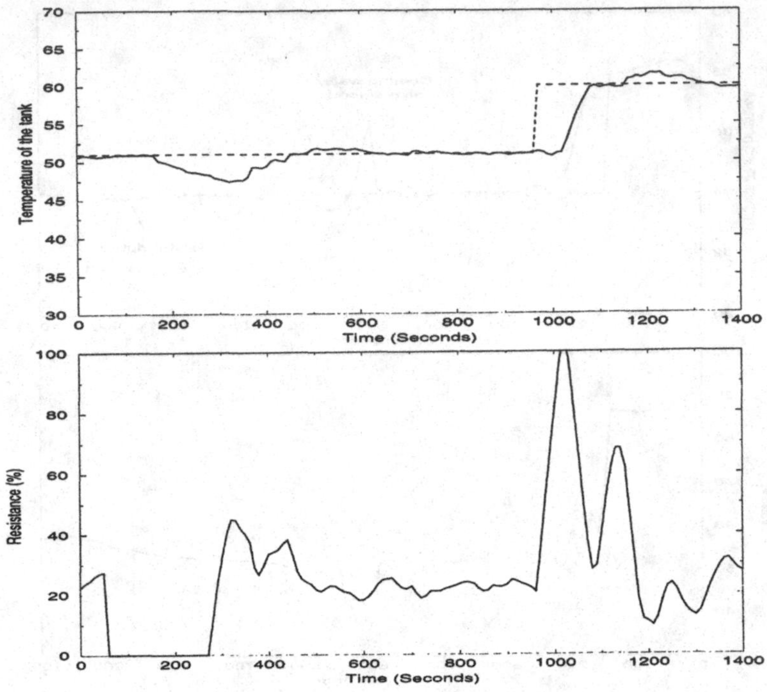

Figure 9.16: Evolution of the tank temperature

of the actuator) was introduced. As can be seen, after the initial drop in the temperature of the tank, caused by the lack of actuation, the control system is able to take the tank temperature to the desired value with a very smooth transient. A change in the set point from 50 $^\circ C$ to 60 $^\circ C$ was then introduced. The temperature of the tank evolves between both set points without big oscillations.

9.2.6 Level Control

Level is another of the most common variables in the process industry. In the Pilot Plant, the level of the tank can be controlled by the input flows (cold or hot water). In this example, the valve V_5 will be used. The system was identified around the nominal operating point (70%) by the reaction curve method. The model transfer function is:

$$G(s) = \frac{1.12}{1 + 87s} e^{-45s}$$

For a sampling time $T = 10$ seconds, the dead time results to be non-integer, thus the controller parameters must be calculated as shown in section

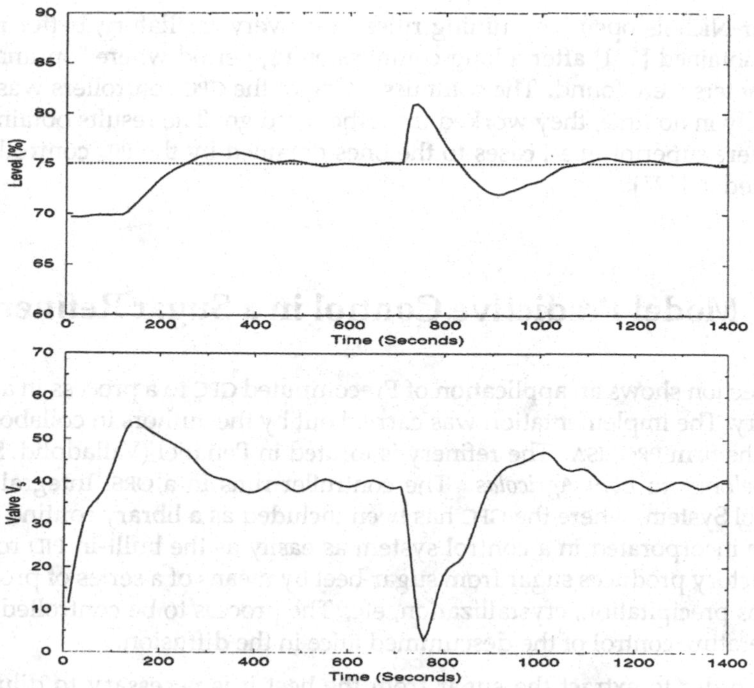

Figure 9.17: Evolution of the level

5.2. The results obtained when working with a weighting factor of 1 and a prediction horizon of 15 are shown in figure 9.17. The set point was changed from 70% to 75%. As can be seen, the level of the tank moves smoothly between both set points. An external perturbation was introduced to test the disturbance rejection capabilities of the GPC. The perturbation consisted of opening the hot water inlet and thus increasing the level of the tank. As can be seen, the perturbation is rejected after a quite well damped transient.

9.2.7 Remarks

The main objective of the control examples presented in this section was to show how easily GPC can be implemented on a commercial distributed control system by using the implementation technique presented in chapter 5. The GPCs were implemented without difficulties using the programming language (ITER) of the Integral Cube distributed control system.

Although comparing the results obtained by GPC with those obtained by using other control techniques was not one of the objectives, GPC has been shown to produce better results than the traditional PID on the examples treated. In all the processes, the results obtained by PID, tuned by the

Ziegler-Nichols open loop tuning rules, were very oscillatory, better results were obtained [121] after a long commissioning period where "optimal" PID parameters were found. The commissioning of the GPC controllers was done virtually in no time, they worked from the word go. The results obtained by GPC were superior in all cases to the ones obtained by the PID controllers as reported in [121].

9.3 Model Predictive Control in a Sugar Refinery

This section shows an application of Precomputed GPC to a process in a sugar refinery. The implementation was carried out by the authors in collaboration with the firm PROCISA. The refinery is located in Peñafiel (Valladolid, Spain) and belongs to *Ebro Agricolas*. The controller runs in a ORSI Integral Cube Control System, where the GPC has been included as a library routine which can be incorporated in a control system as easily as the built-in PID routine. The factory produces sugar from sugar-beet by means of a series of processes such as precipitation, crystallization, etc. The process to be controlled is the temperature control of the descummed juice in the diffusion.

In order to extract the sugar from the beet it is necessary to dilute the saccharose contained in the tuber tissue in water in order to form a juice from which sugar for consumption is obtained.

The juice is obtained in a process known as diffusion. Once the beet has been cut into pieces (called chunks) to increase the interchangeable surface, it enters into the macerator (which revolves at a velocity of 1 r.p.m.) where it is mixed with part of the juice coming from the diffusion process (see figure 9.18). Part of the juice inside the macerator is recirculated in order to be heated by means of steam and in this way it maintains the appropriate temperature for maceration. The juice from the maceration process passes into the diffusor (a slowly revolving pipe 25 m long and with a diameter of 6 m) where it is mixed with water and all the available sugar content is extracted, leaving the pulp as a sub-product. The juice coming out of the diffusor is recirculated to the macerator, from which the juice already prepared is extracted for the next process.

In order for the diffusor to work correctly it is necessary to supply thermal energy to the juice during maceration. In order to obtain this objective, part of the juice from the macerator (150 m^3/h) is made to recirculate through a battery of exchangers; within these the steam proceeding from the general services of the factory provides the heat needed to obtain optimum maceration. Therefore the controller must adjust the steam valve (u) in order to achieve a determined return temperature to the macerator (y).

The system response is seriously disturbed by changes in the steam pressure, which are frequent because the steam used in the exchangers has to be

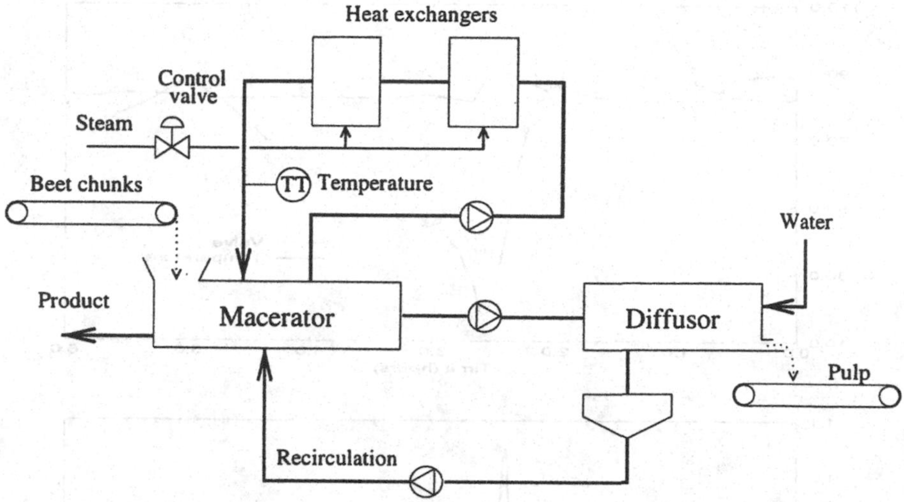

Figure 9.18: Diffusion Process in a Sugar Refinery

shared with other processes which can function in a non-continuous manner.

The process is basically a thermal exchange between the steam and the juice in the pipes of the exchanger, with overdamped behaviour and delay associated to the transportation time of the juice through pipes about 200 meters long. These considerations, together with the observation of the development of the system in certain situations, justify the use of a first order model with delay.

A model was identified by its step response. Starting from the conditions of 82.42 °C and the valve at 57 %, the valve was closed to 37 % in order to observe the evolution; the new stationary state is obtained at 78.61 °C. The values of gain, time constant and delay can easily be obtained from the response (as seen in previous examples):

$$K = \frac{82.42 - 78.61}{57 - 37} = 0.1905 \frac{^0C}{\%} \qquad \tau = 5 \text{ min} \qquad \tau_d = 1 \text{ min } 45 \text{ s}$$

However, it is seen that the system reacts differently when heated to when cooled, the delay being quite a lot greater in the first case. A similar test changing the valve to 57 % again provides values of

$$K = 0.15 \qquad \tau = 5 \text{ min } 20 \text{ s} \qquad \tau_d = 4 \text{ min } 50 \text{ s}$$

Although an adaptive strategy could be used (with the consequent computational cost), a fixed parameter controller was employed, showing, at the same time, the robustness of the method when using the T-polynomial in presence of modelling errors. The error in the delay, which is the most dangerous, appears in this case. The following values of the model were chosen for this:

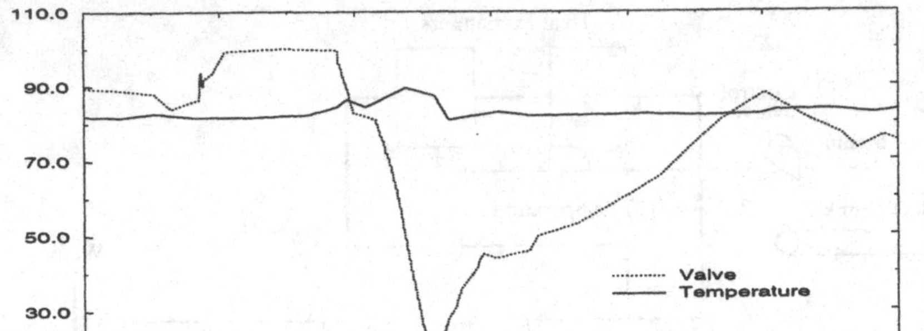

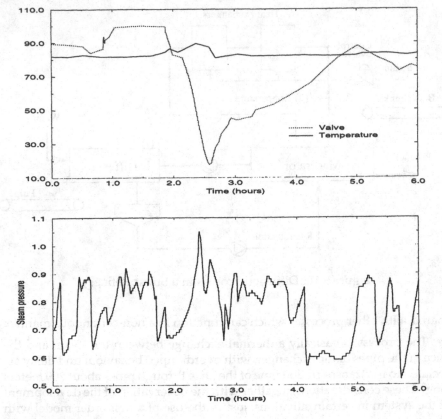

Figure 9.19: System response in the presence of external disturbances

$$K = 0.18 \qquad \tau = 300 \text{ s} \qquad \tau_d = 190 \text{ s}$$

and sampling time of T= 60 s.

It should be noticed that there are great variations in the delay (that produced on heating is about three times greater than that on cooling), due to which it is necessary to introduce a filter $T(z^{-1})$ as suggested in [128] in order to increase the robustness.

The following figures show various moments in operating the temperature control. The behaviour of the controller rejecting the disturbances (sudden variations in the steam pressure and macerator load) can be seen in figure 9.19. On the other hand, figure 9.20 shows the response to a setpoint change in the juice temperature.

It should be emphasised that this controller worked satisfactorily and without interruption until the end of the year's campaign, being handled without difficulty by the plant operators who continuously changed the values

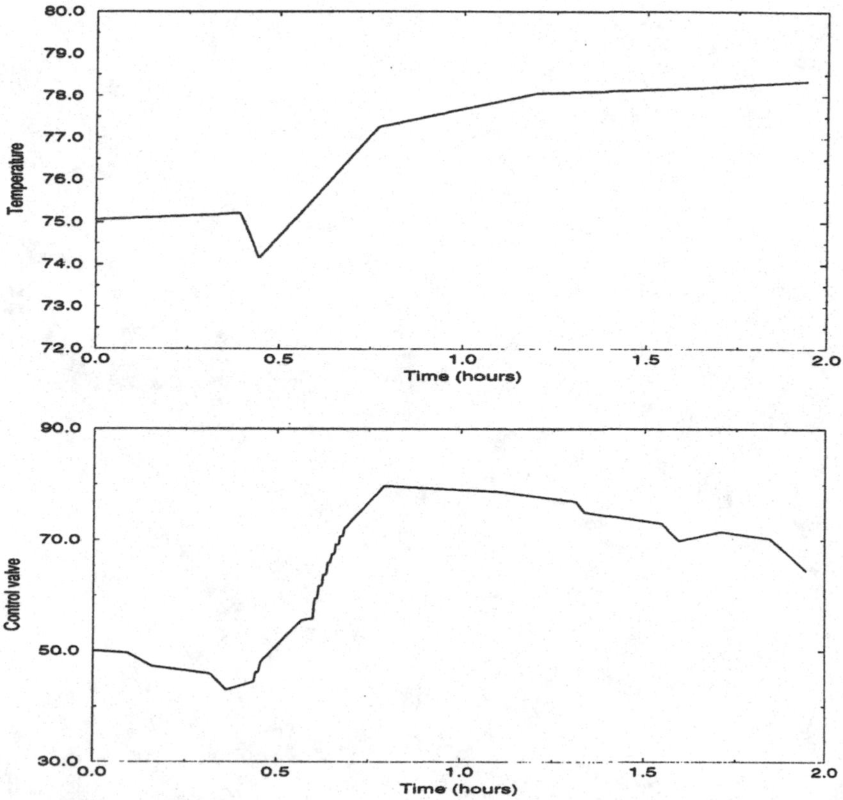

Figure 9.20: Setpoint change

of the model throughout the operation time. Following many operational days the operators themselves concluded that a satisfactory model was given by:

$$K = 0.25 \qquad \tau = 250 \text{ s} \qquad \tau_d = 220 \text{ s} \qquad \lambda = 0.1$$

The sampling time was set to 50 seconds and the following robust filter was used:

$$T(z^{-1}) = A(z^{-1})(1 - 0.8az^{-1})$$

whit a equal to the discrete process pole.

Appendix A

Revision of the Simplex method

The Simplex method [69] is the algorithm most used for solving linear programming problems, such as the ones appearing when using an 1-norm MPC. The Simplex algorithm finds successive and better feasible solutions at the extreme points of the feasible region until it stops, in a finite number of steps, either at the optimum or finds that the optimal solution is not bound by the feasible region. This appendix is dedicated to revising the basic ideas behind the Simplex method.

A.1 Equality Constraints

The problem of minimizing a linear function subject to equality constraints will be considered first:

$$\begin{array}{ll} \text{Minimize} & \mathbf{c}^t\mathbf{x} \\ \text{subject to} & A\mathbf{x} = \mathbf{b} \\ & \mathbf{x} \geq \mathbf{0} \end{array} \tag{A.1}$$

where A is a $q \times p$ real matrix with $q < p$ and full rank.

If the equality constraint equation is multiplied by a matrix T and the columns of A and corresponding components of $\mathbf{x}$ are interchanged in such a way that $T\,A = [I_{q\times q}\ N]$ and $T\,\mathbf{b} = \overline{\mathbf{b}}$, the point $\mathbf{x} = [\mathbf{x}_b^t\ \mathbf{x}_n]^t = [\overline{\mathbf{b}}^t\ 0]$ is a basic feasible solution. The components $\mathbf{x}_b$ are called *basic variables* whereas the remaining components (corresponding to N) are called *nonbasic variables.* Note that this can be done by applying elementary row transformations to matrix A and interchanging the columns of A (and the corresponding $\mathbf{x}$ variables) to take matrix A to the form $[I\ N]$. If the same transformations are applied to I

and b, matrix T and vector $\overline{\mathbf{b}}$ are obtained.

The objective function takes the value $z_0 = \mathbf{c}_b^t\mathbf{x}_b + \mathbf{c}_n^t\mathbf{x}_n = \mathbf{c}_b^t\mathbf{x}_b$. The basic variables can be expressed as a function of the nonbasic variables from the transformed constraint equation:

$$\mathbf{x}_b = \overline{\mathbf{b}} - N\mathbf{x}_n$$

by substituting in the cost function:

$$z = \mathbf{c}^t\mathbf{x} = \mathbf{c}_b^t(\overline{\mathbf{b}} - N\mathbf{x}_n) + \mathbf{c}_n^t\mathbf{x}_n = z_0 + (\mathbf{c}_n^t - \mathbf{c}_b^t N)\mathbf{x}_n$$

As $\mathbf{x}_n \geq 0$, the objective function decreases if any component of$(\mathbf{c}_n^t - \mathbf{c}_b^t N)_i$ is negative and the corresponding nonbasic variable $\mathbf{x}_{n_i}$ increases. This gives an indication of how to obtain a better feasible solution and is the basic idea behind the algorithm. The problem is determining which of the nonbasic variables should be increased (become basic) and which of the basic variables should leave the basis.

This is done as follows:

1. Find an initial basic solution.
2. Form the following tableau:

	$\mathbf{x}_b^t$	$\mathbf{x}_n^t$	
$\mathbf{x}_b$	$\mathbf{I}$	N	$\overline{\mathbf{b}}$
J	$\mathbf{0}$	$\mathbf{c}_n - \mathbf{c}_b^t N$	$\mathbf{c}_b^t\overline{\mathbf{b}}$

3. If $\mathbf{c}_n - \mathbf{c}_b^t N \geq 0$ then stop, the actual basic solution is optimal.
4. Choose one of the negative elements (say j^{th}) of row $\mathbf{c}_n - \mathbf{c}_b^t N$, (the most negative is usually chosen).
5. Choose i such that the ratio $\overline{\mathbf{b}}_i/N_{ij}$ is the minimum for all $N_{ij} > 0$. If there are no non negative elements in that column of the tableau, the problem is not bounded.
6. Make $\mathbf{x}_{n_j}$ a basic variable and $\mathbf{x}_{b_i}$ a non basic variable by *pivoting*:
   (a) Divide i^{th}-row of the tableau by N_{ij}
   (b) Make zero the remaining elements of the jth-column of the nonbasic variable block by multiplying the resulting i^{th}-row by N_{kj} and subtracting it from the kth-row.
7. go to step 2

A.2 Finding an Initial Solution

The Simplex method starts from an initial feasible extreme point. An initial point can be found by applying elementary row transformations to matrix A

and vector b and interchanging the columns of A (and the corresponding $\mathbf{x}$ variables) to take matrix A to the form $[I\ N]$.

A solution can be obtained by using the Simplex algorithms in different ways. One way, known as the two phase method, consists of solving the following augmented system:

$$\begin{array}{ll} \text{Minimize} & \mathbf{1}^t\mathbf{x}_a \\ \text{subject to} & \mathbf{x}_a + A\mathbf{x} = \mathbf{b} \\ & \mathbf{x} \geq \mathbf{0}, \quad \mathbf{x}_a \geq \mathbf{0} \end{array} \tag{A.2}$$

Note that the constraint matrix is now $[IA]$ and that the obvious solution $\mathbf{x} = 0$ and $\mathbf{x}_a = \mathbf{b}$ is an extreme point of the augmented problem. The variables $\mathbf{x}_a$ are called *artificial* variables and are all in the basis for the initial solution. If the algorithm does not find a solution with $\mathbf{x}_a = 0$, the problem is not feasible. Otherwise, the solution constitutes an initial solution to the original problem and the second phase of the algorithm can be started for the original problem with this solution.

Another way of dealing with initial conditions, known as the big-M method [10], solves the whole problem in only one phase. Artificial variables are also introduced as in the two phase method but a term is added to the objective function penalizing the artificial variables with high weighting factors in order to force artificial variables out of the basis. The problem is now stated as:

$$\begin{array}{ll} \text{Minimize} & \mathbf{c}^t\mathbf{x} + \mathbf{m}^t\mathbf{x}_a \\ \text{subject to} & \mathbf{x}_a + A\mathbf{x} = \mathbf{b} \\ & \mathbf{x} \geq \mathbf{0}, \quad \mathbf{x}_a \geq \mathbf{0} \end{array} \tag{A.3}$$

If all the artificial variables are out of the basis at termination, a solution to the optimization problem has been found. Otherwise, if the variable entering the basis is the one with the most positive cost coefficient, we can conclude that the problem has no feasible solution.

A.3 Inequality Constraints

Consider the problem:

$$\begin{array}{ll} \text{Minimize} & \mathbf{c}^t\mathbf{x} \\ \text{subject to} & A\mathbf{x} \leq \mathbf{b} \\ & \mathbf{x} \geq \mathbf{0} \end{array} \tag{A.4}$$

The Simplex method can be used to solve inequality constraints problems by transforming the problem into the standard format by introducing a vector of variables, $\mathbf{x}_s \geq 0$, called *slack* variables, such that the inequality constraint $A\mathbf{x} \leq \mathbf{b}$ is transformed into the equality constraint $A\mathbf{x} + \mathbf{x}_s = \mathbf{b}$. The problem can now be stated in the standard form as:

$$\begin{array}{ll} \text{Minimize} & \mathbf{c}^t\mathbf{x} \\ \text{subject to} & [A\ I]\begin{bmatrix}\mathbf{x}\\ \mathbf{x}_s\end{bmatrix} = \mathbf{b} \\ & \begin{bmatrix}\mathbf{x}\\ \mathbf{x}_s\end{bmatrix} \geq \mathbf{0} \end{array} \tag{A.5}$$

The number of variables is now $q+p$ and the number of constraints is q. Notice that the form of the equality constraint matrix is $[A\ I]$ and that the point $[\mathbf{x}^t\ \mathbf{x}_s^t] = [\mathbf{0}\ \mathbf{b}^t]$ is an initial basic solution to this problem.

Duality

The number of *pivoting* operations needed to solve a standard linear programming problem is in the order of q [69], while the number of floating point operations needed for each pivoting operation is in the order of $q \times p$. The number of floating points operations needed for solving a standard LP problem is therefore in the order of $(q^2 + q \times p)$. For the inequality constraint problem, the number of variables is increased by the number of slack variables which is equal to the number of inequality constraints. In the linear programming problems resulting from robust MPC, all of the constraints are inequality constraints, so the number of operations needed is in the order of $q(q \times (q+p)) = q^3 + q^2p$. That is, linear in the number of variables and cubic in the number of inequality constraints. The problem can be transformed into an LP problem with a different and more convenient structure using duality.

Given the problem (*primal*):

$$\begin{array}{ll} \text{Minimize} & \mathbf{c}^t\mathbf{x} \\ \text{subject to} & A\mathbf{x} \leq \mathbf{b} \\ & \mathbf{x} \geq \mathbf{0} \end{array} \tag{A.6}$$

The *dual* problem is defined as [69]:

$$\begin{array}{ll} \text{Minimize} & -\mathbf{b}^t\lambda \\ \text{subject to} & -A^t\lambda \leq -\mathbf{c} \\ & \lambda \geq \mathbf{0} \end{array} \tag{A.7}$$

The number of operations required by the *dual* problem is in the order of $p(p \times (p+q)) = p^3 + p^2q$. That is, cubic in the number of variables and linear in the number of inequality constraints. So, for problems with more inequality constraints than variables, solving the *dual* problem will require less computation than the primal.

The solutions to both problems are obviously related (Bazaraa and Shetty [10], theorem 6.6.1.). If $\mathbf{x}_o$ and λ_o are the optimal of the *primal* and *dual* problem then, $\mathbf{c}^t\mathbf{x}_o = \mathbf{b}^t\lambda_o$ (the cost is identical) and $(\mathbf{c}^t - \lambda_o^t A)\mathbf{x}_o = 0$. If the primal problem has a feasible solution and has an unbounded objective value the *dual* problem is unfeasible. If the *dual* problem has a feasible solution and has an unbounded objective value the *primal* problem is unfeasible.

References

[1] P. Albertos and R. Ortega. On Generalized Predictive Control: Two Alternative Formulations. *Automatica*, 25(5):753–755, 1989.

[2] A.Y. Allidina and F.M. Hughes. Generalised Self-tuning Controller with Pole Assignment. *Proceedings IEE, Part D*, 127:13–18, 1980.

[3] J.C. Allwright. *Advances in Model-Based Predictive Control*, chapter On min-max Model-Based Predictive Control. Oxford University Press, 1994.

[4] T. Alvarez and C. Prada. Handling Infeasibility in Predictive Control. *Computers and Chemical Engineering*, 21:577–582, 1997.

[5] T. Alvarez, M. Sanzo, and C. Prada. Identification and Constrained Multivariable Predictive Control of Chemical Reactors. In 4th *IEEE Conference on Control Applications, Albany*, pages 663–664, 1995.

[6] P. Ansay and V. Wertz. Model Uncertainties in GPC: a Sistematic two-step design. In *Proceedings of the* 3rd *European Control Conference, Brussels*, 1997.

[7] L.V.R. Arruda, R. Lüders, W.C. Amaral, and F.A.C Gomide. An Object-Oriented Environment for Control Systems in Oil Industry. In *Proceedings of the* 3rd *Conference on Control Applications, Glasgow, UK*, pages 1353–1358, 1994.

[8] K.J. Aström and B. Wittenmark. *Computer Controlled Systems. Theory and Design*. Prentice-Hall. Englewood Cliffs, NJ, 1984.

[9] K.J. Aström and B. Wittenmark. *Adaptive Control*. Addison-Wesley, 1989.

[10] M.S. Bazaraa and C.M. Shetty. *Nonlinear Programming*. Wiley, 1979.

[11] M. Berenguel, E.F. Camacho, and F.R. Rubio. *Simulation Software Package for the Acurex Field*. Departamento de Ingeniería de Sistemas y Automática, ESI Sevilla (Spain), Internal Report, 1994.

[12] R.R. Bitmead, M. Gevers, and V. Wertz. *Adaptive Optimal Control. The Thinking Man's GPC*. Prentice-Hall, 1990.

[13] C. Bordons. *Control Predictivo Generalizado de Procesos Industriales: Formulaciones Aproximadas*. PhD thesis, Universidad de Sevilla, 1994.

[14] S. Boyd, L. El Ghaouni, E. Feron, and V. Balakrishnan. *Linear Matrix Inequalities in Systems and Control Theory*. SIAM books, 1994.

[15] E.H. Bristol. On a new measure of interaction for multivariable process control. *IEEE Trans. on Automatic Control*, 11(1):133–4, 1966.

[16] E.F. Camacho. Constrained Generalized Predictive Control. *IEEE Trans. on Automatic Control*, 38(2):327–332, 1993.

[17] E.F. Camacho and M. Berenguel. *Advances in Model-Based Predictive Control*, chapter Application of Generalized Predictive Control to a Solar Power Plant. Oxford University Press, 1994.

[18] E.F. Camacho, M. Berenguel, and C. Bordons. Adaptive Generalized Predictive Control of a Distributed Collector Field. *IEEE Trans. on Control Systems Technology*, 2:462–468, 1994.

[19] E.F. Camacho, M. Berenguel, and F.R. Rubio. Application of a Gain Scheduling Generalized Predictive Controller to a Solar Power Plant. *Control Engineering Practice*, 2(2):227–238, 1994.

[20] E.F. Camacho, M. Berenguel, and F.R. Rubio. *Advanced Control of Solar Power Plants*. Springer-Verlag, London, 1997.

[21] E.F. Camacho and C. Bordons. Implementation of Self Tuning Generalized Predictive Controllers for the Process Industry. *Int. Journal of Adaptive Control and Signal Processing*, 7:63–73, 1993.

[22] E.F. Camacho and C. Bordons. *Model Predictive Control in the Process Industry*. Springer-Verlag, 1995.

[23] E.F. Camacho, F.R. Rubio, and F.M. Hughes. Self-tuning Control of a Solar Power Plant with a Distributed Collector Field. *IEEE Control Systems Magazine*, pages 72–78, 1992.

[24] R.G. Cameron. The design of Multivarible Systems. In *V Curso de Automática en la Industria*, La Rábida, Huelva, Spain, 1985.

[25] P.J. Campo and M. Morari. Robust Model Predictive Control. In *American Control Conference, Minneapolis, Minnesota*, 1987.

[26] H. Chen and F. Allgower. A Quasi-Infinite Horizon Nonlinear Model Predictive Control Scheme for Constrained Nonlinear Systems. In *Proceedings 16^{t}h Chinese Control Conference*, Qindao, 1996.

[27] T.L. Chia and C.B. Brosilow. Modular Multivariable Control of a Fractionator. *Hydrocarbon Processing*, pages 61–66, 1991.

[28] C.M. Chow and D.W. Clarke. *Advances in Model-Based Predictive Control*, chapter Actuator nonlinearities in predictive control. Oxford University Press, 1994.

[29] D. W. Clarke and P. J. Gawthrop. Self-tuning Controller. *Proceedings IEE*, 122:929–934, 1975.

[30] D. W. Clarke and R. Scattolini. Constrained Receding-horizon Predictive Control. *Proceedings IEE*, 138(4):347–354, july 1991.

[31] D.W. Clarke. Application of Generalized Predictive Control to Industrial Processes. *IEEE Control Systems Magazine*, 122:49–55, 1988.

[32] D.W. Clarke and P.J. Gawthrop. Self-tuning Control. *Proceedings IEEE*, 123:633–640, 1979.

[33] D.W. Clarke and C. Mohtadi. Properties of Generalized Predictive Control. *Automatica*, 25(6):859–875, 1989.

[34] D.W. Clarke, C. Mohtadi, and P.S. Tuffs. Generalized Predictive Control. Part I. The Basic Algorithm. *Automatica*, 23(2):137–148, 1987.

[35] D.W. Clarke, C. Mohtadi, and P.S. Tuffs. Generalized Predictive Control. Part II. Extensions and Interpretations. *Automatica*, 23(2):149–160, 1987.

[36] D.W. Clarke, E. Mosca, and R. Scattolini. Robustness of an Adaptive Predictive Controller. In *Proceedings of the 30th Conference on Decision and Control*, pages 979–984, Brighton, England, 1991.

[37] C.R. Cutler and B.C. Ramaker. Dynamic Matrix Control- A Computer Control Algorithm. In *Automatic Control Conference, San Francisco*, 1980.

[38] P.B. Deshpande and R.H. Ash. *Elements of Computer Process Control*. ISA, 1981.

[39] J. C. Doyle and G. Stein. Multivariable Feedback Design: Concepts for a Classical/Modern Synthesis. *IEEE Trans. on Automatic Control*, 36(1):4–16, 1981.

[40] G. Ferretti, C. Manffezzoni, and R. Scattolini. Recursive Estimation of Time Delay in Sampled Systems. *Automatica*, 27(4):653–661, 1991.

[41] Y.K. Foo and Y.C. Soh. Robust Stability Bounds for Systems with Structured and Unstructured Perturbations. *IEEE Trans. on Automatic Control*, 38(7), 1993.

[42] J.B. Froisy and T. Matsko. IDCOM-M Application to the Shell Fundamental Control Problem. In *AIChE Annual Meeting*, 1990.

[43] C.E. García, D.M. Prett, and M. Morari. Model Predictive Control: Theory and Practice-a Survey. *Automatica*, 25(3):335–348, 1989.

[44] G. Goodwin and K. Sin. *Adaptive Filtering, Predicition and Control*. Prentice-Hall, 1984.

[45] F. Gordillo and F.R. Rubio. Self-tuning Controller with LQG/LTR Structure. In *Proceedings* 1st *European Control Conference, Grenoble*, pages 2159–2163, july 1991.

[46] J.R. Gossner, B. Kouvaritakis, and J.A. Rossiter. Stable Generalized Predictive Control with Constraints and Bounded Disturbances. *Automatica*, 33:551–568, 1997.

[47] C. Greco, G. Menga, E. Mosca, and G. Zappa. Performance Improvement of Self Tuning Controllers by Multistep Horizons: the MUSMAR approach. *Automatica*, 20:681–700, 1984.

[48] P. Grosdidier, J.B. Froisy, and M. Hamman. *IFAC workshop on Model Based Process Control*, chapter The IDCOM-M controller. Pergamon Press, Oxford, 1988.

[49] R. Isermann. *Digital Control Systems*. Springer-Verlag, 1981.

[50] J. Jiang. Optimal Gain Scheduling Controller for a Diesel Engine. *IEEE Control Systems Magazine*, pages 42–48, 1994.

[51] A. Kalt. Distributed Collector System Plant Construction Report. *IEA/SSPS Operating agent DFVLR, Cologne*, 1982.

[52] M.R. Katebi and M.A. Johnson. Predictive Control Design for Large Scale Systems. In *IFAC Conference on Integrated System Engineering*, pages 17–22, Baden-Baden, Germany, 1994.

[53] R. De Keyser. A Gentle Introduction to Model Based Predictive Control. In *PADI2 International Conference on Control Engineering and Signal Processing, Piura, Peru*, 1998.

[54] R.M.C. De Keyser. Basic Principles of Model Based Predictive Control. In 1st *European Control Conference, Grenoble*, pages 1753–1758, july 1991.

[55] R.M.C. De Keyser and A.R. Van Cuawenberghe. Extended Prediction Self-adaptive Control. In *IFAC Symp. on Identification and System Parameter Estimation, York,UK*, pages 1317–1322, 1985.

[56] R.M.C. De Keyser, Ph.G.A. Van de Velde, and F.G.A. Dumortier. A Comparative Study of Self-adaptive Long-range Predictive Control Methods. *Automatica*, 24(2):149–163, 1988.

[57] M.V. Kothare, V. Balakrishnan, and M. Morari. Robust Constrained Predictive Control using Linear Matrix Inequalities . *Automatica*, 32:1361–1379, 1996.

[58] B. Kouvaritakis, J.A. Rossiter, and A.O.T Chang. Stable Generalized Predictive Control: An Algorithm with Guaranteed Stability. *Proceedings IEE, Part D*, 139(4):349–362, 1992.

[59] K. Krämer and H. Ubehauen. Predictive Adaptive Control. Comparison of Main Algorithms. In *Proceedings 1st European Control Conference, Grenoble*, pages 327–332, julio 1991.

[60] A.G. Kutnetsov and D.W. Clarke. *Advances in Model-Based Predictive Control*, chapter Application of constrained GPC for improving performance of controlled plants. Oxford University Press, 1994.

[61] W.H. Kwon and A.E. Pearson. On Feedback Stabilization of Time-Varying Discrete Linear Systems. *IEEE Trans. on Automatic Control*, 23:479–481, 1979.

[62] J.H. Lee, M. Morari, and C.E. García. State-space Interpretation of Model Predictive Control. *Automatica*, 30(4):707–717, 1994.

[63] M.A. Lelic and P.E. Wellstead. Generalized Pole Placement Self Tuning Controller. Part 1. Basic Algorithm. *Int. J. Control*, 46(2):547–568, 1987.

[64] M.A. Lelic and M.B. Zarrop. Generalized Polc Placement Self Tuning Controller. Part 2. Application to Robot Manipulator Control. *Int. J. Control*, 46(2):569–601, 1987.

[65] C.E. Lemke. *Mathematics of the Decision Sciences*, chapter On Complementary Pivot Theory. G.B. Dantzig and A.F. Veinott (Eds.), 1968.

[66] J.M. Lemos and E. Mosca. A Multipredictor-based LQ self-tuning Controller. In *IFAC Symp. on Identification and System Parameter Estimation, York,UK*, pages 137–141, 1985.

[67] D.A. Linkers and M. Mahfonf. *Advances in Model-Based Predictive Control*, chapter Generalized Predictive Control in Clinical Anaesthesia. Oxford University Press, 1994

[68] L. Ljung. *System Identification. Theory for the user*. Prentice-Hall, 1987.

[69] D.E. Luenberger. *Linear and Nonlinear Programming*. Addison-Wesley, 1984.

[70] J. Lunze. *Robust Multivariable Feedback Control*. Prentice-Hall, 1988.

[71] J.M. Martin-Sanchez and J. Rodellar. *Adaptive Predictive Control. From the concepts to plant optimization*. Prentice -Hall International (UK), 1996.

[72] D. Mayne and H. Michalska. Receding Horizon Control of Nonlinear Systems. *IEEE Trans. on Automatic Control*, 35:814–824, 1990.

[73] D. Mayne and H. Michalska. Robust Receding Horizon Control of Constrained Nonlinear Systems. *IEEE Trans. on Automatic Control*, 38:1623–1633, 1993.

[74] C. Mohtadi. *Advanced self-tuning algorithms*. PhD thesis, Oxford University, U.K., 1986.

[75] C. Mohtadi, S.L. Shah, and D.G. Fisher. Frequency Response Characteristics of MIMO GPC. In *Proceedings* 1st *European Control Conference, Grenoble*, pages 1845–1850, july 1991.

[76] M. Morari. *Advances in Model-Based Predictive Control*, chapter Model Predictive Control: Multivariable Control Technique of Choice in the 1990s? Oxford University Press, 1994.

[77] M. Morari and E. Zafiriou. *Robust Process Control*. Prentice-Hall, 1989.

[78] E. Mosca. *Optimal, Predictive and Adaptive Control*. Prentice Hall, 1995.

[79] E. Mosca, J.M. Lemos, and J. Zhang. Stabilizing I/O Receding Horizon Control. In *IEEE Conference on Decision and Control*, 1990.

[80] E. Mosca and J. Zhang. Stable Redesign of Predictive Control. *Automatica*, 28:1229–1233, 1992.

[81] K.R. Muske, E.S. Meadows, and J.B. Rawlings. The Stability of Constrained Receding Horizon Control with State Estimation. In *Proceedings of the American Control Conference*, pages 2837–2841, Baltimore, USA, 1994.

[82] K.R. Muske and J. Rawlings. Model Predictive Control with Linear Models. *AIChE Journal*, 39:262–287, 1993.

[83] R.B. Newell and P.L. Lee. *Applied Process Control. A Case Study*. Prentice-Hall, 1989.

[84] G. De Nicolao and R. Scattolini. *Advances in Model-Based Predictive Control*, chapter Stability and Output Terminal Constraints in Predictive Control. Oxford University Press, 1994.

[85] J.E. Normey, E.F. Camacho, and C. Bordons. Robustness Analysis of Generalized Predictive Controllers for Industrial Processes. In *Proceedings of the 2nd Portuguese Conference on Automatic Control*, pages 309–314, Porto, Portugal, 1996.

[86] J.E. Normey-Rico, C. Bordons, and E.F. Camacho. Improving the Robustness of Dead-Time Compensating PI Controllers. *Control Engineering Practice*, 5(6):801–810, 1997.

[87] J.E. Normey-Rico and E.F. Camacho. A Smith Predictor Based Generalized Predictive Controller. *Internal Report GAR 1996/02. University of Sevilla,* 1996.

[88] J.E. Normey-Rico and E.F. Camacho. Robustness Effect of a Prefilter in Generalized Predictive Control. *IEE Proc. on Control Theory and Applications (in press),* 1999.

[89] J.E. Normey-Rico, J. Gomez-Ortega, and E.F. Camacho. A Smith Predictor Based Generalized Predictive Controller for Mobile Robot Path-Tracking. In 3rd *IFAC Symposium on Intelligent Autonomous Vehicles,* pages 471–476, Madrid, Spain, 1998.

[90] M. Ohshima, I. Hshimoto, T. Takamatsu, and H. Ohno. Robust Stability of Model Predictive Control. *Int. Chem. Eng.,* 31(1), 1991.

[91] A.W. Ordys and D.W. Clarke. A State-space Description for GPC Controllers. *Int. Journal of System Science,* 24(9):1727–1744, 1993.

[92] A.W. Ordys and M.J. Grimble. *Advances in Model-Based Predictive Control,* chapter Evaluation of stochastic characteristics for a constrained GPC algorithm. Oxford University Press, 1994.

[93] G.C. Papavasilicu and J.C. Allwright. A Descendent Algorithm for a Min-Max Problem in Model Predictive Control. In *Proceedings of the 30th Conference on Decision and Control,* Brighton, England, 1991.

[94] V. Peterka. Predictor-based Self-tuning Control. *Automatica,* 20(1):39–50, 1984.

[95] D.M. Prett and M. Morari. *Shell Process Control Workshop.* Butterworths, 1987.

[96] D.M. Prett and R.D. Morari. Optimization and constrained multivariable control of a catalytic cracking unit. In *Proceedings of the Joint Automatic Control Conference.,* 1980.

[97] A.I. Propoi. Use of LP methods for synthesizing sampled-data automatic systems. *Automn Remote Control,* 24, 1963.

[98] S.J. Qin and T.A. Badgwell. An Overview of Industrial Model Predictive Control Technology. In Chemical Process Control: Assessment and New Directions for Research. In *AIChE Symposium Series 316, 93. Jeffrey C. Kantor, Carlos E. Garcia and Brice Carnahan Eds.* 232-256, 1997.

[99] J.M. Quero and E.F. Camacho. Neural Generalized Predictive Controllers. In *Proc. of the IEEE International Conference on System Engineering, Pittsburg,* 1990.

[100] J. Rawlings and K. Muske. The Stability of Constrained Receding Horizon Control. *IEEE Trans. on Automatic Control,* 38:1512–1516, 1993.

[101] J. Richalet. *Practique de la commande predictive.* Hermes, 1992.

[102] J. Richalet. Industrial Applications of Model Based Predictive Control. *Automatica,* 29(5):1251–1274, 1993.

[103] J. Richalet, S. Abu el Ata-Doss, C. Arber, H.B. Kuntze, A. Jacubash, and W. Schill. Predictive Functional Control. Application to fast and accurate robots. In *Proc. 10th IFAC Congress, Munich,* 1987.

[104] J. Richalet, A. Rault, J.L. Testud, and J. Papon. Algorithmic Control of Industrial Processes. In 4th IFAC *Symposium on Identification and System Parameter Estimation. Tbilisi* URSS, 1976.

[105] J. Richalet, A. Rault, J.L. Testud, and J. Papon. Model Predictive Heuristic Control: Application to Industrial Processes. *Automatica,* 14(2):413–428, 1978.

[106] B.D. Robinson and D.W. Clarke. Robustness effects of a prefilter in Generalized Predictive Control. *Proceedings IEE, Part D,* 138:2–8, 1991.

[107] A. Rossiter, J.R. Gossner, and B. Kouvaritakis. Infinite Horizon Stable Predictive Control. *IEEE Trans. on Automatic Control,* 41(10), 1996.

[108] J.A. Rossiter and B. Kouvaritakis. Constrained Stable Generalized Predictive Control. *Proceedings IEE, Part D,* 140(4), 1993.

[109] J.A. Rossiter and B. Kouvaritakis. *Advances in Model-Based Predictive Control,* chapter Advances in Generalized and Constrained Predictive Control. Oxford University Press, 1994.

[110] R. Rouhani and R.K. Mehra. Model Algorithmic Control; Basic Theoretical Properties. *Automatica,* 18(4):401–414, 1982.

[111] P. Scokaert and D.W. Clarke. *Advances in Model-Based Predictive Control,* chapter Stability and feasibility in coinstrained predictive control. Oxford University Press, 1994.

[112] S.L. Shah, C. Mohtadi, and D.W. Clarke. Multivariable adaptive control without a prior knowledge of the delay matrix. *Systems and Control Letters,* 9:295–306, 1987.

[113] I. Skrjanc and D. Matko. *Advances in Model-Based Predictive Control,* chapter Fuzzy Predictive Controller with Adaptive Gain. Oxford University Press, 1994.

[114] O.J.M. Smith. Close Control of Loops with Deadtime. *Chem. Eng. Prog.,* 53(5):217, 1957.

[115] R. Söeterboek. *Predictive Control. A unified approach.* Prentice-Hall, 1992.

[116] E. Srinivasa and M. Chidambaram. Robust Control of a distillation column by the method of inequalities. *Journal of Process Control*, 1(3):171–176, 1993.

[117] G.W. Stewart. *Introduction to Matrix Computations*. Academis Press, Inc., 1973.

[118] Y. Tan and R. De Keyser. *Advances in Model-Based Predictive Control*, chapter Neural Network Based Predictive Control. Oxford University Press, 1994.

[119] C.A. Tsiligiannis and S.A Svoronos. Multivariable self-tuning control via the right interactor matrix. *IEE Trans. Aut. Control*, 31:987–989, 1986.

[120] M.L. Tyler and M. Morari. Propositional Logic in Control and Monitoring Problems. *Technical Report AUT96-15, Institut f r Automatik, ETH-Swiss Federal Institute of Technology, Zurich, Switzerland*, 1996.

[121] E.R. Velasco. *Control y Supervisión de Planta Piloto mediante Sistema de Control Distribuido*. P.F.C. Univesidad de Sevilla, 1994.

[122] R.A.J. De Vries and H.B. Verbruggen. *Advances in Model-Based Predictive Control*, chapter Multivariable Unified Predictive Control. Oxford University Press, 1994.

[123] P.E. Wellstead, D. Prager, and P. Zanker. A Pole Assignment Selt Tuning Regulator. *Proceedings IEE, Part D*, 126:781–787, 1978.

[124] T. H. Yang and E. Polak. Moving Horizon Control of Nonlinear Systems with Input Saturations, Disturbances and Plant Uncertainties. *Int. Journal of Control*, pages 875–903, 1993.

[125] B.E. Ydstie. Extended Horizon Adaptive Control. In *Proc. 9th IFAC World Congress, Budapest, Hungary*, 1984.

[126] T.W. Yoon and D.W. Clarke. Prefiltering in Receding-Horizon Predictive Control. *Internal Report 1995/93, University of Oxford, Department of Engineering Science*, 1993.

[127] T.W. Yoon and D.W. Clarke. *Advances in Model-Based Predictive Control*, chapter Towards Robust Adaptive Predictive Control, pages 402–414. Oxford University Press, 1994.

[128] T.W. Yoon and D.W. Clarke. Observer Design in Receding-Horizon Control. *International Journal of Control*, 2:151–171, 1995.

[129] L.A. Zadeh and B.H. Whalen. On Optimal Control and Linear Programming. *IRE Trans. Aut. Control*, 7(4), 1962.

[130] A. Zheng and M. Morari. Stability of Model Predictive Control with Soft Constraints. *Internal Report. California Institute of Technology*, 1994.

[131] Y. Zhu and T. Backx. *Identification of Multivariable Industrial Processes.* Springer-Verlag, 1993.

Index

Printed in the United States
By Bookmasters